THE ORIGIN OF SPECIES

THE ORIGIN OF SPECIES

BY MEANS OF NATURAL SELECTION OR THE PRESERVATION OF FAVOURED RACES IN THE STRUGGLE FOR LIFE

BY CHARLES DARWIN, M.A.,
LL.D., F.R.S.

WITH PORTRAIT

CASTLE BOOKS

This edition published in 2004 by
CASTLE BOOKS ®
A division of BOOK SALES, INC.
114 Northfield Avenue
Edison, New Jersey 08837

ISBN: 0-7858-1911-8

Printed in the United States of America

CONTENTS.

CHAPTER I.

VARIATION UNDER DOMESTICATION.

CHAPTER II.

VARIATION UNDER NATURE.

CONTENTS.

CONTENTS.

CHAPTER VI.

DIFFICULTIES OF THE THEORY.

CHAPTER VII.

MISCELLANEOUS OBJECTIONS TO THE THEORY OF NATURAL SELECTION.

CONTENTS.

CHAPTER VIII.

INSTINCT.

CHAPTER IX.

HYBRIDISM.

CHAPTER X.

ON THE IMPERFECTION OF THE GEOLOGICAL RECORD.

CONTENTS.

CONTENTS.

CHAPTER XIV.

MUTUAL AFFINITIES OF ORGANIC BEINGS : MORPHOLOGY : EMBRYOLOGY : RUDIMENTARY ORGANS.

CHAPTER XV.

RECAPITULATION AND CONCLUSION.

INSTRUCTION TO BINDER.

The Diagram to front page 140, and to face the latter part of the Volume.

ADDITIONS AND CORRECTIONS

TO THE SIXTH EDITION.

———◦◦◦———

NUMEROUS small corrections have been made in the last and present editions on various subjects, according as the evidence has become somewhat stronger or weaker. The more important corrections and some additions in the present volume are tabulated on the following page, for the convenience of those interested in the subject, and who possess the fifth edition. The second edition was little more than a reprint of the first. The third edition was largely corrected and added to, and the fourth and fifth still more largely. As copies of the present work will be sent abroad, it may be of use if I specify the state of the foreign editions. The third French and second German editions were from the third English, with some few of the additions given in the fourth edition. A new fourth French edition has been translated by Colonel Moulinié; of which the first half is from the fifth English, and the latter half from the present edition. A third German edition, under the superintendence of Professor Victor Carus, was from the fourth English edition; a fifth is now preparing by the same author from the present volume. The second American edition was from the English second, with a few of the additions given in the third; and a third American edition has been printed from the fifth English edition. The Italian is from the third, the Dutch and three Russian editions from the second English edition, and the Swedish from the fifth English edition.

Fifth Edition.	Sixth Edition.	Chief Additions and Corrections.
Page	Page	
100	106	Influence of fortuitous destruction on natural selection.
158	156	On the convergence of specific forms.
220	221	Account of the Ground-Woodpecker of La Plata modified.
225	227	On the modification of the eye.
230	233	Transitions through the acceleration or retardation of the period of reproduction.
231	234	The account of the electric organ of fishes added to.
233	237	Analogical resemblance between the eyes of Cephalopods and Vertebrates.
234	239	Claparède on the analogical resemblance of the hair-claspers of the Acaridæ.
248	254	The probable use of the rattle to the Rattle-snake.
248	254	Helmholtz on the imperfection of the human eye.
255	262	The first part of this new chapter consists of portions, in a much modified state, taken from chap. iv. of the former editions. The latter and larger part is new, and relates chiefly to the supposed incompetency of natural selection to account for the incipient stages of useful structures. There is also a discussion on the causes which prevent in many cases the acquisition through natural selection of useful structures. Lastly, reasons are given for disbelieving in great and sudden modifications. Gradations of character, often accompanied by changes of function, are likewise here incidentally considered.
268	333	The statement with respect to young cuckoos ejecting their foster-brothers confirmed.
270	334	On the cuckoo-like habits of the Molothrus.
307	373	On fertile hybrid moths.
319	386	The discussion on the fertility of hybrids not having been acquired through natural selection condensed and modified.

Fifth Edition.	Sixth Edition.	Chief Additions and Corrections.
Page	Page	
326	392	On the causes of sterility of hybrids, added to and corrected.
377	445	Pyrgoma found in the chalk.
402	471	Extinct forms serving to connect existing groups.
440	512	On earth adhering to the feet of migratory birds.
463	536	On the wide geographical range of a species of Galaxias, a fresh-water fish.
505	582	Discussion on analogical resemblances, enlarged and modified.
516	596	Homological structure of the feet of certain marsupial animals.
518	600	On serial homologies, corrected.
520	601	Mr. E. Ray Lankester on morphology.
521	604	On the asexual reproduction of Chironomus.
541	626	On the origin of rudimentary parts, corrected.
547	633	Recapitulation on the sterility of hybrids, corrected.
552	639	Recapitulation on the absence of fossils beneath the Cambrian system, corrected.
568	657	Natural selection not the exclusive agency in the modification of species, as always maintained in this work.
572	661	The belief in the separate creation of species generally held by naturalists, until a recent period.

AN HISTORICAL SKETCH

OF THE PROGRESS OF OPINION ON THE ORIGIN OF SPECIES,

PREVIOUSLY TO THE PUBLICATION OF THE FIRST EDITION OF THIS WORK.

I WILL here give a brief sketch of the progress of opinion on the Origin of Species. Until recently the great majority of naturalists believed that species were immutable productions, and had been separately created. This view has been ably maintained by many authors. Some few naturalists, on the other hand, have believed that species undergo modification, and that the existing forms of life are the descendants by true generation of pre-existing forms. Passing over allusions to the subject in the classical writers,* the first author who in

---

* Aristotle, in his 'Physicæ Auscultationes' (lib. 2, cap. 8, s. 2), after remarking that rain does not fall in order to make the corn grow, any more than it falls to spoil the farmer's corn when threshed out of doors, applies the same argument to organisation; and adds (as translated by Mr. Clair Grece, who first pointed out the passage to me), "So what hinders the different parts [of the body] from having this merely accidental relation in nature? as the teeth, for example, grow by necessity, the front ones sharp, adapted for dividing, and the grinders flat, and serviceable for masticating the food; since they were not made for the sake of this, but it was the result of accident. And in like manner as to the other parts in which there appears to exist an adaptation to an end. Wheresoever, therefore, all things together (that is all the parts of one whole) happened like as if they were made for the sake of something, these were preserved, having been appropriately constituted

modern times has treated it in a scientific spirit was Buffon. But as his opinions fluctuated greatly at different periods, and as he does not enter on the causes or means of the transformation of species, I need not here enter on details.

Lamarck was the first man whose conclusions on the subject excited much attention. This justly-celebrated naturalist first published his views in 1801; he much enlarged them in 1809 in his 'Philosophie Zoologique,' and subsequently, in 1815, in the Introduction to his 'Hist. Nat. des Animaux sans Vertébres.' In these works he upholds the doctrine that all species, including man, are descended from other species. He first did the eminent service of arousing attention to the probability of all change in the organic, as well as in the inorganic world, being the result of law, and not of miraculous interposition. Lamarck seems to have been chiefly led to his conclusion on the gradual change of species, by the difficulty of distinguishing species and varieties, by the almost perfect gradation of forms in certain groups, and by the analogy of domestic productions. With respect to the means of modification, he attributed something to the direct action of the physical conditions of life, something to the crossing of already existing forms, and much to use and disuse, that is, to the effects of habit. To this latter agency he seems to attribute all the beautiful adaptations in nature;—such as the long neck of the giraffe for browsing on the

---

by an internal spontaneity; and whatsoever things were not thus constituted, perished, and still perish." We here see the principle of natural selection shadowed forth, but how little Aristotle fully comprehended the principle, is shown by his remarks on the formation of the teeth.

branches of trees. But he likewise believed in a law of progressive development; and as all the forms of life thus tend to progress, in order to account for the existence at the present day of simple productions, he maintains that such forms are now spontaneously generated.*

Geoffroy Saint-Hilaire, as is stated in his 'Life,' written by his son, suspected, as early as 1795, that what we call species are various degenerations of the same type. It was not until 1828 that he published his conviction that the same forms have not been perpetuated since the origin of all things. Geoffroy seems to have relied chiefly on the conditions of life, or the "*monde ambiant*" as the cause of change. He was cautious in drawing conclusions, and did not believe that existing species are now undergoing modification; and, as his son adds, " C'est donc un problème à réserver entièrement à l'avenir, supposé même que l'avenir doive avoir prise sur lui."

In 1813, Dr. W. C. Wells read before the Royal

---

* I have taken the date of the first publication of Lamarck from Isid. Geoffroy Saint-Hilaire's ('Hist. Nat. Générale,' tom. ii. p. 405, 1859) excellent history of opinion on this subject. In this work a full account is given of Buffon's conclusions on the same subject. It is curious how largely my grandfather, Dr. Erasmus Darwin, anticipated the views and erroneous grounds of opinion of Lamarck in his 'Zoonomia' (vol. i. pp. 500–510), published in 1794. According to Isid. Geoffroy there is no doubt that Goethe was an extreme partisan of similar views, as shown in the Introduction to a work written in 1794 and 1795, but not published till long afterwards : he has pointedly remarked (' Goethe als Naturforscher,' von Dr. Karl Meding, s. 34) that the future question for naturalists will be how, for instance, cattle got their horns, and not for what they are used. It is rather a singular instance of the manner in which similar views arise at about the same time, that Goethe in Germany, Dr. Darwin in England, and Geoffroy Saint-Hilaire (as we shall immediately see) in France, came to the same conclusion on the origin of species, in the years 1794–5.

Society ' An Account of a White female, part of whose skin resembles that of a Negro '; but his paper was not published until his famous ' Two Essays upon Dew and Single Vision ' appeared. in 1818. In this paper he distinctly recognises the principle of natural selection, and this is the first recognition which has been indicated; but he applies it only to the races of man, and to certain characters alone. After remarking that negroes and mulattoes enjoy an immunity from certain tropical diseases, he observes, firstly, that all animals tend to vary in some degree, and, secondly, that agriculturists improve their domesticated animals by selection; and then, he adds, but what is done in this latter case " by art, seems to be done with equal efficacy, though more slowly, by nature, in the formation of varieties of mankind, fitted for the country which they inhabit. Of the accidental varieties of man, which would occur among the first few and scattered inhabitants of the middle regions of Africa, some one would be better fitted than the others to bear the diseases of the country. This race would consequently multiply, while the others would decrease; not only from their inability to sustain the attacks of disease, but from their incapacity of contending with their more vigorous neighbours. The colour of this vigorous race I take for granted, from what has been already said, would be dark. But the same disposition to form varieties still existing, a darker and a darker race would in the course of time occur: and as the darkest would be the best fitted for the climate, this would at length become the most prevalent, if not the only race, in the particular country in which it had originated." He then extends these same views to the white inhabitants of colder climates. I am indebted to Mr. Rowley, of the United States, for having called my attention, through Mr. Brace, to the above passage in Dr. Wells' work.

The Hon. and Rev. W. Herbert, afterwards Dean of Manchester, in the fourth volume of the 'Horticultural Transactions,' 1822, and in his work on the 'Amaryllidaceæ' (1837, pp. 19, 339), declares that "horticultural experiments have established, beyond the possibility of refutation, that botanical species are only a higher and more permanent class of varieties." He extends the same view to animals. The Dean believes that single species of each genus were created in an originally highly plastic condition, and that these have produced, chiefly by intercrossing, but likewise by variation, all our existing species.

In 1826 Professor Grant, in the concluding paragraph in his well-known paper ('Edinburgh Philosophical Journal,' vol. xiv. p. 283) on the Spongilla, clearly declares his belief that species are descended from other species, and that they become improved in the course of modification. This same view was given in his 55th Lecture, published in the 'Lancet' in 1834.

In 1831 Mr. Patrick Matthew published his work on 'Naval Timber and Arboriculture,' in which he gives precisely the same view on the origin of species as that (presently to be alluded to) propounded by Mr. Wallace and myself in the 'Linnean Journal,' and as that enlarged in the present volume. Unfortunately the view was given by Mr. Matthew very briefly in scattered passages in an Appendix to a work on a different subject, so that it remained unnoticed until Mr. Matthew himself drew attention to it in the 'Gardener's Chronicle,' on April 7th, 1860. The differences of Mr.

Matthew's view from mine are not of much importance : he seems to consider that the world was nearly de-populated at successive periods, and then re-stocked ; and he gives as an alternative, that new forms may be generated "without the presence of any mould or germ of former aggregates." I am not sure that I understand some passages ; but it seems that he attributes much influence to the direct action of the conditions of life. He clearly saw, however, the full force of the principle of natural selection.

The celebrated geologist and naturalist, Von Buch, in his excellent 'Description Physique des Isles Canaries' (1836, p. 147), clearly expresses his belief that varieties slowly become changed into permanent species, which are no longer capable of intercrossing.

Rafinesque, in his 'New Flora of North America,' published in 1836, wrote (p. 6) as follows :—"All species might have been varieties once, and many varieties are gradually becoming species by assuming constant and peculiar characters ;" but farther on (p. 18) he adds, " except the original types or ancestors of the genus."

In 1843–44 Professor Haldeman ('Boston Journal of Nat. Hist. U. States,' vol. iv. p. 468) has ably given the arguments for and against the hypothesis of the develop-ment and modification of species : he seems to lean towards the side of change.

The 'Vestiges of Creation' appeared in 1844. In the tenth and much improved edition (1853) the anonymous author says (p. 155):—"The proposition determined on after much consideration is, that the several series of animated beings, from the simplest and oldest up to the highest and most recent, are, under the providence of God, the results, *first*, of an impulse which has been imparted to the forms of life, advancing them, in definite times, by generation, through grades of organisation terminating in the highest dicotyledons and vertebrata, these grades being few in number, and generally marked by intervals of organic character, which we find to be a practical difficulty in ascertaining affinities ; *second*, of another impulse connected with the vital forces, tending, in the course of generations, to modify organic structures in accordance with external circumstances, as food, the nature of the habitat, and the meteoric agencies, these being the 'adaptations' of the natural theologian." The author apparently believes that organisation progresses by sudden leaps, but that the effects produced by the conditions of life are gradual. He argues with much force on general grounds that species are not immutable productions. But I cannot see how the two supposed "impulses" account in a scientific sense for the numerous and beautiful co-adaptations which we see throughout nature; I cannot see that we thus gain any insight how, for instance, a woodpecker has become adapted to its peculiar habits of life. The work, from its powerful and brilliant style, though displaying in the earlier editions little accurate knowledge and a great want of scientific caution, immediately had a very wide circulation. In my opinion it has done excellent service in this country in calling attention to the subject, in removing prejudice, and in thus preparing the ground for the reception of analogous views.

In 1846 the veteran geologist M. J. d'Omalius d'Halloy published in an excellent though short paper ('Bulletins de l'Acad. Roy. Bruxelles,' tom. xiii. p. 581) his opinion that it is more probable that new species have been produced by descent with modification than that they have been separately created: the author first promulgated this opinion in 1831.

Professor Owen, in 1849 ('Nature of Limbs,' p. 86), wrote as follows:—"The archetypal idea was manifested in the flesh under diverse such modifications, upon this planet, long prior to the existence of those animal species that actually exemplify it. To what natural laws or secondary causes the orderly succession and progression of such organic phenomena may have been committed, we, as yet, are ignorant." In his Address to the British Association, in 1858, he speaks (p. li.) of "the axiom of the continuous operation of creative power, or of the ordained becoming of living things." Farther on (p. xc.), after referring to geographical distribution, he adds, "These phenomena shake our confidence in the conclusion that the Apteryx of New Zealand and the Red Grouse of England were distinct creations in and for those islands respectively. Always, also, it may be well to bear in mind that by the word 'creation' the zoologist means 'a process he knows not what.'" He amplifies this idea by adding that when such cases as that of the Red Grouse are "enumerated by the zoologist as evidence of distinct creation of the bird in and for such islands, he chiefly expresses that he knows not how the Red Grouse came to be there, and there exclusively; signifying also, by this mode of expressing such ignorance, his belief that both the bird and the islands owed their origin to a great first Creative Cause." If we interpret these sentences given in the same Address, one by the other, it appears that this eminent philosopher felt in 1858 his confidence shaken that the Apteryx and the Red Grouse first appeared in their respective homes, " he knew not how," or by some process " he knew not what."

This Address was delivered after the papers by Mr. Wallace and myself on the Origin of Species, presently to be referred to, had been read before the Linnean Society. When the first edition of this work was published, I was so completely deceived, as were many others, by such expressions as " the continuous operation of creative power," that I included Professor Owen with other palæontologists as being firmly convinced of the immutability of species; but it appears ('Anat. of Vertebrates,' vol. iii. p. 796) that this was on my part a preposterous error. In the last edition of this work I inferred, and the inference still seems to me perfectly just, from a passage beginning with the words " no doubt the type-form," &c. (Ibid. vol. i. p. xxxv.), that Professor Owen admitted that natural selection may have done something in the formation of a new species; but this it appears (Ibid. vol. iii. p. 798) is inaccurate and without evidence. I also gave some extracts from a correspondence between Professor Owen and the Editor of the ' London Review,' from which it appeared manifest to the Editor as well as to myself, that Professor Owen claimed to have promulgated the theory of natural selection before I had done so; and I expressed my surprise and satisfaction at this announcement; but as far as it is possible to understand certain recently published passages (Ibid. vol. iii. p. 798) I have either partially or wholly again fallen into error. It is consolatory to me that others find Professor Owen's controversial writings as difficult to understand and to reconcile with each other, as I do. As far as the mere

enunciation of the principle of natural selection is concerned, it is quite immaterial whether or not Professor
Owen preceded me, for both of us, as shown in this
historical sketch, were long ago preceded by Dr. Wells
and Mr. Matthews.

M. Isidore Geoffroy Saint-Hilaire, in his lectures
delivered in 1850 (of which a Résumé appeared in the
' Revue et Mag. de Zoolog.,' Jan. 1851), briefly gives
his reason for believing that specific characters " sont
fixés, pour chaque espèce, tant qu'elle se perpétue au
milieu des mêmes circonstances : ils se modifient, si les
circonstances ambiantes viennent à changer." " En
résumé, *l'observation* des animaux sauvages démontre
déjà la variabilité *limitée* des espèces. Les *expériences*
sur les animaux sauvages devenus domestiques, et sur
les animaux domestiques redevenus sauvages, la démon-
trent plus clairement encore. Ces mêmes expériences
prouvent, de plus, que les différences produites peuvent
être de *valeur générique*." In his ' Hist. Nat. Générale '
(tom. ii. p. 430, 1859) he amplifies analogous conclusions.

From a circular lately issued it appears that Dr.
Freke, in 1851 (' Dublin Medical Press,' p. 322), propounded the doctrine that all organic beings have
descended from one primordial form. His grounds of
belief and treatment of the subject are wholly different
from mine ; but as Dr. Freke has now (1861) published
his Essay on the ' Origin of Species by means of
Organic Affinity,' the difficult attempt to give any idea
of his views would be superfluous on my part.

Mr. Herbert Spencer, in an Essay (originally published in the ' Leader,' March, 1852, and republished in
his ' Essays,' in 1858), has contrasted the theories of the
Creation and the Development of organic beings with

remarkable skill and force. He argues from the analogy of domestic productions, from the changes which the embryos of many species undergo, from the difficulty of distinguishing species and varieties, and from the principle of general gradation, that species have been modified; and he attributes the modification to the change of circumstances. The author (1855) has also treated Psychology on the principle of the necessary acquirement of each mental power and capacity by gradation.

In 1852 M. Naudin, a distinguished botanist, expressly stated, in an admirable paper on the Origin of Species ('Revue Horticole,' p. 102; since partly republished in the 'Nouvelles Archives du Muséum,' tom. i. p. 171), his belief that species are formed in an analogous manner as varieties are under cultivation; and the latter process he attributes to man's power of selection. But he does not show how selection acts under nature. He believes, like Dean Herbert, that species, when nascent, were more plastic than at present. He lays weight on what he calls the principle of finality, "puissance mystérieuse, indéterminée; fatalité pour les uns; pour les autres, volonté providentielle, dont l'action incessante sur les êtres vivants détermine, à toutes les époques de l'existence du monde, la forme, le volume, et la durée de chacun d'eux, en raison de sa destinée dans l'ordre de choses dont il fait partie. C'est cette puissance qui harmonise chaque membre à l'ensemble, en l'appropriant à la fonction qu'il doit remplir dans l'organisme général de la nature, fonction qui est pour lui sa raison d'être." *

* From references in Bronn's 'Untersuchungen über die Entwickelungs-Gesetze,' it appears that the celebrated botanist and

In 1853 a celebrated geologist, Count Keyserling ('Bulletin de la Soc. Géolog.,' 2nd Ser., tom. x. p. 357), suggested that as new diseases, supposed to have been caused by some miasma, have arisen and spread over the world, so at certain periods the germs of existing species may have been chemically affected by circum-ambient molecules of a particular nature, and thus have given rise to new forms.

In this same year, 1853, Dr. Schaaffhausen published an excellent pamphlet ('Verhand. des Naturhist. Vereins der Preuss. Rheinlands,' &c.), in which he maintains the development of organic forms on the earth. He infers that many species have kept true for long periods, whereas a few have become modified. The distinction of species he explains by the destruction of intermediate graduated forms. "Thus living plants and animals are not separated from the extinct by new creations, but are to be regarded as their descendants through continued reproduction."

A well-known French botanist, M. Lecoq, writes in 1854 ('Etudes sur Géograph. Bot.,' tom. i. p. 250), "On voit que nos recherches sur la fixité ou la variation de l'espèce, nous conduisent directement aux idées émises,

---

palæontologist Unger published, in 1852, his belief that species undergo development and modification. Dalton, likewise, in Pander and Dalton's work on Fossil Sloths, expressed, in 1821, a similar belief. Similar views have, as is well known, been maintained by Oken in his mystical 'Natur-Philosophie.' From other references in Godron's work 'Sur l'Espèce,' it seems that Bory St. Vincent, Burdach, Poiret, and Fries, have all admitted that new species are continually being produced.

I may add, that of the thirty-four authors named in this Historical Sketch, who believe in the modification of species, or at least disbelieve in separate acts of creation, twenty-seven have written on special branches of natural history or geology.

par deux hommes justement célèbres, Geoffroy Saint-Hilaire et Gœthe." Some other passages scattered through M. Lecoq's large work, make it a little doubtful how far he extends his views on the modification of species.

The 'Philosophy of Creation' has been treated in a masterly manner by the Rev. Baden Powell, in his 'Essays on the Unity of Worlds,' 1855. Nothing can be more striking than the manner in which he shows that the introduction of new species is " a regular, not a casual phenomenon," or, as Sir John Herschel expresses it, " a natural in contradistinction to a miraculous process."

The third volume of the 'Journal of the Linnean Society' contains papers, read July 1st, 1858, by Mr. Wallace and myself, in which, as stated in the introductory remarks to this volume, the theory of Natural Selection is promulgated by Mr. Wallace with admirable force and clearness.

Von Baer, towards whom all zoologists feel so profound a respect, expressed about the year 1859 (see Prof. Rudolph Wagner, 'Zoologisch-Anthropologische Untersuchungen,' 1861, s. 51) his conviction, chiefly grounded on the laws of geographical distribution, that forms now perfectly distinct have descended from a single parent-form.

In June, 1859, Professor Huxley gave a lecture before the Royal Institution on the 'Persistent Types of Animal Life.' Referring to such cases, he remarks, "It is difficult to comprehend the meaning of such facts as these, if we suppose that each species of animal and plant, or each great type of organisation, was formed and placed upon the surface of the globe at long intervals by a distinct act of creative power; and it is well to recollect that such an assumption is as unsupported by tradition or revelation as it is opposed to the general analogy of nature. If, on the other hand, we view 'Persistent Types' in relation to that hypothesis which supposes the species living at any time to be the result of the gradual modification of pre-existing species a hypothesis which, though un-proven, and sadly damaged by some of its supporters, is yet the only one to which physiology lends any countenance; their existence would seem to show that the amount of modification which living beings have undergone during geological time is but very small in relation to the whole series of changes which they have suffered."

In December, 1859, Dr. Hooker published his 'Intro-duction to the Australian Flora.' In the first part of this great work he admits the truth of the descent and modification of species, and supports this doctrine by many original observations.

The first edition of this work was published on November 24th, 1859, and the second edition on January 7th, 1860.

ORIGIN OF SPECIES.

INTRODUCTION.

WHEN on board H.M.S. 'Beagle,' as naturalist, I was much struck with certain facts in the distribution of the organic beings inhabiting South America, and in the geological relations of the present to the past inhabitants of that continent. These facts, as will be seen in the latter chapters of this volume, seemed to throw some light on the origin of species—that mystery of mysteries, as it has been called by one of our greatest philosophers. On my return home, it occurred to me, in 1837, that something might perhaps be made out on this question by patiently accumulating and reflecting on all sorts of facts which could possibly have any bearing on it. After five years' work I allowed myself to speculate on the subject, and drew up some short notes; these I enlarged in 1844 into a sketch of the conclusions, which then seemed to me probable: from that period to the present day I have steadily pursued the same object. I hope that I may be excused for entering on these personal details, as I give them to show that I have not been hasty in coming to a decision.

My work is now (1859) nearly finished; but as it will take me many more years to complete it, and as my health is far from strong, I have been urged to publish this Abstract. I have more especially been induced to do this, as Mr. Wallace, who is now studying the natural history of the Malay archipelago, has arrived at almost exactly the same general conclusions that I have on the origin of species. In 1858 he sent me a memoir on this subject, with a request that I would forward it to Sir Charles Lyell, who sent it to the Linnean Society, and it is published in the third volume of the Journal of that Society. Sir C. Lyell and Dr. Hooker, who both knew of my work—the latter having read my sketch of 1844—honoured me by thinking it advisable to publish, with Mr. Wallace's excellent memoir, some brief extracts from my manuscripts.

This Abstract, which I now publish, must necessarily be imperfect. I cannot here give references and authorities for my several statements; and I must trust to the reader reposing some confidence in my accuracy. No doubt errors will have crept in, though I hope I have always been cautious in trusting to good authorities alone. I can here give only the general conclusions at which I have arrived, with a few facts in illustration, but which, I hope, in most cases will suffice. No one can feel more sensible than I do of the necessity of hereafter publishing in detail all the facts, with references, on which my conclusions have been grounded; and I hope in a future work to do this. For I am well aware that scarcely a single point is discussed in this volume on which facts cannot be adduced, often apparently leading to conclusions directly opposite to those at which I have arrived. A fair result can be obtained only by fully stating and balancing the facts and arguments on both sides of each question; and this is here impossible.

I much regret that want of space prevents my having the satisfaction of acknowledging the generous assistance which I have received from very many naturalists, some of them personally unknown to me. I cannot, however, let this opportunity pass without expressing my deep obligations to Dr. Hooker, who, for the last fifteen years, has aided me in every possible way by his large stores of knowledge and his excellent judgment.

In considering the Origin of Species, it is quite conceivable that a naturalist, reflecting on the mutual affinities of organic beings, on their embryological relations, their geographical distribution, geological succession, and other such facts, might come to the conclusion that species had not been independently created, but had descended, like varieties, from other species. Nevertheless, such a conclusion, even if well founded, would be unsatisfactory, until it could be shown how the innumerable species inhabiting this world have been modified, so as to acquire that perfection of structure and coadaptation which justly excites our admiration. Naturalists continually refer to external conditions, such as climate, food, &c., as the only possible cause of variation. In one limited sense, as we shall hereafter see, this may be true; but it is preposterous to attribute to mere external conditions, the structure, for instance, of the woodpecker, with its feet, tail, beak, and tongue, so admirably adapted to catch insects under the bark of trees. In the case of the mistletoe, which draws its nourishment from certain trees, which has seeds that must be transported by certain birds, and which has flowers with separate sexes absolutely requiring the agency of certain insects to bring pollen from one flower to the other, it is equally preposterous to account for the structure of this parasite, with its relations to several distinct organic beings, by the effects of external conditions, or of habit, or of the volition of the plant itself.

It is, therefore, of the highest importance to gain a clear insight into the means of modification and co-adaptation. At the commencement of my observations it seemed to me probable that a careful study of domesticated animals and of cultivated plants would offer the best chance of making out this obscure problem. Nor have I been disappointed; in this and in all other perplexing cases I have invariably found that our knowledge, imperfect though it be, of variation under domestication, afforded the best and safest clue. I may venture to express my conviction of the high value of such studies, although they have been very commonly neglected by naturalists.

From these considerations, I shall devote the first chapter of this Abstract to Variation under Domestication. We shall thus see that a large amount of hereditary modification is at least possible; and, what is equally or more important, we shall see how great is the power of man in accumulating by his Selection successive slight variations. I will then pass on to the variability of species in a state of nature; but I shall, unfortunately, be compelled to treat this subject far too briefly, as it can be treated properly only by giving long catalogues of facts. We shall, however, be enabled to discuss what circumstances are most favourable to variation. In the next chapter the Struggle for Existence amongst all organic beings throughout the world, which inevitably follows from the high geometrical ratio of their increase, will be considered. This is the doctrine of Malthus, applied to the whole animal and vegetable kingdoms. As many more individuals of each species are born than can possibly survive; and as, consequently, there is a frequently recurring struggle for existence, it follows that any being, if it vary however slightly in any manner profitable to itself, under the complex and sometimes varying conditions of life, will have a better chance of surviving, and thus be *naturally selected*. From the strong principle of inheritance, any selected variety will tend to propagate its new and modified form.

This fundamental subject of Natural Selection will be treated at some length in the fourth chapter; and we shall then see how Natural Selection almost inevitably causes much Extinction of the less improved forms of life, and leads to what I have called Divergence of Character. In the next chapter I shall discuss the complex and little known laws of variation. In the five succeeding chapters, the most apparent and gravest difficulties in accepting the theory will be given: namely, first, the difficulties of transitions, or how a simple being or a simple organ can be changed and perfected into a highly developed being or into an elaborately constructed organ; secondly, the subject of Instinct, or the mental powers of animals; thirdly, Hybridism, or the infertility of species and the fertility of varieties when intercrossed; and fourthly, the imperfection of the Geological Record. In the next chapter I shall consider the geological succession of organic beings throughout time; in the twelfth and thirteenth, their geographical distribution throughout space; in the fourteenth, their classification or mutual affinities, both when mature and in an embryonic condition. In the last chapter I shall give a brief recapitulation of the whole work, and a few concluding remarks.

No one ought to feel surprise at much remaining as yet unexplained in regard to the origin of species and varieties, if he make due allowance for our profound ignorance in regard to the mutual relations of the many beings which live around us. Who can explain why one species ranges widely and is very numerous, and why another allied species has a narrow range and is rare? Yet these relations are of the highest importance, for they determine the present welfare and, as I believe, the future success and modification of every inhabitant of this world. Still less do we know of the mutual relations of the innumerable inhabitants of the world during the many past geological epochs in its history. Although much remains obscure, and will long remain obscure, I can entertain no doubt, after the most deliberate study and dispassionate judgment of which I am capable, that the view which most naturalists until recently entertained, and which I formerly entertained—namely, that each species has been independently created—is erroneous. I am fully convinced that species are not immutable; but that those belonging to what are called the same genera are lineal descendants of some other and generally extinct species, in the same manner as the acknowledged varieties of any one species are the descendants of that species. Furthermore, I am convinced that Natural Selection has been the most important, but not the exclusive, means of modification.

CHAPTER I.

VARIATION UNDER DOMESTICATION.

Causes of Variability—Effects of Habit and the use or disuse of
Parts—Correlated Variation—Inheritance—Character of Domestic Varieties—Difficulty of distinguishing between Varieties
and Species—Origin of Domestic Varieties from one or more
Species—Domestic Pigeons, their Differences and Origin—
Principles of Selection, anciently followed, their Effects—
Methodical and Unconscious Selection—Unknown Origin of
our Domestic Productions—Circumstances favourable to Man's
power of Selection.

Causes of Variability.

WHEN we compare the individuals of the same variety
or sub-variety of our older cultivated plants and animals,
one of the first points which strikes us is, that they
generally differ more from each other than do the
individuals of any one species or variety in a state of
nature. And if we reflect on the vast diversity of the
plants and animals which have been cultivated, and
which have varied during all ages under the most
different climates and treatment, we are driven to
conclude that this great variability is due to our
domestic productions having been raised under conditions of life not so uniform as, and somewhat
different from, those to which the parent species had
been exposed under nature. There is, also, some pro-

bability in the view propounded by Andrew Knight,
that this variability may be partly connected with
excess of food. It seems clear that organic beings
must be exposed during several generations to new
conditions to cause any great amount of variation; and
that, when the organisation has once begun to vary, it
generally continues varying for many generations. No
case is on record of a variable organism ceasing to vary
under cultivation. Our oldest cultivated plants, such
as wheat, still yield new varieties: our oldest domesticated animals are still capable of rapid improv-
ment or modification.

As far as I am able to judge, after long attending to
the subject, the conditions of life appear to act in two
ways,—directly on the whole organisation or on certain
parts alone, and indirectly by affecting the reproductive
system. With respect to the direct action, we must
bear in mind that in every case, as Professor Weismann has lately insisted, and as I have incidentally
shown in my work on 'Variation under Domestication,'
there are two factors: namely, the nature of the organism, and the nature of the conditions. The former seems
to be much the more important; for nearly similar variations sometimes arise under, as far as we can judge,
dissimilar conditions; and, on the other hand, dissimilar
variations arise under conditions which appear to be
nearly uniform. The effects on the offspring are either
definite or indefinite. They may be considered as definite when all or nearly all the offspring of individuals
exposed to certain conditions during several generations
are modified in the same manner. It is extremely
difficult to come to any conclusion in regard to the
extent of the changes which have been thus definitely

induced. There can, however, be little doubt about many slight changes,—such as size from the amount of food, colour from the nature of the food, thickness of the skin and hair from climate, &c. Each of the endless variations which we see in the plumage of our fowls must have had some efficient cause; and if the same cause were to act uniformly during a long series of generations on many individuals, all probably would be modified in the same manner. Such facts as the complex and extraordinary out-growths which variably follow from the insertion of a minute drop of poison by a gall-producing insect, show us what singular modifications might result in the case of plants from a chemical change in the nature of the sap.

Indefinite variability is a much more common result of changed conditions than definite variability, and has probably played a more important part in the formation of our domestic races. We see indefinite variability in the endless slight peculiarities which distinguish the individuals of the same species, and which cannot be accounted for by inheritance from either parent or from some more remote ancestor. Even strongly-marked differences occasionally appear in the young of the same litter, and in seedlings from the same seed-capsule. At long intervals of time, out of millions of individuals reared in the same country and fed on nearly the same food, deviations of structure so strongly pronounced as to deserve to be called monstrosities arise; but monstrosities cannot be separated by any distinct line from slighter variations. All such changes of structure, whether extremely slight or strongly marked, which appear amongst many individuals living together, may be considered as the indefinite effects of the conditions of life on each individual organism, in nearly the same manner as the chill affects different men in an indefinite manner, according to their state of body or constitution, causing coughs or colds, rheumatism, or inflammation of various organs.

With respect to what I have called the indirect action of changed conditions, namely, through the reproductive system of being affected, we may infer that variability is thus induced, partly from the fact of this system being extremely sensitive to any change in the conditions, and partly from the similarity, as Kölreuter and others have remarked, between the variability which follows from the crossing of distinct species, and that which may be observed with plants and animals when reared under new or unnatural conditions. Many facts clearly show how eminently susceptible the reproductive system is to very slight changes in the surrounding conditions. Nothing is more easy than to tame an animal, and few things more difficult than to get it to breed freely under confinement, even when the male and female unite. How many animals there are which will not breed, though kept in an almost free state in their native country! This is generally, but erroneously, attributed to vitiated instincts. Many cultivated plants display the utmost vigour, and yet rarely or never seed! In some few cases it has been discovered that a very trifling change, such as a little more or less water at some particular period of growth, will determine whether or not a plant will produce seeds. I cannot here give the details which I have collected and elsewhere published on this curious subject; but to show how singular the laws are which determine the reproduction of animals under confinement, I may mention that carnivorous animals, even from the tropics, breed in this country pretty freely under confinement, with the exception of the plantigrades or bear family, which seldom produce young; whereas carnivorous birds, with the rarest exceptions, hardly ever lay fertile eggs. Many exotic plants have pollen utterly worthless, in the same condition as in the most sterile hybrids. When, on the one hand, we see domesticated animals and plants, though often weak and sickly, breeding freely under confinement; and when, on the other hand, we see individuals, though taken young from a state of nature perfectly tamed, long-lived and healthy (of which I could give numerous instances), yet having their reproductive system so seriously affected by unperceived causes as to fail to act, we need not be surprised at this system, when it does act under confinement, acting irregularly, and producing offspring somewhat unlike their parents. I may add, that as some organisms breed freely under the most unnatural conditions (for instance, rabbits and ferrets kept in hutches), showing that their reproductive organs are not easily affected; so will some animals and plants withstand domestication or cultivation, and vary very slightly—perhaps hardly more than in a state of nature.

Some naturalists have maintained that all variations are connected with the act of sexual reproduction; but this is certainly an error; for I have given in another work a long list of " sporting plants," as they are called by gardeners;—that is, of plants which have suddenly produced a single bud with a new and sometimes widely different character from that of the other buds on the same plant. These bud variations, as they may be named, can be propagated by grafts, offsets, &c., and sometimes by seed. They occur rarely under nature, but are far from rare under culture. As a single bud out of the many thousands, produced year after year on the same tree under uniform conditions, has been known suddenly to assume a new character; and as buds on distinct trees, growing under different conditions, have sometimes yielded nearly the same variety —for instance, buds on peach-trees producing nectarines, and buds on common roses producing moss-roses—we clearly see that the nature of the conditions is of subordinate importance in comparison with the nature of the organism in determining each particular form of variation;—perhaps of not more importance than the nature of the spark, by which a mass of combustible matter is ignited, has in determining the nature of the flames.

Effects of Habit and of the Use or Disuse of Parts; Correlated Variation; Inheritance.

Changed habits produce an inherited effect, as in the period of the flowering of plants when transported from one climate to another. With animals the increased use or disuse of parts has had a more marked influence; thus I find in the domestic duck that the bones of the wing weigh less and the bones of the leg more, in proportion to the whole skeleton, than do the same bones in the wild-duck; and this change may be safely attributed to the domestic duck flying much less, and walking more, than its wild parents. The great and inherited development of the udders in cows and goats in countries where they are habitually milked, in comparison with these organs in other countries, is probably another instance of the effects of use. Not one of our domestic animals can be named which has not in some country drooping ears; and the view which has been suggested that the drooping is due to disuse of the muscles of the ear, from the animals being seldom much alarmed, seems probable.

Many laws regulate variation, some few of which can be dimly seen, and will hereafter be briefly discussed. I will here only allude to what may be called correlated variation. Important changes in the embryo or larva will probably entail changes in the mature animal. In monstrosities, the correlations between quite distinct parts are very curious; and many instances are given in Isidore Geoffroy St. Hilaire's great work on this subject. Breeders believe that long limbs are almost always accompanied by an elongated head. Some instances of correlation are quite whimsical: thus cats which are entirely white and have blue eyes are generally deaf; but it has been lately stated by Mr. Tait that this is confined to the males. Colour and constitutional peculiarities go together, of which many remarkable cases could be given amongst animals and plants. From facts collected by Heusinger, it appears that white sheep and pigs are injured by certain plants, whilst dark-coloured individuals escape: Professor Wyman has recently communicated to me a good illustration of this fact; on asking some farmers in Virginia how it was that all their pigs were black, they informed him that the pigs ate the paint-root (Lachnanthes), which coloured their bones pink, and which caused the hoofs of all but the black varieties to drop off; and one of the "crackers" (*i.e.* Virginia squatters) added, "we select the black members of a litter for raising, as they alone have a good chance of living." Hairless dogs have imperfect teeth ; long-haired and coarse-haired animals are apt to have, as is asserted, long or many horns ; pigeons with feathered feet have skin between their outer toes ; pigeons with short beaks have small feet, and those with long beaks large feet. Hence if man goes on selecting, and thus augmenting, any peculiarity, he will almost certainly modify unintentionally other parts of the structure, owing to the mysterious laws of correlation.

The results of the various, unknown, or but dimly understood laws of variation are infinitely complex and diversified. It is well worth while carefully to study the several treatises on some of our old cultivated plants, as on the hyacinth, potato, even the dahlia, &c. ; and it is really surprising to note the endless points of structure and constitution in which the varieties and sub-varieties differ slightly from each other. The whole organisation seems to have become plastic, and departs in a slight degree from that of the parental type.

Any variation which is not inherited is unimportant for us. But the number and diversity of inheritable deviations of structure, both those of slight and those of considerable physiological importance, are endless. Dr. Prosper Lucas's treatise, in two large volumes, is the fullest and the best on this subject. No breeder doubts how strong is the tendency to inheritance ; that like produces like is his fundamental belief : doubts have been thrown on this principle only by theoretical writers. When any deviation of structure often ap-

pears, and we see it in the father and child, we cannot
tell whether it may not be due to the same cause
having acted on both; but when amongst individuals,
apparently exposed to the same conditions, any very
rare deviation, due to some extraordinary combination
of circumstances, appears in the parent—say, once
amongst several million individuals—and it reappears
in the child, the mere doctrine of chances almost com-
pels us to attribute its reappearance to inheritance.
Every one must have heard of cases of albinism,
prickly skin, hairy bodies, &c., appearing in several
members of the same family. If strange and rare
deviations of structure are really inherited, less strange
and commoner deviations may be freely admitted to be
inheritable. Perhaps the correct way of viewing the
whole subject would be, to look at the inheritance of
every character whatever as the rule, and non-in-
heritance as the anomaly.

The laws governing inheritance are for the most part
unknown. No one can say why the same peculiarity
in different individuals of the same species, or in
different species, is sometimes inherited and sometimes
not so; why the child often reverts in certain characters
to its grandfather or grandmother or more remote
ancestor; why a peculiarity is often transmitted from
one sex to both sexes, or to one sex alone, more commonly but not exclusively to the like sex. It is a fact
of some importance to us, that peculiarities appearing
in the males of our domestic breeds are often transmitted, either exclusively or in a much greater degree,
to the males alone. A much more important rule,
which I think may be trusted, is that, at whatever
period of life a peculiarity first appears, it tends to

reappear in the offspring at a corresponding age, though sometimes earlier. In many cases this could not be otherwise; thus the inherited peculiarities in the horns of cattle could appear only in the offspring when nearly mature; peculiarities in the silkworm are known to appear at the corresponding caterpillar or cocoon stage. But hereditary diseases and some other facts make me believe that the rule has a wider extension, and that, when there is no apparent reason why a peculiarity should appear at any particular age, yet that it does tend to appear in the offspring at the same period at which it first appeared in the parent. I believe this rule to be of the highest importance in explaining the laws of embryology. These remarks are of course confined to the first *appearance* of the peculiarity, and not to the primary cause which may have acted on the ovules or on the male element; in nearly the same manner as the increased length of the horns in the offspring from a short-horned cow by a long-horned bull, though appearing late in life, is clearly due to the male element.

Having alluded to the subject of reversion, I may here refer to a statement often made by naturalists—namely, that our domestic varieties, when run wild, gradually but invariably revert in character to their aboriginal stocks. Hence it has been argued that no deductions can be drawn from domestic races to species in a state of nature. I have in vain endeavoured to discover on what decisive facts the above statement has so often and so boldy been made. There would be great difficulty in proving its truth: we may safely conclude that very many of the most strongly marked domestic varieties could not possibly live in a wild state. In many cases we do not know what the aboriginal stock was, and so could not tell whether or not nearly perfect reversion had ensued. It would be necessary, in order to prevent the effects of intercrossing, that only a single variety should have been turned loose in its new home. Nevertheless, as our varieties certainly do occasionally revert in some of their characters to ancestral forms, it seems to me not improbable that if we could succeed in naturalising, or were to cultivate, during many generations, the several races, for instance, of the cabbage, in very poor soil (in which case, however, some effect would have to be attributed to the *definite* action of the poor soil), that they would, to a large extent, or even wholly, revert to the wild aboriginal stock. Whether or not the experiment would succeed, is not of great importance for our line of argument; for by the experiment itself the conditions of life are changed. If it could be shown that our domestic varieties manifested a strong tendency to reversion,—that is, to lose their acquired characters, whilst kept under the same conditions, and whilst kept in a considerable body, so that free intercrossing might check, by blending together, any slight deviations in their structure, in such case, I grant that we could deduce nothing from domestic varieties in regard to species. But there is not a shadow of evidence in favour of this view: to assert that we could not breed our cart and race-horses, long and short-horned cattle, and poultry of various breeds, and esculent vegetables, for an unlimited number of generations, would be opposed to all experience.

Character of Domestic Varieties; difficulty of distinguishing between Varieties and Species; origin of Domestic Varieties from one or more Species.

When we look to the hereditary varieties or races of our domestic animals and plants, and compare them with closely allied species, we generally perceive in each domestic race, as already remarked, less uniformity of character than in true species. Domestic races often have a somewhat monstrous character; by which I mean, that, although differing from each other, and from other species of the same genus, in several trifling respects, they often differ in an extreme degree in some one part, both when compared one with another, and more especially when compared with the species under nature to which they are nearest allied. With these exceptions (and with that of the perfect fertility of varieties when crossed,—a subject hereafter to be discussed), domestic races of the same species differ from each other in the same manner as do the closely-allied species of the same genus in a state of nature, but the differences in most cases are less in degree. This must be admitted as true, for the domestic races of many animals and plants have been ranked by some competent judges as the descendants of aboriginally distinct species, and by other competent judges as mere varieties. If any well marked distinction existed between a domestic race and a species, this source of doubt would not so perpetually recur. It has often been stated that domestic races do not differ from each other in characters of generic value. It can be shown that this statement is not correct; but naturalists differ much in determining what characters are of generic value; all such valuations being at present empirical. When it is explained how genera originate under nature, it will be seen that we have no right to expect often to find a generic amount of difference in our domesticated races.

In attempting to estimate the amount of structural difference between allied domestic races, we are soon involved in doubt, from not knowing whether they are descended from one or several parent species. This point, if it could be cleared up, would be interesting; if, for instance, it could be shown that the greyhound, bloodhound, terrier, spaniel, and bull-dog, which we all know propagate their kind truly, were the offspring of any single species, then such facts would have great weight in making us doubt about the immutability of the many closely allied natural species—for instance, of the many foxes—inhabiting different quarters of the world. I do not believe, as we shall presently see, that the whole amount of difference between the several breeds of the dog has been produced under domestication; I believe that a small part of the difference is due to their being descended from distinct species. In the case of strongly marked races of some other domesticated species, there is presumptive or even strong evidence, that all are descended from a single wild stock.

It has often been assumed that man has chosen for domestication animals and plants having an extraordinary inherent tendency to vary, and likewise to withstand diverse climates. I do not dispute that these capacities have added largely to the value of most of our domesticated productions · but how could a savage possibly know, when he first tamed an animal, whether it would vary in succeeding generations, and whether it would endure other climates? Has the little variability of the ass and goose, or the small power of endurance of warmth by the reindeer, or of cold by the common camel, prevented their domestication? I cannot doubt that if other animals and plants, equal in number to our domesticated productions, and belonging to equally diverse classes and countries, were taken from a state of nature, and could be made to breed for an equal number of generations under domestication, they would on a average vary as largely as the parent species of our existing domesticated productions have varied.

In the case of most of our anciently domesticated animals and plants, it is not possible to come to any definite conclusion, whether they are descended from one or several wild species. The argument mainly relied on by those who believe in the multiple origin of our domestic animals is, that we find in the most ancient times, on the monuments of Egypt, and in the lake-habitations of Switzerland, much diversity in the breeds; and that some of these ancient breeds closely resemble, or are even identical with, those still existing. But this only throws far backwards the history of civilisation, and shows that animals were domesticated at a much earlier period than has hitherto been supposed. The lake-inhabitants of Switzerland cultivated several kinds of wheat and barley, the pea, the poppy for oil, and flax; and they possessed several domesticated animals. They also carried on commerce with other nations. All this clearly shows, as Heer has remarked, that they had at this early age progressed considerably in civilisation; and this again implies a long continued previous period of less advanced civilisation, during which the domesticated animals, kept by different tribes in different districts, might have varied and given rise to distinct races.   Since the discovery of flint tools in the superficial formations of many parts of the world, all geologists believe that barbarian man existed at an enormously remote period; and we know that at the present day there is hardly a tribe so barbarous, as not to have domesticated at least the dog.

The origin of most of our domestic animals will probably for ever remain vague.   But I may here state, that, looking to the domestic dogs of the whole world, I have, after a laborious collection of all known facts, come to the conclusion that several wild species of Canidæ have been tamed, and that their blood, in some cases mingled together, flows in the veins of our domestic breeds.   In regard to sheep and goats I can form no decided opinion.   From facts communicated to me by Mr. Blyth, on the habits, voice, constitution, and structure of the humped Indian cattle, it is almost certain that they are descended from a different aboriginal stock from our European cattle; and some competent judges believe that these latter have had two or three wild progenitors,—whether or not these deserve to be called species.   This conclusion, as well as that of the specific distinction between the humped and common cattle, may, indeed, be looked upon as established by the admirable researches of Professor Rütimeyer.   With respect to horses, from reasons which I cannot here give, I am doubtfully inclined to believe, in opposition to several authors, that all the races belong to the same species.   Having kept nearly all the English breeds of the fowl alive, having

bred and crossed them, and examined their skeletons,
it appears to me almost certain that all are the
descendants of the wild Indian fowl, Gallus bankiva;
and this is the conclusion of Mr. Blyth, and of others
who have studied this bird in India. In regard to
ducks and rabbits, some breeds of which differ much
from each other, the evidence is clear that they are all
descended from the common wild duck and rabbit.

The doctrine of the origin of our several domestic
races from several aboriginal stocks, has been carried to
an absurd extreme by some authors. They believe that
every race which breeds true, let the distinctive
characters be ever so slight, has had its wild prototype.
At this rate there must have existed at least a score of
species of wild cattle, as many sheep, and several goats,
in Europe alone, and several even within Great Britain.
One author believes that there formerly existed eleven
wild species of sheep peculiar to Great Britain! When
we bear in mind that Britain has now not one peculiar
mammal, and France but few distinct from those of
Germany, and so with Hungary, Spain, &c., but that
each of these kingdoms possesses several peculiar breeds
of cattle, sheep, &c., we must admit that many domestic
breeds must have originated in Europe; for whence
otherwise could they have been derived? So it is in
India. Even in the case of the breeds of the domestic
dog throughout the world, which I admit are descended
from several wild species, it cannot be doubted that
there has been an immense amount of inherited
variation; for who will believe that animals closely
resembling the Italian greyhound, the bloodhound, the
bull-dog, pug-dog, or Blenheim spaniel, &c.—so unlike
all wild Canidæ—ever existed in a state of nature? It

has often been loosely said that all our races of dogs have been produced by the crossing of a few aboriginal species; but by crossing we can only get forms in some degree intermediate between their parents; and if we account for our several domestic races by this process, we must admit the former existence of the most extreme forms, as the Italian greyhound, bloodhound, bull-dog, &c., in the wild state. Moreover, the possibility of making distinct races by crossing has been greatly exaggerated. Many cases are on record, showing that a race may be modified by occasional crosses, if aided by the careful selection of the individuals which present the desired character; but to obtain a race intermediate between two quite distinct races, would be very difficult. Sir J. Sebright expressly experimented with this object and failed. The offspring from the first cross between two pure breeds is tolerably and sometimes (as I have found with pigeons) quite uniform in character, and everything seems simple enough; but when these mongrels are crossed one with another for several generations, hardly two of them are alike, and then the difficulty of the task becomes manifest.

Breeds of the Domestic Pigeon, their Differences and Origin.

Believing that it is always best to study some special group, I have,, after deliberation, taken up domestic pigeons. I have kept every breed which I could purchase or obtain, and have been most kindly favoured with skins from several quarters of the world, more especially by the Hon. W. Elliot from India, and by the Hon. C. Murray from Persia. Many treatises in

different languages have been published on pigeons,
and some of them are very important, as being of
considerable antiquity. I have associated with several
eminent fanciers, and have been permitted to join two
of the London Pigeon Clubs. The diversity of the
breeds is something astonishing. Compare the English
carrier and the short-faced tumbler, and see the
wonderful difference in their beaks, entailing corresponding differences in their skulls. The carrier, more
especially the male bird, is also remarkable from the
wonderful development of the carunculated skin about
the head; and this is accompanied by greatly elongated
eyelids, very large external orifices to the nostrils, and
a wide gape of mouth. The short-faced tumbler has a
beak in outline almost like that of a finch; and the
common tumbler has the singular inherited habit of
flying at a great height in a compact flock, and tumbling
in the air head over heels. The runt is a bird of great
size, with long massive beak and large feet; some of the
sub-breeds of runts have very long necks, others very
long wings and tails, others singularly short tails. The
barb is allied to the carrier, but, instead of a long beak
has a very short and broad one. The pouter has a
much elongated body, wings, and legs; and its
enormously developed crop, which it glories in inflating,
may well excite astonishment and even laughter. The
turbit has a short and conical beak, with a line of
reversed feathers down the breast; and it has the habit
of continually expanding, slightly, the upper part of the
œsophagus. The Jacobin has the feathers so much
reversed along the back of the neck that they form a
hood; and it has, proportionally to its size, elongated
wing and tail feathers. The trumpeter and laugher, as

their names express, utter a very different coo from the
other breeds. The fantail has thirty or even forty tail-
feathers, instead of twelve or fourteen—the normal
number in all the members of the great pigeon family:
these feathers are kept expanded, and are carried so
erect, that in good birds the head and tail touch: the
oil-gland is quite aborted. Several other less distinct
breeds might be specified.

In the skeletons of the several breeds, the development of the bones of the face in length and breadth and
curvature differs enormously. The shape, as well as the
breadth and length of the ramus of the lower jaw, varies
in a highly remarkable manner. The caudal and sacral
vertebræ vary in number; as does the number of the
ribs, together with their relative breadth and the presence of processes. The size and shape of the apertures
in the sternum are highly variable; so is the degree
of divergence and relative size of the two arms of
the furcula. The proportional width of the gape of
mouth, the proportional length of the eyelids, of the
orifice of the nostrils, of the tongue (not always in strict
correlation with the length of beak), the size of the crop
and of the upper part of the œsophagus; the development and abortion of the oil-gland; the number of the
primary wing and caudal feathers; the relative length
of the wing and tail to each other and to the body; the
relative length of the leg and foot; the number of
scutellæ on the toes, the development of skin between
the toes, are all points of structure which are variable.
The period at which the perfect plumage is acquired
varies, as does the state of the down with which the
nestling birds are clothed when hatched. The shape
and size of the eggs vary. The manner of flight, and in

some breeds the voice and disposition, differ remarkably. Lastly, in certain breeds, the males and females have come to differ in a slight degree from each other.

Altogether at least a score of pigeons might be chosen, which, if shown to an ornithologist, and he were told that they were wild birds, would certainly be ranked by him as well-defined species. Moreover, I do not believe that any ornithologist would in this case place the English carrier, the short-faced tumbler, the runt, the barb, pouter, and fantail in the same genus; more especially as in each of these breeds several truly-inherited sub-breeds, or species, as he would call them, could be shown him.

Great as are the differences between the breeds of the pigeon, I am fully convinced that the common opinion of naturalists is correct, namely, that all are descended from the rock-pigeon (Columba livia), including under this term several geographical races or sub-species, which differ from each other in the most trifling respects. As several of the reasons which have led me to this belief are in some degree applicable in other cases, I will here briefly give them. If the several breeds are not varieties, and have not proceeded from the rock-pigeon, they must have descended from at least seven or eight aboriginal stocks; for it is impossible to make the present domestic breeds by the crossing of any lesser number: how, for instance, could a pouter be produced by crossing two breeds unless one of the parent-stocks possessed the characteristic enormous crop? The supposed aboriginal stocks must all have been rock-pigeons, that is, they did not breed or willingly perch on trees. But besides C. livia, with its geographical sub-species, only two or three other

species of rock-pigeons are known; and these have
not any of the characters of the domestic breeds.
Hence the supposed aboriginal stocks must either
still exist in the countries where they were originally
domesticated, and yet be unknown to ornithologists;
and this, considering their size, habits, and remarkable
characters, seems improbable; or they must have
become extinct in the wild state. But birds breeding
on precipices, and good fliers, are unlikely to be exterminated; and the common rock-pigeon, which has
the same habits with the domestic breeds, has not
been exterminated even on several of the smaller
British islets, or on the shores of the Mediterranean.
Hence the supposed extermination of so many species
having similar habits with the rock-pigeon seems a
very rash assumption. Moreover, the several above-
named domesticated breeds have been transported to
all parts of the world, and, therefore, some of them
must have been carried back again into their native
country; but not one has become wild or feral, though
the dovecot-pigeon, which is the rock-pigeon in a very
slightly altered state, has become feral in several places.
Again, all recent experience shows that it is difficult
to get wild animals to breed freely under domestication; yet on the hypothesis of the multiple origin
of our pigeons, it must be assumed that at least seven
or eight species were so thoroughly domesticated in
ancient times by half-civilised man, as to be quite
prolific under confinement.

An argument of great weight, and applicable in
several other cases, is, that the above-specified breeds,
though agreeing generally with the wild rock-pigeon
in constitution, habits, voice, colouring, and in most

parts of their structure, yet are certainly highly abnormal in other parts; we may look in vain through the whole great family of Columbidæ for a beak like that of the English carrier, or that of the short-faced tumbler, or barb; for reversed feathers like those of the Jacobin; for a crop like that of the pouter; for tail-feathers like those of the fantail. Hence it must be assumed not only that half-civilised man succeeded in thoroughly domesticating several species, but that he intentionally or by chance picked out extraordinarily abnormal species; and further, that these very species have since all become extinct or unknown. So many strange contingencies are improbable in the highest degree.

Some facts in regard to the colouring of pigeons well deserve consideration. The rock-pigeon is of a slaty-blue, with white loins; but the Indian sub-species, C. intermedia of Strickland, has this part bluish. The tail has a terminal dark bar, with the outer feathers externally edged at the base with white. The wings have two black bars. Some semi-domestic breeds, and some truly wild breeds, have, besides the two black bars, the wings chequered with black. These several marks do not occur together in any other species of the whole family. Now, in every one of the domestic breeds, taking thoroughly well-bred birds, all the above marks, even to the white edging of the outer tail-feathers, sometimes concur perfectly developed. Moreover, when birds belonging to two or more distinct breeds are crossed, none of which are blue or have any of the above-specified marks, the mongrel offspring are very apt suddenly to acquire these characters. To give one instance out of several which I have observed:—I crossed some white fantails, which breed

very true, with some black barbs—and it so happens
that blue varieties of barbs are so rare that I never
heard of an instance in England; and the mongrels
were black, brown, and mottled. I also crossed a barb
with a spot, which is a white bird with a red tail and
red spot on the forehead, and which notoriously breeds
very true; the mongrels were dusky and mottled. I
then crossed one of the mongrel barb-fantails with a
mongrel barb-spot, and they produced a bird of as
beautiful a blue colour, with the white loins, double
black wing-bar, and barred and white-edged tail-
feathers, as any wild rock-pigeon! We can understand these facts, on the well-known principle of
reversion to ancestral characters, if all the domestic
breeds are descended from the rock-pigeon. But if we
deny this, we must make one of the two following
highly improbable suppositions. Either, first, that all
the several imagined aboriginal stocks were coloured
and marked like the rock-pigeon, although no other
existing species is thus coloured and marked, so that
in each separate breed there might be a tendency to
revert to the very same colours and markings. Or,
secondly, that each breed, even the purest, has within
a dozen, or at most within a score, of generations, been
crossed by the rock-pigeon: I say within a dozen or
twenty generations, for no instance is known of crossed
descendants reverting to an ancestor of foreign blood,
removed by a greater number of generations. In a
breed which has been crossed only once, the tendency
to revert to any character derived from such a cross
will naturally become less and less, as in each succeeding generation there will be less of the foreign
blood; but when there has been no cross, and there

is a tendency in the breed to revert to a character which was lost during some former generation, this tendency, for all that we can see to the contrary, may be transmitted undiminished for an indefinite number of generations. These two distinct cases of reversion are often confounded together by those who have written on inheritance.

Lastly, the hybrids or mongrels from between all the breeds of the pigeon are perfectly fertile, as I can state from my own observations, purposely made, on the most distinct breeds. Now, hardly any cases have been ascertained with certainty of hybrids from two quite distinct species of animals being perfectly fertile. Some authors believe that long-continued domestication eliminates this strong tendency to sterility in species. From the history of the dog, and of some other domestic animals, this conclusion is probably quite correct, if applied to species closely related to each other. But to extend it so far as to suppose that species, aboriginally as distinct as carriers, tumblers, pouters, and fantails now are, should yield offspring perfectly fertile *inter se*, would be rash in the extreme.

From these several reasons, namely,—the improbability of man having formerly made seven or eight supposed species of pigeons to breed freely under domestication;—these supposed species being quite unknown in a wild state, and their not having become anywhere feral;—these species presenting certain very abnormal characters, as compared with all other Columbidæ, though so like the rock-pigeon in most respects;—the occasional re-appearance of the blue colour and various black marks in all the breeds, both when kept pure and when crossed;—and lastly, the mongrel offspring being perfectly fertile ;—from these several reasons, taken together, we may safely conclude that all our domestic breeds are descended from the rock-pigeon or Columba livia with its geographical sub-species.

In favour of this view, I may add, firstly, that the wild C. livia has been found capable of domestication in Europe and in India ; and that it agrees in habits and in a great number of points of structure with all the domestic breeds.   Secondly, that, although an English carrier or a short-faced tumbler differs immensely in certain characters from the rock-pigeon, yet that, by comparing the several sub-breeds of these two races, more especially those brought from distant countries, we can make, between them and the rock-pigeon, an almost perfect series ; so we can in some other cases, but not with all the breeds.   Thirdly, those characters which are mainly distinctive of each breed are in each eminently variable, for instance the wattle and length of beak of the carrier, the shortness of that of the tumbler, and the number of tail-feathers in the fantail ; and the explanation of this fact will be obvious when we treat of Selection.   Fourthly, pigeons have been watched and tended with the utmost care, and loved by many people.   They have been domesticated for thousands of years in several quarters of the world ; the earliest known record of pigeons is in the fifth Ægyptian dynasty, about 3000 B.C., as was pointed out to me by Professor Lepsius ; but Mr. Birch informs me that pigeons are given in a bill of fare in the previous dynasty.   In the time of the Romans, as we hear from Pliny, immense prices were given for pigeons ; " nay, they are come to this pass, that they

can reckon up their pedigree and race." Pigeons were
much valued by Akber Khan in India, about the year
1600; never less than 20,000 pigeons were taken with
the court. "The monarchs of Iran and Turan sent him
some very rare birds;" and, continues the courtly
historian, "His Majesty by crossing the breeds, which
method was never practised before, has improved them
astonishingly." About this same period the Dutch were
as eager about pigeons as were the old Romans. The
paramount importance of these considerations in explaining the immense amount of variation which pigeons
have undergone, will likewise be obvious when we treat
of Selection. We shall then, also, see how it is that the
several breeds so often have a somewhat monstrous
character. It is also a most favourable circumstance
for the production of distinct breeds, that male and
female pigeons can be easily mated for life; and thus
different breeds can be kept together in the same
aviary.

I have discussed the probable origin of domestic
pigeons at some, yet quite insufficient, length; because
when I first kept pigeons and watched the several
kinds, well knowing how truly they breed, I felt fully
as much difficulty in believing that since they had
been domesticated they had all proceeded from a
common parent, as any naturalist could in coming to
a similar conclusion in regard to the many species of
finches, or other groups of birds, in nature. One
circumstance has struck me much; namely, that
nearly all the breeders of the various domestic
animals and the cultivators of plants, with whom I
have conversed, or whose treatises I have read, are
firmly convinced that the several breeds to which each

has attended, are descended from so many aboriginally distinct species. Ask, as I have asked, a celebrated raiser of Hereford cattle, whether his cattle might not have descended from Long-horns, or both from a common parent-stock, and he will laugh you to scorn. I have never met a pigeon, or poultry, or duck, or rabbit fancier, who was not fully convinced that each main breed was descended from a distinct species. Van Mons, in his treatise on pears and apples, shows how utterly he disbelieves that the several sorts, for instance a Ribston-pippin or Codlin-apple, could ever have proceeded from the seeds of the same tree. Innumerable other examples could be given. The explanation, I think, is simple: from long-continued study they are strongly impressed with the differences between the several races; and though they well know that each race varies slightly, for they win their prizes by selecting such slight differences, yet they ignore all general arguments, and refuse to sum up in their minds slight differences accumulated during many successive generations. May not those naturalists who, knowing far less of the laws of inheritance than does the breeder, and knowing no more than he does of the intermediate links in the long lines of descent, yet admit that many of our domestic races are descended from the same parents—may they not learn a lesson of caution, when they deride the idea of species in a state of nature being lineal descendants of other species?

Principles of Selection anciently followed, and their Effects.

Let us now briefly consider the steps by which domestic races have been produced, either from one or from several allied species. Some effect may be attributed to the direct and definite action of the external conditions of life, and some to habit; but he would be a bold man who would account by such agencies for the differences between a dray- and race-horse, a greyhound and bloodhound, a carrier and tumbler pigeon. One of the most remarkable features in our domesticated races is that we see in them adaptation, not indeed to the animal's or plant's own good, but to man's use or fancy. Some variations useful to him have probably arisen suddenly, or by one step; many botanists, for instance, believe that the fuller's teasel, with its hooks, which cannot be rivalled by any mechanical contrivance, is only a variety of the wild Dipsacus; and this amount of change may have suddenly arisen in a seedling. So it has probably been with the turnspit dog; and this is known to have been the case with the ancon sheep. But when we compare the dray-horse and race-horse, the dromedary and camel, the various breeds of sheep fitted either for cultivated land or mountain pasture, with the wool of one breed good for one purpose, and that of another breed for another purpose; when we compare the many breeds of dogs, each good for man in different ways; when we compare the game-cock, so pertinacious in battle, with other breeds so little quarrelsome, with "everlasting layers" which never desire to sit, and with the bantam so small and elegant; when we compare the host of agricultural, culinary, orchard, and flower-garden races of plants, most useful to man at different seasons and for different purposes, or so beautiful in his eyes, we must, I think, look further than to mere variability. We cannot suppose that all the breeds were suddenly produced as perfect and as useful as we now see them; indeed, in many cases, we know that this has not been their history. The key is man's power of accumulative selection: nature gives successive variations; man adds them up in certain directions useful to him. In this sense he may be said to have made for himself useful breeds.

The great power of this principle of selection is not hypothetical. It is certain that several of our eminent breeders have, even within a single lifetime, modified to a large extent their breeds of cattle and sheep. In order fully to realise what they have done, it is almost necessary to read several of the many treatises devoted to this subject, and to inspect the animals. Breeders habitually speak of an animal's organisation as something plastic, which they can model almost as they please. If I had space I could quote numerous passages to this effect from highly competent authorities. Youatt, who was probably better acquainted with the works of agriculturists than almost any other individual, and who was himself a very good judge of animals, speaks of the principle of selection as "that which enables the agriculturist, not only to modify the character of his flock, but to change it altogether. It is the magician's wand, by means of which he may summon into life whatever form and mould he pleases." Lord Somerville, speaking of what breeders have done for sheep, says:—"It would seem as if they had chalked out upon a wall a form perfect in itself, and then had given it existence." In Saxony the importance of the principle of selection in regard to merino sheep is so fully recognised, that men follow it as a trade: the sheep are placed on a table and are studied, like a picture by a connoisseur; this is done three times at intervals of months, and the sheep are each time marked and classed, so that the very best may ultimately be selected for breeding.

What English breeders have actually effected is proved by the enormous prices given for animals with a good pedigree; and these have been exported to almost every quarter of the world. The improvement is by no means generally due to crossing different breeds; all the best breeders are strongly opposed to this practice, except sometimes amongst closely allied sub-breeds. And when a cross has been made, the closest selection is far more indispensable even than in ordinary cases. If selection consisted merely in separating some very distinct variety, and breeding from it, the principle would be so obvious as hardly to be worth notice; but its importance consists in the great effect produced by the accumulation in one direction, during successive generations, of differences absolutely inappreciable by an uneducated eye—differences which I for one have vainly attempted to appreciate. Not one man in a thousand has accuracy of eye and judgment sufficient to become an eminent breeder. If gifted with these qualities, and he studies his subject for years, and devotes his lifetime to it with indomitable perseverance, he will succeed, and may make great improvements; if he wants any of these qualities, he will assuredly fail. Few would readily believe in the natural capacity and years of practice requisite to become even a skilful pigeon-fancier.

The same principles are followed by horticulturists; but the variations are here often more abrupt. No one supposes that our choicest productions have been produced by a single variation from the aboriginal stock. We have proofs that this has not been so in several cases in which exact records have been kept; thus, to give a very trifling instance, the steadily-increasing size of the common gooseberry may be quoted. We see an astonishing improvement in many florists' flowers, when the flowers of the present day are compared with drawings made only twenty or thirty years ago. When a race of plants is once pretty well established, the seed-raisers do not pick out the best plants, but merely go over their seed-beds, and pull up the "rogues," as they call the plants that deviate from the proper standard. With animals this kind of selection is, in fact, likewise followed; for hardly any one is so careless as to breed from his worst animals.

In regard to plants, there is another means of observing the accumulated effects of selection—namely, by comparing the diversity of flowers in the different varieties of the same species in the flower-garden; the diversity of leaves, pods, or tubers, or whatever part is valued, in the kitchen-garden, in comparison with the flowers of the same varieties; and the diversity of fruit of the same species in the orchard, in comparison with the leaves and flowers of the same set of varieties. See how different the leaves of the cabbage are, and how extremely alike the flowers; how unlike the flowers of the heartsease are, and how alike the leaves; how much the fruit of the different kinds of goose-

berries differ in size, colour, shape, and hairiness, and
yet the flowers present very slight differences.   It is
not that the varieties which differ largely in some one
point do not differ at all in other points ; this is hardly
ever,—I speak after careful observation,—perhaps
never, the case.   The law of correlated variation, the
importance of which should never be overlooked, will
ensure some differences ; but, as a general rule, it
cannot be doubted that the continued selection of
slight variations, either in the leaves, the flowers, or
the fruit, will produce races differing from each other
chiefly in these characters.

It may be objected that the principle of selection has
been reduced to methodical practice for scarcely more
than three-quarters of a century ; it has certainly been
more attended to of late years, and many treatises have
been published on the subject ; and the result has been,
in a corresponding degree, rapid and important.   But
it is very far from true that the principle is a modern
discovery.   I could give several references to works of
high antiquity, in which the full importance of the
principle is acknowledged.   In rude and barbarous
periods of English history choice animals were often
imported, and laws were passed to prevent their ex-
portation : the destruction of horses under a certain
size was ordered, and this may be compared to the
" roguing " of plants by nurserymen.   The principle of
selection I find distinctly given in an ancient Chinese
encyclopædia.   Explicit rules are laid down by some
of the Roman classical writers.   From passages in
Genesis, it is clear that the colour of domestic animals
was at that early period attended to.   Savages now
sometimes cross their dogs with wild canine animals,

to improve the breed, and they formerly did so, as is
attested by passages in Pliny. The savages in South
Africa match their draught cattle by colour, as do some
of the Esquimaux their teams of dogs. Livingstone
states that good domestic breeds are highly valued by
the negroes in the interior of Africa who have not
associated with Europeans. Some of these facts do not
show actual selection, but they show that the breeding
of domestic animals was carefully attended to in
ancient times, and is now attended to by the lowest
savages. It would, indeed, have been a strange fact,
had attention not been paid to breeding, for the inheritance of good and bad qualities is so obvious.

Unconscious Selection.

At the present time, eminent breeders try by
methodical selection, with a distinct object in view, to
make a new strain or sub-breed, superior to anything of
the kind in the country. But, for our purpose, a form
of Selection, which may be called Unconscious, and
which results from every one trying to possess and
breed from the best individual animals, is more
important. Thus, a man who intends keeping pointers
naturally tries to get as good dogs as he can, and
afterwards breeds from his own best dogs, but he has no
wish or expectation of permanently altering the breed.
Nevertheless we may infer that this process, continued
during centuries, would improve and modify any breed,
in the same way as Bakewell, Collins, &c., by this very
same process, only carried on more methodically, did
greatly modify, even during their lifetimes, the forms
and qualities of their cattle. Slow and insensible

changes of this kind can never be recognised unless
actual measurements or careful drawings of the breeds
in question have been made long ago, which may serve
for comparison. In some cases, however, unchanged, or
but little changed individuals of the same breed exist in
less civilised districts, where the breed has been less
improved. There is reason to believe that King
Charles's spaniel has been unconsciously modified to a
large extent since the time of that monarch. Some
highly competent authorities are convinced that the
setter is directly derived from the spaniel, and has probably been slowly altered from it. It is known that the
English pointer has been greatly changed within the last
century, and in this case the change has, it is believed,
been chiefly effected by crosses with the foxhound : but
what concerns us is, that the change has been effected
unconsciously and gradually, and yet so effectually,
that, though the old Spanish pointer certainly came
from Spain, Mr. Borrow has not seen, as I am informed
by him, any native dog in Spain like our pointer.

By a similar process of selection, and by careful
training, English racehorses have come to surpass in
fleetness and size the parent Arabs, so that the latter, by
the regulations for the Goodwood Races, are favoured in
the weights which they carry. Lord Spencer and others
have shown how the cattle of England have increased in
weight and in early maturity, compared with the stock
formerly kept in this country. By comparing the
accounts given in various old treatises of the former and
present state of carrier and tumbler pigeons in Britain,
India, and Persia, we can trace the stages through
which they have insensibly passed, and come to differ
so greatly from the rock-pigeon.

Youatt gives an excellent illustration of the effects of a course of selection, which may be considered as unconscious, in so far that the breeders could never have expected, or even wished, to produce the result which ensued—namely, the production of two distinct strains. The two flocks of Leicester sheep kept by Mr. Buckley and Mr. Burgess, as Mr. Youatt remarks, "have been purely bred from the original stock of Mr. Bakewell for upwards of fifty years. There is not a suspicion existing in the mind of any one at all acquainted with the subject, that the owner of either of them has deviated in any one instance from the pure blood of Mr. Bakewell's flock, and yet the difference between the sheep possessed by these two gentlemen is so great that they have the appearance of being quite different varieties."

If there exist savages so barbarous as never to think of the inherited character of the offspring of their domestic animals, yet any one animal particularly useful to them, for any special purpose, would be carefully preserved during famines and other accidents, to which savages are so liable, and such choice animals would thus generally leave more offspring than the inferior ones; so that in this case there would be a kind of unconscious selection going on. We see the value set on animals even by the barbarians of Tierra del Fuego, by their killing and devouring their old women, in times of dearth, as of less value than their dogs.

In plants the same gradual process of improvement, through the occasional preservation of the best individuals, whether or not sufficiently distinct to be ranked at their first appearance as distinct varieties, and whether or not two or more species or races have

become blended together by crossing, may plainly be
recognised in the increased size and beauty which
we now see in the varieties of the heartsease, rose,
pelargonium, dahlia, and other plants, when compared
with the older varieties or with their parent-stocks.
No one would ever expect to get a first-rate heartsease
or dahlia from the seed of a wild plant. No one would
expect to raise a first-rate melting pear from the seed of
the wild pear, though he might succeed from a poor
seedling growing wild, if it had come from a garden-
stock. The pear though cultivated in classical times,
appears, from Pliny's description, to have been a fruit
of very inferior quality. I have seen great surprise
expressed in horticultural works at the wonderful skill
of gardeners, in having produced such splendid results
from such poor materials; but the art has been simple,
and, as far as the final result is concerned, has been
followed almost unconsciously. It has consisted in
always cultivating the best known variety, sowing its
seeds, and, when a slightly better variety chanced to
appear, selecting it, and so onwards. But the gardeners
of the classical period, who cultivated the best pears
which they could procure, never thought what splendid
fruit we should eat; though we owe our excellent fruit
in some small degree, to their having naturally chosen
and preserved the best varieties they could anywhere
find.

A large amount of change, thus slowly and unconsciously accumulated, explains, as I believe, the
well-known fact, that in a number of cases we cannot
recognise, and therefore do not know, the wild parent-
stocks of the plants which have been longest cultivated
in our flower and kitchen gardens. If it has taken

centuries or thousands of years to improve or modify most of our plants up to their present standard of usefulness to man, we can understand how it is that neither Australia, the Cape of Good Hope, nor any other region inhabited by quite uncivilised man, has afforded us a single plant worth culture. It is not that these countries, so rich in species, do not by a strange chance possess the aboriginal stocks of any useful plants, but that the native plants have not been improved by continued selection up to a standard of perfection comparable with that acquired by the plants in countries anciently civilised.

In regard to the domestic animals kept by uncivilised man, it should not be overlooked that they almost always have to struggle for their own food, at least during certain seasons. And in two countries very differently circumstanced, individuals of the same species, having slightly different constitutions or structure, would often succeed better in the one country than in the other; and thus by a process of "natural selection," as will hereafter be more fully explained, two sub-breeds might be formed. This, perhaps, partly explains why the varieties kept by savages, as has been remarked by some authors, have more of the character of true species than the varieties kept in civilised countries.

On the view here given of the important part which selection by man has played, it becomes at once obvious, how it is that our domestic races show adaptation in their structure or in their habits to man's wants or fancies. We can, I think, further understand the frequently abnormal character of our domestic races, and likewise their differences being so great in external characters, and relatively so slight in internal parts or organs. Man can hardly select, or only with much difficulty, any deviation of structure excepting such as is externally visible; and indeed he rarely cares for what is internal. He can never act by selection, excepting on variations which are first given to him in some slight degree by nature. No man would ever try to make a fantail till he saw a pigeon with a tail developed in some slight degree in an unusual manner, or a pouter till he saw a pigeon with a crop of somewhat unusual size; and the more abnormal or unusual any character was when it first appeared, the more likely it would be to catch his attention. But to use such an expression as trying to make a fantail, is, I have no doubt, in most cases, utterly incorrect. The man who first selected a pigeon with a slightly larger tail, never dreamed what the descendants of that pigeon would become through long-continued, partly un-conscious and partly methodical, selection. Perhaps the parent-bird of all fantails had only fourteen tail-feathers somewhat expanded, like the present Java fantail, or like individuals of other and distinct breeds, in which as many as seventeen tail-feathers have been counted. Perhaps the first pouter-pigeon did not inflate its crop much more than the turbit now does the upper part of its œsophagus,—a habit which is disregarded by all fanciers, as it is not one of the points of the breed.

Nor let it be thought that some great deviation of structure would be necessary to catch the fancier's eye: he perceives extremely small differences, and it is in human nature to value any novelty, however slight, in one's own possession. Nor must the value which would formerly have been set on any slight differences in the individuals of the same species, be judged of by the value which is now set on them, after several breeds have fairly been established. It is known that with pigeons many slight variations now occasionally appear, but these are rejected as faults or deviations from the standard of perfection in each breed. The common goose has not given rise to any marked varieties; hence the Toulouse and the common breed, which differ only in colour, that most fleeting of characters, have lately been exhibited as distinct at our poultry-shows.

These views appear to explain what has sometimes been noticed—namely, that we know hardly anything about the origin or history of any of our domestic breeds. But, in fact, a breed, like a dialect of a language, can hardly be said to have a distinct origin. A man preserves and breeds from an individual with some slight deviation of structure, or takes more care than usual in matching his best animals, and thus improves them, and the improved animals slowly spread in the immediate neighbourhood. But they will as yet hardly have a distinct name, and from being only slightly valued, their history will have been disregarded. When further improved by the same slow and gradual process, they will spread more widely, and will be recognised as something distinct and valuable, and will then probably first receive a provincial name. In semi-civilised countries, with little free communication, the spreading of a new sub-breed would be a slow process. As soon as the points of value are once acknowledged, the principle, as I have called it, of unconscious selection will always tend,—perhaps more at one period than at another, as the breed rises or falls in fashion,—perhaps more in one district than in another, according to the state of civilisation of the inhabitants,—slowly to add to the characteristic features of the breed, whatever they may be. But the chance will be infinitely small of any record having been preserved of such slow, varying, and insensible changes.

Circumstances favourable to Man's Power of Selection.

I will now say a few words on the circumstances, favourable, or the reverse, to man's power of selection. A high degree of variability is obviously favourable, as freely giving the materials for selection to work on; not that mere individual differences are not amply sufficient, with extreme care, to allow of the accumulation of a large amount of modification in almost any desired direction. But as variations manifestly useful or pleasing to man appear only occasionally, the chance of their appearance will be much increased by a large number of individuals being kept. Hence, number is of the highest importance for success. On this principle Marshall formerly remarked, with respect to the sheep of parts of Yorkshire, " as they generally belong to poor people, and are mostly *in small lots*, they never can be improved." On the other hand, nurserymen, from keeping large stocks of the same plant, are generally far more successful than amateurs in raising new and valuable varieties. A large number of individuals of an animal or plant can be reared only where the conditions for its propagation are favourable. When the individuals are scanty, all will be allowed to breed, whatever their quality may be, and this will effectually prevent selection. But probably the most important element is that the animal or plant should be so highly valued by man, that the closest attention is paid to even the slightest deviations in its qualities or structure. Unless such attention be paid nothing can be effected. I have seen it gravely remarked, that it was most fortunate that the strawberry began to vary just when gardeners began to attend to this plant. No doubt the strawberry had always varied since it was cultivated, but the slight varieties had been neglected. As soon, however, as gardeners picked out individual plants with slightly larger, earlier, or better fruit, and raised seedlings from them, and again picked out the best seedlings and bred from them, then (with some aid by crossing distinct species) those many admirable varieties of the strawberry were raised which have appeared during the last half-century.

With animals, facility in preventing crosses is an important element in the formation of new races,—at least, in a country which is already stocked with other races. In this respect enclosure of the land plays a part. Wandering savages or the inhabitants of open plains rarely possess more than one breed of the same species. Pigeons can be mated for life, and this is a great convenience to the fancier, for thus many races may be improved and kept true, though mingled in the same aviary; and this circumstance must have largely favoured the formation of new breeds. Pigeons, I may add, can be propagated in great numbers and at a very quick rate, and inferior birds may be freely rejected, as when killed they serve for food. On the other hand, cats, from their nocturnal rambling habits, cannot be easily matched, and, although so much valued by women and children, we rarely see a distinct breed long kept up; such breeds as we do sometimes see are almost always imported from some other country. Although I do not doubt that some domestic animals vary less than others, yet the rarity or absence of distinct breeds of the cat, the donkey, peacock, goose, &c., may be attributed in main part to selection not having been brought into play : in cats, from the difficulty in pairing them ; in donkeys, from only a few being kept by poor people, and little attention paid to their breeding ; for recently in certain parts of Spain and of the United States this animal has been surprisingly modified and improved by careful selection : in peacocks, from not being very easily reared and a large stock not kept : in geese, from being valuable only for two purposes, food and feathers, and more especially from no pleasure having been felt in the display of distinct breeds ; but the goose, under the conditions to which it is exposed when domesticated, seems to have a singularly inflexible organisation, though it has varied to a slight extent, as I have elsewhere described.

Some authors have maintained that the amount of variation in our domestic productions is soon reached, and can never afterwards be exceeded. It would be somewhat rash to assert that the limit has been attained in any one case; for almost all our animals and plants have been greatly improved in many ways within a recent period; and this implies variation. It would be equally rash to assert that characters now increased to their utmost limit, could not, after remaining fixed for many centuries, again vary under new conditions of life. No doubt, as Mr. Wallace has remarked with much truth, a limit will be at last reached. For instance, there must be a limit to the fleetness of any terrestrial animal, as this will be determined by the friction to be overcome, the weight of body to be carried, and the power of contraction in the muscular fibres. But what concerns us is that the domestic varieties of the same species differ from each other in almost every character, which man has attended to and selected, more than do the distinct species of the same genera. Isidore Geoffroy St. Hilaire has proved this in regard to size, and so it is with colour and probably with the length of hair. With respect to fleetness, which depends on many bodily characters, Eclipse was far fleeter, and a dray-horse is incomparably stronger than any two natural species belonging to the same genus. So with plants, the seeds of the different varieties of the bean or maize probably differ more in size, than do the seeds of the distinct species in any one genus in the same two families. The same remark holds good in regard to the fruit of the several varieties of the plum, and still more strongly with the melon, as well as in many other analogous cases.

To sum up on the origin of our domestic races of animals and plants. Changed conditions of life are of the highest importance in causing variability, both by acting directly on the organisation, and indirectly by affecting the reproductive system. It is not probable that variability is an inherent and necessary contingent, under all circumstances. The greater or less force of inheritance and reversion determine whether variations shall endure. Variability is governed by many unknown laws, of which correlated growth is probably the most important. Something, but how much we do not know, may be attributed to the definite action of the conditions of life. Some, perhaps a great, effect may

E

be attributed to the increased use or disuse of parts. The final result is thus rendered infinitely complex. In some cases the intercrossing of aboriginally distinct species appears to have played an important part in the origin of our breeds. When several breeds have once been formed in any country, their occasional intercrossing, with the aid of selection, has, no doubt, largely aided in the formation of new sub-breeds; but the importance of crossing has been much exaggerated, both in regard to animals and to those plants which are propagated by seed. With plants which are temporarily propagated by cuttings, buds, &c., the importance of crossing is immense; for the cultivator may here disregard the extreme variability both of hybrids and of mongrels, and the sterility of hybrids; but plants not propagated by seed are of little importance to us, for their endurance is only temporary. Over all these causes of Change, the accumulative action of Selection, whether applied methodically and quickly, or unconsciously and slowly but more efficiently seems to have been the predominant Power.

CHAPTER II.

VARIATION UNDER NATURE.

Variability—Individual differences—Doubtful species—Wide ranging, much diffused, and common species, vary most—Species of
the larger genera in each country vary more frequently than
the species of the smaller genera—Many of the species of the
larger genera resemble varieties in being very closely, but unequally, related to each other, and in having restricted ranges.

BEFORE applying the principles arrived at in the last
chapter to organic beings in a state of nature, we must
briefly discuss whether these latter are subject to any
variation.   To treat this subject properly, a long
catalogue of dry facts ought to be given; but these
I shall reserve for a future work.   Nor shall I here
discuss the various definitions which have been given
of the term species.   No one definition has satisfied all
naturalists; yet every naturalist knows vaguely what
he means when he speaks of a species.   Generally the
term includes the unknown element of a distinct act of
creation.   The term "variety" is almost equally
difficult to define; but here community of descent
is almost universally implied, though it can rarely be
proved.   We have also what are called monstrosities;
but they graduate into varieties   By a monstrosity I
presume is meant some considerable deviation of
structure, generally injurious, or not useful to the
species.   Some authors use the term "variation" in a

technical sense, as implying a modification directly due
to the physical conditions of life; and "variations" in
this sense are supposed not to be inherited; but who
can say that the dwarfed condition of shells in the
brackish waters of the Baltic, or dwarfed plants on
Alpine summits, or the thicker fur of an animal from
far northwards, would not in some cases be inherited
for at least a few generations? and in this case I
presume that the form would be called a variety.

It may be doubted whether sudden and considerable
deviations of structure such as we occasionally see in
our domestic productions, more especially with plants,
are ever permanently propagated in a state of nature.
Almost every part of every organic being is so beautifully related to its complex conditions of life that it
seems as improbable that any part should have been
suddenly produced perfect, as that a complex machine
should have been invented by man in a perfect state.
Under domestication monstrosities sometimes occur
which resemble normal structures in widely different
animals.   Thus pigs have occasionally been born
with a sort of proboscis, and if any wild species of the
same genus had naturally possessed a proboscis, it
might have been argued that this had appeared as a
monstrosity; but I have as yet failed to find, after
diligent search, cases of monstrosities resembling
normal structures in nearly allied forms, and these
alone bear on the question. If monstrous forms of
this kind ever do appear in a state of nature and are
capable of reproduction (which is not always the case),
as they occur rarely and singly, their preservation
would depend on unusually favourable circumstances.
They would, also, during the first and succeeding

generations cross with the ordinary form, and thus their abnormal character would almost inevitably be lost. But I shall have to return in a future chapter to the preservation and perpetuation of single or occasional variations.

Individual Differences.

The many slight differences which appear in the offspring from the same parents, or which it may be presumed have thus arisen, from being observed in the individuals of the same species inhabiting the same confined locality, may be called individual differences. No one supposes that all the individuals of the same species are cast in the same actual mould. These individual differences are of the highest importance for us, for they are often inherited, as must be familiar to every one; and they thus afford materials for natural selection to act on and accumulate, in the same manner as man accumulates in any given direction individual differences in his domesticated productions. These individual differences generally affect what naturalists consider unimportant parts; but I could show by a long catalogue of facts, that parts which must be called important, whether viewed under a physiological or classificatory point of view, sometimes vary in the individuals of the same species. I am convinced that the most experienced naturalist would be surprised at the number of the cases of variability, even in important parts of structure, which he could collect on good authority, as I have collected, during a course of years. It should be remembered that systematists are far from being pleased at finding variability in important characters, and that there are not many

men who will laboriously examine internal and
important organs, and compare them in many specimens
of the same species. It would never have been expected that the branching of the main nerves close to
the great central ganglion of an insect would have been
variable in the same species; it might have been
thought that changes of this nature could have been
effected only by slow degrees; yet Sir J. Lubbock has
shown a degree of variability in these main nerves in
Coccus, which may almost be compared to the irregular
branching of the stem of a tree. This philosophical
naturalist, I may add, has also shown that the muscles
in the larvæ of certain insects are far from uniform.
Authors sometimes argue in a circle when they state
that important organs never vary; for these same
authors practically rank those parts as important (as
some few naturalists have honestly confessed) which do
not vary; and, under this point of view, no instance
will ever be found of an important part varying; but
under any other point of view many instances assuredly
can be given.

There is one point connected with individual
differences, which is extremely perplexing: I refer to
those genera which have been called "protean" or
"polymorphic," in which the species present an
inordinate amount of variation. With respect to many
of these forms, hardly two naturalists agree whether to
rank them as species or as varieties. We may instance
Rubus, Rosa, and Hieracium amongst plants, several
genera of insects and of Brachiopod shells. In most
polymorphic genera some of the species have fixed and
definite characters. Genera which are polymorphic
in one country seem to be, with a few exceptions,

polymorphic in other countries, and likewise, judging from Brachiopod shells, at former periods of time. These facts are very perplexing, for they seem to show that this kind of variability is independent of the conditions of life. I am inclined to suspect that we see, at least in some of these polymorphic genera, variations which are of no service or disservice to the species, and which consequently have not been seized on and rendered definite by natural selection, as hereafter to be explained.

Individuals of the same species often present, as is known to every one, great differences of structure, independently of variation, as in the two sexes of various animals, in the two or three castes of sterile females or workers amongst insects, and in the immature and larval states of many of the lower animals. There are, also, cases of dimorphism and trimorphism, both with animals and plants. Thus, Mr. Wallace, who has lately called attention to the subject, has shown that the females of certain species of butterflies, in the Malayan archipelago, regularly appear under two or even three conspicuously distinct forms, not connected by intermediate varieties. Fritz Müller has described analogous but more extraordinary cases with the males of certain Brazilian Crustaceans: thus, the male of a Tanais regularly occurs under two distinct forms; one of these has strong and differently shaped pincers, and the other has antennæ much more abundantly furnished with smelling-hairs. Although in most of these cases, the two or three forms, both with animals and plants, are not now connected by intermediate gradations, it is probable that they were once thus connected. Mr. Wallace, for instance,

describes a certain butterfly which presents in the
same island a great range of varieties connected by
intermediate links, and the extreme links of the chain
closely resemble the two forms of an allied dimorphic
species inhabiting another part of the Malay archipelago.
Thus also with ants, the several worker-castes are
generally quite distinct; but in some cases, as we
shall hereafter see, the castes are connected together by
finely graduated varieties. So it is, as I have myself
observed, with some dimorphic plants. It certainly at
first appears a highly remarkable fact that the same
female butterfly should have the power of producing at
the same time three distinct female forms and a male;
and that an hermaphrodite plant should produce from
the same seed-capsule three distinct hermaphrodite
forms, bearing three different kinds of females and
three or even six different kinds of males. Nevertheless these cases are only exaggerations of the
common fact that the female produces offspring of
two sexes which sometimes differ from each other in a
wonderful manner.

Doubtful Species.

The forms which possess in some considerable degree
the character of species, but which are so closely similar
to other forms, or are so closely linked to them by
intermediate gradations, that naturalists do not like to
rank them as distinct species, are in several respects the
most important for us. We have every reason to
believe that many of these doubtful and closely allied
forms have permanently retained their characters for a
long time; for as long, as far as we know, as have good
and true species. Practically, when a naturalist can

unite by means of intermediate links any two forms, he treats the one as a variety of the other; ranking the most common, but sometimes the one first described, as the species, and the other as the variety. But cases of great difficulty, which I will not here enumerate, sometimes arise in deciding whether or not to rank one form as a variety of another, even when they are closely connected by intermediate links; nor will the commonly-assumed hybrid nature of the intermediate forms always remove the difficulty. In very many cases, however, one form is ranked as a variety of another, not because the intermediate links have actually been found, but because analogy leads the observer to suppose either that they do now somewhere exist, or may formerly have existed; and here a wide door for the entry of doubt and conjecture is opened.

Hence, in determining whether a form should be ranked as a species or a variety, the opinion of naturalists having sound judgment and wide experience seems the only guide to follow. We must, however, in many cases, decide by a majority of naturalists, for few well-marked and well-known varieties can be named which have not been ranked as species by at least some competent judges.

That varieties of this doubtful nature are far from uncommon cannot be disputed. Compare the several floras of Great Britain, of France, or of the United States, drawn up by different botanists, and see what a surprising number of forms have been ranked by one botanist as good species, and by another as mere varieties. Mr. H. C. Watson, to whom I lie under deep obligation for assistance of all kinds, has marked for me 182 British plants, which are generally con-

sidered as varieties, but which have all been ranked by
botanists as species; and in making this list he has
omitted many trifling varieties. but which nevertheless
have been ranked by some botanists as species, and he
has entirely omitted several highly polymorphic genera.
Under genera, including the most polymorphic forms,
Mr. Babington gives 251 species, whereas Mr. Bentham
gives only 112,—a difference of 139 doubtful forms!
Amongst animals which unite for each birth, and which
are highly locomotive, doubtful forms, ranked by one
zoologist as a species and by another as a variety, can
rarely be found within the same country, but are
common in separated areas.   How many of the birds
and insects in North America and Europe, which differ
very slightly from each other, have been ranked by
one eminent naturalist as undoubted species, and by
another as varieties, or, as they are often called,
geographical races!   Mr. Wallace, in several valuable
papers on the various animals, especially on the
Lepidoptera, inhabiting the islands of the great
Malayan archipelago, shows that they may be classed
under four heads, namely, as variable forms, as local
forms, as geographical races or sub-species, and as true
representative species.   The first or variable forms vary
much within the limits of the same island.   The local
forms are moderately constant and distinct in each
separate island; but when all from the several islands
are compared together, the differences are seen to be so
slight and graduated, that it is impossible to define or
describe them, though at the same time the extreme
forms are sufficiently distinct.   The geographical races
or sub-species are local forms completely fixed and
isolated; but as they do not differ from each other by

strongly marked and important characters, "there is no possible test but individual opinion to determine which of them shall be considered as species and which as varieties." Lastly, representative species fill the same place in the natural economy of each island as do the local forms and sub-species; but as they are distinguished from each other by a greater amount of difference than that between the local forms and sub-species, they are almost universally ranked by naturalists as true species. Nevertheless, no certain criterion can possibly be given by which variable forms, local forms, sub-species, and representative species can be recognised.

Many years ago, when comparing, and seeing others compare, the birds from the closely neighbouring islands of the Galapagos archipelago, one with another, and with those from the American mainland, I was much struck how entirely vague and arbitrary is the distinction between species and varieties. On the islets of the little Madeira group there are many insects which are characterised as varieties in Mr. Wollaston's admirable work, but which would certainly be ranked as distinct species by many entomologists. Even Ireland has a few animals, now generally regarded as varieties, but which have been ranked as species by same zoologists. Several experienced ornithologists consider our British red grouse as only a strongly-marked race of a Norwegian species, whereas the greater number rank it as an undoubted species peculiar to Great Britain. A wide distance between the homes of two doubtful forms leads many naturalists to rank them as distinct species; but what distance, it has been well asked, will suffice; if that between

America and Europe is ample, will that between Europe and the Azores, or Madeira, or the Canaries, or between the several islets of these small archipelagos, be sufficient?

Mr. B. D. Walsh, a distinguished entomologist of the United States, has described what he calls Phytophagic varieties and Phytophagic species. Most vegetable-feeding insects live on one kind of plant or on one group of plants; some feed indiscriminately on many kinds, but do not in consequence vary. In several cases, however, insects found living on different plants, have been observed by Mr. Walsh to present in their larval or mature state, or in both states, slight, though constant differences in colour, size, or in the nature of their secretions. In some instances the males alone, in other instances both males and females, have been observed thus to differ in a slight degree. When the differences are rather more strongly marked, and when both sexes and all ages are affected, the forms are ranked by all entomologists as good species. But no observer can determine for another, even if he can do so for himself, which of these Phytophagic forms ought to be called species and which varieties. Mr. Walsh ranks the forms which it may be supposed would freely intercross, as varieties; and those which appear to have lost this power, as species. As the differences depend on the insects having long fed on distinct plants, it cannot be expected that intermediate links connecting the several forms should now be found. The naturalist thus loses his best guide in determining whether to rank doubtful forms as varieties or species. This likewise necessarily occurs with closely allied organisms, which inhabit distinct continents or islands. When, on the other

hand, an animal or plant ranges over the same
continent, or inhabits many islands in the same
archipelago, and presents different forms in the different
areas, there is always a good chance that intermediate
forms will be discovered which will link together the
extreme states; and these are then degraded to the
rank of varieties.

Some few naturalists maintain that animals never
present varieties; but then these same naturalists rank
the slightest difference as of specific value; and when
the same identical form is met with in two distant
countries, or in two geological formations, they believe
that two distinct species are hidden under the same
dress. The term species thus comes to be a mere
useless abstraction, implying and assuming a separate
act of creation. It is certain that many forms,
considered by highly-competent judges to be varieties,
resemble species so completely in character, that they
have been thus ranked by other highly-competent
judges. But to discuss whether they ought to be called
species or varieties, before any definition of these terms
has been generally accepted, is vainly to beat the air.

Many of the cases of strongly-marked varieties or
doubtful species well deserve consideration; for several
interesting lines of argument, from geographical distribution, analogical variation, hybridism, &c., have
been brought to bear in the attempt to determine their
rank; but space does not here permit me to discuss
them. Close investigation, in many cases, will no
doubt bring naturalists to agree how to rank doubtful
forms. Yet it must be confessed that it is in the best
known countries that we find the greatest number of
them. I have been struck with the fact, that if any

animal or plant in a state of nature be highly useful to
man, or from any cause closely attracts his attention,
varieties of it will almost universally be found recorded.
These varieties, moreover, will often be ranked by some
authors as species. Look at the common oak, how
closely it has been studied; yet a German author
makes more than a dozen species out of forms, which
are almost universally considered by other botanists to
be varieties; and in this country the highest botanical
authorities and practical men can be quoted to show
that the sessile and pedunculated oaks are either good
and distinct species or mere varieties.

I may here allude to a remarkable memoir lately
published by A. de Candolle, on the oaks of the whole
world. No one ever had more ample materials for the
discrimination of the species, or could have worked on
them with more zeal and sagacity. He first gives in
detail all the many points of structure which vary in
the several species, and estimates numerically the
relative frequency of the variations. He specifies
above a dozen characters which may be found varying
even on the same branch, sometimes according to age
or development, sometimes without any assignable
reason. Such characters are not of course of specific
value, but they are, as Asa Gray has remarked in
commenting on this memoir, such as generally enter
into specific definitions. De Candolle then goes on to
say that he gives the rank of species to the forms that
differ by characters never varying on the same tree, and
never found connected by intermediate states. After
this discussion, the result of so much labour, he
emphatically remarks: "They are mistaken, who
repeat that the greater part of our species are clearly

limited, and that the doubtful species are in a feeble minority. This seemed to be true, so long as a genus was imperfectly known, and its species were founded upon a few specimens, that is to say, were provisional. Just as we come to know them better, intermediate forms flow in, and doubts as to specific limits augment." He also adds that it is the best known species which present the greatest number of spontaneous varieties and sub-varieties. Thus Quercus robur has twenty-eight varieties, all of which, excepting six, are clustered round three sub-species, namely, Q. pedunculata, sessiliflora, and pubescens. The forms which connect these three sub-species are comparatively rare; and, as Asa Gray again remarks, if these connecting forms which are now rare, were to become wholly extinct, the three sub-species would hold exactly the same relation to each other, as do the four or five provisionally admitted species which closely surround the typical Quercus robur. Finally, De Candolle admits that out of the 300 species, which will be enumerated in his Prodromus as belonging to the oak family, at least two-thirds are provisional species, that is, are not known strictly to fulfil the definition above given of a true species. It should be added that De Candolle no longer believes that species are immutable creations, but concludes that the derivative theory is the most natural one, " and the most accordant with the known facts in palæontology, geographical botany and zoology, of anatomical structure and classification."

When a young naturalist commences the study of a group of organisms quite unknown to him, he is at first much perplexed in determining what differences to consider as specific, and what as varietal; for he knows nothing of the amount and kind of variation to which the group is subject; and this shows, at least, how very generally there is some variation. But if he confine his attention to one class within one country, he will soon make up his mind how to rank most of the doubtful forms. His general tendency will be to make many species, for he will become impressed, just like the pigeon or poultry fancier before alluded to, with the amount of difference in the forms which he is continually studying; and he has little general knowledge of analogical variation in other groups and in other countries, by which to correct his first impressions. As he extends the range of his observations, he will meet with more cases of difficulty; for he will encounter a greater number of closely-allied forms. But if his observations be widely extended, he will in the end generally be able to make up his own mind: but he will succeed in this at the expense of admitting much variation,—and the truth of this admission will often be disputed by other naturalists. When he comes to study allied forms brought from countries not now continuous, in which case he cannot hope to find intermediate links, he will be compelled to trust almost entirely to analogy, and his difficulties will rise to a climax.

Certainly no clear line of demarcation has as yet been drawn between species and sub-species—that is, the forms which in the opinion of some naturalists come very near to, but do not quite arrive at, the rank of species: or, again, between sub-species and well-marked varieties, or between lesser varieties and individual differences. These differences blend into each other by an insensible series; and a series impresses the mind with the idea of an actual passage.

Hence I look at individual differences, though of small interest to the systematist, as of the highest importance for us, as being the first steps towards such slight varieties as are barely thought worth recording in works on natural history. And I look at varieties which are in any degree more distinct and permanent, as steps towards more strongly-marked and permanent varieties; and at the latter, as leading to sub-species, and then to species. The passage from one stage of difference to another may, in many cases, be the simple result of the nature of the organism and of the different physical conditions to which it has long been exposed; but with respect to the more important and adaptive characters, the passage from one stage of difference to another, may be safely attributed to the cumulative action of natural selection, hereafter to be explained, and to the effects of the increased use or disuse of parts. A well-marked variety may therefore be called an incipient species; but whether this belief is justifiable must be judged by the weight of the various facts and considerations to be given throughout this work.

It need not be supposed that all varieties or incipient species attain the rank of species. They may become extinct, or they may endure as varieties for very long periods, as has been shown to be the case by Mr. Wollaston with the varieties of certain fossil land-shells in Madeira, and with plants by Gaston de Saporta. If a variety were to flourish so as to exceed in numbers the parent species, it would then rank as the species, and the species as the variety; or it might come to supplant and exterminate the parent species; or both

F

might co-exist, and both rank as independent species. But we shall hereafter return to this subject.

From these remarks it will be seen that I look at the term species as one arbitrarily given, for the sake of convenience, to a set of individuals closely resembling each other, and that it does not essentially differ from the term variety, which is given to less distinct and more fluctuating forms. The term variety, again, in comparison with mere individual differences, is also applied arbitrarily, for convenience' sake.

Wide-ranging, much diffused, and common Species vary most.

Guided by theoretical considerations, I thought that some interesting results might be obtained in regard to the nature and relations of the species which vary most, by tabulating all the varieties in several well-worked floras. At first this seemed a simple task; but Mr. H. C. Watson, to whom I am much indebted for valuable advice and assistance on this subject, soon convinced me that there were many difficulties, as did subsequently Dr. Hooker, even in stronger terms. I shall reserve for a future work the discussion of these difficulties, and the tables of the proportional numbers of the varying species. Dr. Hooker permits me to add that after having carefully read my manuscript, and examined the tables, he thinks that the following statements are fairly well established. The whole subject, however, treated as it necessarily here is with much brevity, is rather perplexing, and allusions cannot be avoided to the "struggle for existence," "divergence of character," and other questions, hereafter to be discussed.

Alphonse de Candolle and others have shown that plants which have very wide ranges generally present varieties ; and this might have been expected, as they are exposed to diverse physical conditions, and as they come into competition (which, as we shall hereafter see, is an equally or more important circumstance) with different sets of organic beings. But my tables further show that, in any limited country, the species which are the most common, that is abound most in individuals, and the species which are most widely diffused within their own country (and this is a different consideration from wide range, and to a certain extent from commonness), oftenest give rise to varieties sufficiently well-marked to have been recorded in botanical works. Hence it is the most flourishing, or, as they may be called, the dominant species,—those which range widely, are the most diffused in their own country, and are the most numerous in individuals,— which oftenest produce well-marked varieties, or, as I consider them, incipient species. And this, perhaps, might have been anticipated ; for, as varieties, in order to become in any degree permanent, necessarily have to struggle with the other inhabitants of the country, the species which are already dominant will be the most likely to yield offspring, which, though in some slight degree modified, still inherit those advantages that enabled their parents to become dominant over their compatriots. In these remarks on predominance, it should be understood that reference is made only to the forms which come into competition with each other, and more especially to the members of the same genus or class having nearly similar habits of life. With respect to the number of individuals or commonness of species, the comparison of course relates only to the members of the same group.   One of the higher plants may be said to be dominant if it be more numerous in individuals and more widely diffused than the other plants of the same country, which live under nearly the same conditions.   A plant of this kind is not the less dominant because some conferva inhabiting the water or some parasitic fungus is infinitely more numerous in individuals, and more widely diffused.   But if the conferva or parasitic fungus exceeds its allies in the above respects, it will then be dominant within its own class.

Species of the Larger Genera in each Country vary more frequently than the Species of the Smaller Genera.

If the plants inhabiting a country, as described in any Flora, be divided into two equal masses, all those in the larger genera (*i.e.*, those including many species) being placed on one side, and all those in the smaller genera on the other side, the former will be found to include a somewhat larger number of the very common and much diffused or dominant species.   This might have been anticipated; for the mere fact of many species of the same genus inhabiting any country, shows that there is something in the organic or inorganic conditions of that country favourable to the genus; and, consequently, we might have expected to have found in the larger genera, or those including many species, a larger proportional number of dominant species.   But so many causes tend to obscure this result, that I am surprised that my tables show even a

small majority on the side of the larger genera. I will
here allude to only two causes of obscurity. Fresh-
water and salt-loving plants generally have very wide
ranges and are much diffused, but this seems to be
connected with the nature of the stations inhabited by
them, and has little or no relation to the size of the
genera to which the species belong. Again, plants low
in the scale of organisation are generally much more
widely diffused than plants higher in the scale; and
here again there is no close relation to the size of the
genera. The cause of lowly-organised plants ranging
widely will be discussed in our chapter on Geographical
Distribution.

From looking at species as only strongly-marked and
well-defined varieties, I was led to anticipate that the
species of the larger genera in each country would
oftener present varieties, than the species of the smaller
genera; for wherever many closely related species (*i.e.*,
species of the same genus) have been formed, many
varieties or incipient species ought, as a general rule, to
be now forming. Where many large trees grow, we
expect to find saplings. Where many species of a
genus have been formed through variation, circumstances have been favourable for variation; and hence
we might expect that the circumstances would generally
be still favourable to variation. On the other hand, if
we look at each species as a special act of creation,
there is no apparent reason why more varieties should
occur in a group having many species, than in one
having few.

To test the truth of this anticipation I have arranged
the plants of twelve countries, and the coleopterous
insects of two districts, into two nearly equal masses,

the species of the larger genera on one side, and those
of the smaller genera on the other side, and it has
invariably proved to be the case that a larger proportion of the species on the side of the larger genera
presented varieties, than on the side of the smaller
genera. Moreover, the species of the large genera
which present any varieties, invariably present a larger
average number of varieties than do the species of the
small genera. Both these results follow when another
division is made, and when all the least genera, with
from only one to four species, are altogether excluded
from the tables. These facts are of plain signification
on the view that species are only strongly-marked and
permanent varieties; for wherever many species of the
same genus have been formed, or where, if we may use
the expression, the manufactory of species has been
active, we ought generally to find the manufactory still
in action, more especially as we have every reason to
believe the process of manufacturing new species to be
a slow one. And this certainly holds true, if varieties
be looked at as incipient species; for my tables clearly
show as a general rule that, wherever many species of a
genus have been formed, the species of that genus
present a number of varieties, that is of incipient
species, beyond the average. It is not that all large
genera are now varying much, and are thus increasing
in the number of their species, or that no small genera
are now varying and increasing; for if this had been
so, it would have been fatal to my theory; inasmuch
as geology plainly tells us that small genera have in
the lapse of time often increased greatly in size; and
that large genera have often come to their maxima,
decline, and disappeared. All that we want to show

is, that, where many species of a genus have been
formed, on an average many are still forming; and this
certainly holds good.

*Many of the Species included within the Larger Genera
resemble Varieties in being very closely, but unequally,
related to each other, and in having restricted ranges.*

There are other relations between the species of
large genera and their recorded varieties which deserve
notice.   We have seen that there is no infallible
criterion by which to distinguish species and well-
marked varieties; and when intermediate links have
not been found between doubtful forms, naturalists are
compelled to come to a determination by the amount
of difference between them, judging by analogy
whether or not the amount suffices to raise one or
both to the rank of species.   Hence the amount of
difference is one very important criterion in settling
whether two forms should be ranked as species or
varieties.   Now Fries has remarked in regard to
plants, and Westwood in regard to insects, that in
large genera the amount of difference between the
species is often exceedingly small.   I have endeavoured
to test this numerically by averages, and, as far as
my imperfect results go, they confirm the view.   I
have also consulted some sagacious and experienced
observers, and, after deliberation, they concur in this
view.   In this respect, therefore, the species of the
larger genera resemble varieties, more than do the
species of the smaller genera.   Or the case may be put
in another way, and it may be said, that in the larger
genera, in which a number of varieties or incipient

species greater than the average are now manufacturing, many of the species already manufactured
still to a certain extent resemble varieties, for they
differ from each other by less than the usual amount of
difference.

Moreover, the species of the larger genera are
related to each other, in the same manner as the
varieties of any one species are related to each other.
No naturalist pretends that all the species of a genus
are equally distinct from each other; they may
generally be divided into sub-genera, or sections, or
lesser groups. As Fries has well remarked, little
groups of species are generally clustered like satellites
around other species. And what are varieties but
groups of forms, unequally related to each other, and
clustered round certain forms—that is, round their
parent-species. Undoubtedly there is one most important point of difference between varieties and
species; namely, that the amount of difference between
varieties, when compared with each other or with their
parent-species, is much less than that between the
species of the same genus. But when we come to
discuss the principle, as I call it, of Divergence of
Character, we shall see how this may be explained,
and how the lesser differences between varieties tend
to increase into the greater differences between species.

There is one other point which is worth notice.
Varieties generally have much restricted ranges: this
statement is indeed scarcely more than a truism, for, if
a variety were found to have a wider range than that
of its supposed parent-species, their denominations
would be reversed. But there is reason to believe
that the species which are very closely allied to other

species, and in so far resemble varieties, often have much restricted ranges. For instance, Mr. H. C. Watson has marked for me in the well-sifted London Catalogue of plants (4th edition) 63 plants which are therein ranked as species, but which he considers as so closely allied to other species as to be of doubtful value : these 63 reputed species range on an average over 6·9 of the provinces into which Mr. Watson has divided Great Britain. Now, in this same Catalogue, 53 acknowledged varieties are recorded, and these range over 7·7 provinces ; whereas, the species to which these varieties belong range over 14·3 provinces. So that the acknowledged varieties have nearly the same restricted average range, as have the closely allied forms, marked for me by Mr. Watson as doubtful species, but which are almost universally ranked by British botanists as good and true species.

Summary.

Finally, varieties cannot be distinguished from species,—except, first, by the discovery of intermediate linking forms ; and, secondly, by a certain indefinite amount of difference between them ; for two forms, if differing very little, are generally ranked as varieties, notwithstanding that they cannot be closely connected ; but the amount of difference considered necessary to give to any two forms the rank of species cannot be defined. In genera having more than the average number of species in any country, the species of these genera have more than the average number of varieties. In large genera the species are apt to be closely, but unequally, allied together, forming little clusters round other species. Species very closely allied to other species apparently have restricted ranges. In all these respects the species of large genera present a strong analogy with varieties. And we can clearly understand these analogies, if species once existed as varieties, and thus originated; whereas, these analogies are utterly inexplicable if species are independent creations.

We have, also, seen that it is the most flourishing or dominant species of the larger genera within each class which on an average yield the greatest number of varieties; and varieties, as we shall hereafter see, tend to become converted into new and distinct species. Thus the larger genera tend to become larger; and throughout nature the forms of life which are now dominant tend to become still more dominant by leaving many modified and dominant descendants. But by steps hereafter to be explained, the larger genera also tend to break up into smaller genera. And thus, the forms of life throughout the universe become divided into groups subordinate to groups.

CHAPTER III.

STRUGGLE FOR EXISTENCE.

Its bearing on natural selection—The term used in a wide sense—
Geometrical ratio of increase—Rapid increase of naturalised
animals and plants—Nature of the checks to increase—Competition universal—Effects of climate—Protection from the
number of individuals—Complex relations of all animals and
plants throughout nature—Struggle for life most severe between
individuals and varieties of the same species: often severe
between species of the same genus—The relation of organism
to organism the most important of all relations.

BEFORE entering on the subject of this chapter, 1 must
make a few preliminary remarks, to show how the
struggle for existence bears on Natural Selection. It
has been seen in the last chapter that amongst organic
beings in a state of nature there is some individual
variability : indeed I am not aware that this has ever
been disputed. It is immaterial for us whether a
multitude of doubtful forms be called species or sub-
species or varieties; what rank, for instance, the two
or three hundred doubtful forms of British plants are
entitled to hold, if the existence of any well-marked
varieties be admitted. But the mere existence of
individual variability and of some few well-marked
varieties, though necessary as the foundation for the
work, helps us but little in understanding how species
arise in nature. How have all those exquisite
adaptations of one part of the organisation to another

part, and to the conditions of life, and of one organic
being to another being, been perfected ?   We see these
beautiful co-adaptations most plainly in the woodpecker
and the mistletoe; and only a little less plainly in the
humblest parasite which clings to the hairs of a
quadruped or feathers of a bird; in the structure of
the beetle which dives through the water : in the
plumed seed which is wafted by the gentlest breeze;
in short, we see beautiful adaptations everywhere and
in every part of the organic world.

Again, it may be asked, how is it that varieties,
which I have called incipient species, become ultimately
converted into good and distinct species, which in most
cases obviously differ from each other far more than do
the varieties of the same species?   How do those
groups of species, which constitute what are called
distinct genera, and which differ from each other more
than do the species of the same genus, arise ?   All
these results, as we shall more fully see in the next
chapter, follow from the struggle for life.   Owing to
this struggle, variations, however slight and from
whatever cause proceeding, if they be in any degree
profitable to the individuals of a species, in their
infinitely complex relations to other organic beings and
to their physical conditions of life, will tend to the
preservation of such individuals, and will generally
be inherited by the offspring.   The offspring, also, will
thus have a better chance of surviving, for, of the many
individuals of any species which are periodically born,
but a small number can survive.   I have called this
principle, by which each slight variation, if useful, is
preserved, by the term Natural Selection, in order to
mark its relation to man's power of selection.   But the

expression often used by Mr. Herbert Spencer of the Survival of the Fittest is more accurate, and is sometimes equally convenient. We have seen that man by selection can certainly produce great results, and can adapt organic beings to his own uses, through the accumulation of slight but useful variations, given to him by the hand of Nature. But Natural Selection, as we shall hereafter see, is a power incessantly ready for action, and is as immeasurably superior to man's feeble efforts, as the works of Nature are to those of Art.

We will now discuss in a little more detail the struggle for existence. In my future work this subject will be treated, as it well deserves, at greater length. The elder De Candolle and Lyell have largely and philosophically shown that all organic beings are exposed to severe competition. In regard to plants, no one has treated this subject with more spirit and ability than W. Herbert, Dean of Manchester, evidently the result of his great horticultural knowledge. Nothing is easier than to admit in words the truth of the universal struggle for life, or more difficult—at least I have found it so—than constantly to bear this conclusion in mind. Yet unless it be thoroughly engrained in the mind, the whole economy of nature, with every fact on distribution, rarity, abundance, extinction, and variation, will be dimly seen or quite misunderstood. We behold the face of nature bright with gladness, we often see superabundance of food; we do not see or we forget, that the birds which are idly singing round us mostly live on insects or seeds, and are thus constantly destroying life; or we forget how largely these songsters, or their eggs, or their nestlings, are destroyed by birds and beasts of prey; we do not always bear in mind, that, though food may be now superabundant, it is not so at all seasons of each recurring year.

The Term, Struggle for Existence, used in a large sense.

I should premise that I use this term in a large and metaphorical sense including dependence of one being on another, and including (which is more important) not only the life of the individual, but success in leaving progeny. Two canine animals, in a time of dearth, may be truly said to struggle with each other which shall get food and live. But a plant on the edge of a desert is said to struggle for life against the drought, though more properly it should be said to be dependent on the moisture. A plant which annually produces a thousand seeds, of which only one of an average comes to maturity, may be more truly said to struggle with the plants of the same and other kinds which already clothe the ground. The mistletoe is dependent on the apple and a few other trees, but can only in a far-fetched sense be said to struggle with these trees, for, if too many of these parasites grow on the same tree, it languishes and dies. But several seedling mistletoes, growing close together on the same branch, may more truly be said to struggle with each other. As the mistletoe is disseminated by birds, its existence depends on them; and it may metaphorically be said to struggle with other fruit-bearing plants, in tempting the birds to devour and thus disseminate its seeds. In these several senses, which pass into each other, I use for convenience' sake the general term of Struggle for Existence.

Geometrical Ratio of Increase.

A struggle for existence inevitably follows from the high rate at which all organic beings tend to increase. Every being, which during its natural lifetime produces several eggs or seeds, must suffer destruction during some period of its life, and during some season or occasional year, otherwise, on the principle of geometrical increase, its numbers would quickly become so inordinately great that no country could support the product. Hence, as more individuals are produced than can possibly survive, there must in every case be a struggle for existence, either one individual with another of the same species, or with the individuals of distinct species, or with the physical conditions of life. It is the doctrine of Malthus applied with manifold force to the whole animal and vegetable kingdoms; for in this case there can be no artificial increase of food, and no prudential restraint from marriage. Although some species may be now increasing, more or less rapidly, in numbers, all cannot do so, for the world would not hold them.

There is no exception to the rule that every organic being naturally increases at so high a rate, that, if not destroyed, the earth would soon be covered by the progeny of a single pair. Even slow-breeding man has doubled in twenty-five years, and at this rate, in less than a thousand years, there would literally not be standing-room for his progeny. Linnæus has calculated that if an annual plant produced only two seeds—and there is no plant so unproductive as this—and their seedlings next year produced two, and so on, then in twenty years there would be a million plants. The elephant is reckoned the slowest breeder of all known animals, and I have taken some pains to estimate its probable minimum rate of natural increase; it will be safest to assume that it begins breeding when thirty years old, and goes on breeding till ninety years old, bringing forth six young in the interval, and surviving till one hundred years old; if this be so, after a period of from 740 to 750 years there would be nearly nineteen million elephants alive, descended from the first pair.

But we have better evidence on this subject than mere theoretical calculations, namely, the numerous recorded cases of the astonishingly rapid increase of various animals in a state of nature, when circumstances have been favourable to them during two or three following seasons. Still more striking is the evidence from our domestic animals of many kinds which have run wild in several parts of the world; if the statements of the rate of increase of slow-breeding cattle and horses in South America, and latterly in Australia, had not been well authenticated, they would have been incredible. So it is with plants; cases could be given of introduced plants which have become common throughout whole islands in a period of less than ten years. Several of the plants, such as the cardoon and a tall thistle, which are now the commonest over the wide plains of La Plata, clothing square leagues of surface almost to the exclusion of every other plant, have been introduced from Europe; and there are plants which now range in India, as I hear from Dr. Falconer, from Cape Comorin to the Himalaya, which have been imported from America since its discovery. In such cases, and endless others could be given, no one supposes, that the fertility of the animals or plants has been suddenly and temporarily increased in any sensible degree. The obvious explanation is that the conditions of life have been highly favourable, and that there has consequently been less destruction of the old and young, and that nearly all the young have been enabled to breed. Their geometrical ratio of increase, the result of which never fails to be surprising, simply explains their extraordinarily rapid increase and wide diffusion in their new homes.

In a state of nature almost every full-grown plant annually produces seed, and amongst animals there are very few which do not annually pair. Hence we may confidently assert, that all plants and animals are tending to increase at a geometrical ratio,—that all would rapidly stock every station in which they could anyhow exist,—and that this geometrical tendency to increase must be checked by destruction at some period of life. Our familiarity with the larger domestic animals tends, I think, to mislead us: we see no great destruction falling on them, but we do not keep in mind that thousands are annually slaughtered for food, and that in a state of nature an equal number would have somehow to be disposed of.

The only difference between organisms which annually produce eggs or seeds by the thousand, and those which produce extremely few, is, that the slow-breeders would require a few more years to people, under favourable conditions, a whole district, let it be ever so large. The condor lays a couple of eggs and the ostrich a score, and yet in the same country the condor may be the more numerous of the two; the Fulmar petrel lays but one egg, yet it is believed to be the most numerous bird in the world. One fly deposits hundreds of eggs, and another, like the hippobosca, a single one; but this difference does not determine how many individuals of the two species can be supported in a district. A large number of eggs is of some importance to those species which depend on a fluctuating amount of food, for it allows them rapidly to increase in number. But the real importance of a large number of eggs or seeds is to make up for much destruction at some period of life; and this period in the great majority of cases is an early one. If an animal can in any way protect its own eggs or young, a small number may be produced; and yet the average stock be fully kept up; but if many eggs or young are destroyed, many must be produced, or the species will become extinct. It would suffice to keep up the full number of a tree, which lived on an average for a thousand years, if a single seed were produced once in a thousand years, supposing that this seed were never destroyed, and could be ensured to germinate in a fitting place. So that, in all cases, the average number of any animal or plant depends only indirectly on the number of its eggs or seeds.

In looking at Nature, it is most necessary to keep the foregoing considerations always in mind—never to forget that every single organic being may be said to be striving to the utmost to increase in numbers; that each lives by a struggle at some period of its life; that heavy destruction inevitably falls either on the young or old, during each generation or at recurrent intervals. Lighten any check, mitigate the destruction ever so little, and the number of the species will almost instantaneously increase to any amount.

Nature of the Checks to Increase.

The causes which check the natural tendency of each species to increase are most obscure. Look at the most vigorous species; by as much as it swarms in numbers, by so much will it tend to increase still further. We know not exactly what the checks are even in a single instance. Nor will this surprise any one who reflects how ignorant we are on this head, even in regard to mankind, although so incomparably better known than any other animal. This subject of the checks to increase has been ably treated by several authors, and I hope in a future work to discuss it at considerable length, more especially in regard to the feral animals of South America. Here I will make only a few remarks, just to recall to the reader's mind some of the chief points. Eggs or very young animals seem generally to suffer most, but this is not invariably the case. With plants there is a vast destruction of seeds, but, from some observations which I have made it appears that the seedlings suffer most from germinating in ground already thickly stocked with other plants. Seedlings, also, are destroyed in vast numbers by various enemies; for instance, on a piece of ground three feet long and two wide, dug and cleared, and where there could be no choking from other plants, I marked all the seedlings of our native weeds as they came up, and out of 357 no less than 295 were destroyed, chiefly by slugs and insects. If turf which has long been mown, and the case would be the same with turf closely browsed by quadrupeds, be let to grow, the more vigorous plants gradually kill the less vigorous, though fully grown plants; thus out of twenty species growing on a little plot of mown turf (three feet by four) nine species perished, from the other species being allowed to grow up freely.

The amount of food for each species of course gives the extreme limit to which each can increase; but very frequently it is not the obtaining food, but the serving as prey to other animals, which determines the average numbers of a species. Thus, there seems to be little doubt that the stock of partridges, grouse, and hares on any large estate depends chiefly on the destruction of vermin. If not one head of game were shot during the next twenty years in England, and, at the same time, if no vermin were destroyed, there would, in all probability, be less game than at present, although hundreds of thousands of game animals are now annually shot. On the other hand, in some cases, as with the elephant, none are destroyed by beasts of prey; for even the tiger in India most rarely dares to attack a young elephant protected by its dam.

Climate plays an important part in determining the average numbers of a species, and periodical seasons of extreme cold or drought seem to be the most effective of all checks. I estimated (chiefly from the greatly reduced numbers of nests in the spring) that the winter of 1854–5 destroyed four-fifths of the birds in my own grounds; and this is a tremendous destruction, when we remember that ten per cent. is an extraordinarily severe mortality from epidemics with man. The action of climate seems at first sight to be quite independent of the struggle for existence; but in so far as climate chiefly acts in reducing food, it bring on the most severe struggle between the individuals, whether of the same or of distinct species, which subsist on the same kind of food. Even when climate, for instance extreme cold, acts directly, it will be the least vigorous individuals, or those which have got least food through the advancing winter, which will suffer most. When we travel from south to north, or from a damp region to a dry, we invariably see some species gradually getting rarer and rarer, and finally disappearing; and the change of climate being conspicuous, we are tempted to attribute the whole effect to its direct action. But this is a false view; we forget that each species, even where it most abounds, is constantly suffering enormous destruction at some period of its life, from enemies or from competitors for the same place and food; and if these enemies or competitors be in the least degree favoured by any slight change of climate, they will increase in numbers; and as each area is already fully stocked with inhabitants, the other species must decrease. When we travel southward and see a species decreasing in numbers, we may feel sure that the cause lies quite as much in other species being favoured, as in this one being hurt. So it is when we travel northward, but in a somewhat lesser degree, for the number of species of all kinds, and therefore of competitors, decreases northwards; hence in going northwards, or in ascending a mountain, we far oftener meet with stunted forms, due to the *directly* injurious action of climate, than we do in proceeding southwards or in descending a mountain. When we reach the Arctic regions, or snow-capped summits, or absolute deserts, the struggle for life is almost exclusively with the elements.

That climate acts in main part indirectly by favouring other species, we clearly see in the prodigious number

of plants which in our gardens can perfectly well
endure our climate, but which never become naturalised,
for they cannot compete with our native plants nor
resist destruction by our native animals.

When a species, owing to highly favourable circumstances, increases inordinately in numbers in a small
tract, epidemics—at least, this seems generally to
occur with our game animals—often ensue; and here
we have a limiting check independent of the struggle
for life. But even some of these so-called epidemics
appear to be due to parasitic worms, which have from
some cause, possibly in part through facility of diffusion
amongst the crowded animals, been disproportionally
favoured: and here comes in a sort of struggle between
the parasite and its prey.

On the other hand, in many cases, a large stock of
individuals of the same species, relatively to the
numbers of its enemies, is absolutely necessary for its
preservation. Thus we can easily raise plenty of corn
and rape-seed, &c., in our fields, because the seeds are
in great excess compared with the number of birds
which feed on them; nor can the birds, though having
a superabundance of food at this one season, increase in
number proportionally to the supply of seed, as their
numbers are checked during winter; but any one who
has tried, knows how troublesome it is to get seed from
a few wheat or other such plants in a garden: I have
in this case lost every single seed. This view of the
necessity of a large stock of the same species for its
preservation, explains, I believe, some singular facts in
nature such as that of very rare plants being sometimes
extremely abundant, in the few spots where they do
exist; and that of some social plants being social, that

is abounding in individuals, even on the extreme verge
of their range.   For in such cases, we may believe,
that a plant could exist only where the conditions of
its life were so favourable that many could exist
together, and thus save the species from utter
destruction.   I should add that the good effects of
intercrossing, and the ill effects of close interbreeding,
no doubt come into play in many of these cases; but I
will not here enlarge on this subject.

*Complex Relations of all Animals and Plants to each
other in the Struggle for Existence.*

Many cases are on record showing how complex and
unexpected are the checks and relations between
organic beings, which have to struggle together in
the same country.   I will give only a single instance,
which, though a simple one, interested me.   In Staffordshire, on the estate of a relation, where I had ample
means of investigation, there was a large and extremely
barren heath, which had never been touched by the
hand of man; but several hundred acres of exactly
the same nature had been enclosed twenty-five years
previously and planted with Scotch fir.   The change in
the native vegetation of the planted part of the heath
was most remarkable, more than is generally seen in
passing from one quite different soil to another: not
only the proportional numbers of the heath-plants
were wholly changed, but twelve species of plants (not
counting grasses and carices) flourished in the plantations, which could not be found on the heath.   The
effect on the insects must have been still greater, for
six insectivorous birds were very common in the

plantations, which were not to be seen on the heath; and the heath was frequented by two or three distinct insectivorous birds.   Here we see how potent has been the effect of the introduction of a single tree, nothing whatever else having been done, with the exception of the land having been enclosed, so that cattle could not enter.   But how important an element enclosure is, I plainly saw near Farnham, in Surrey.   Here there are extensive heaths, with a few clumps of old Scotch firs on the distant hill-tops: within the last ten years large spaces have been enclosed, and self-sown firs are now springing up in multitudes, so close together that all cannot live.   When I ascertained that these young trees had not been sown or planted, I was so much surprised at their numbers that I went to several points of view, whence I could examine hundreds of acres of the unenclosed heath, and literally I could not see a single Scotch fir, except the old planted clumps.   But on looking closely between the stems of the heath, I found a multitude of seedlings and little trees which had been perpetually browsed down by the cattle.   In one square yard, at a point some hundred yards distant from one of the old clumps, I counted thirty-two little trees; and one of them, with twenty-six rings of growth, had, during many years tried to raise its head above the stems of the heath, and had failed.   No wonder that, as soon as the land was enclosed, it became thickly clothed with vigorously growing young firs.   Yet the heath was so extremely barren and so extensive that no one would ever have imagined that cattle would have so closely and effectually searched it for food.

Here we see that cattle absolutely determine the

existence of the Scotch fir; but in several parts of the world insects determine the existence of cattle. Perhaps Paraguay offers the most curious instance of this; for here neither cattle nor horses nor dogs have ever run wild, though they swarm southward and northward in a feral state; and Azara and Rengger have shown that this is caused by the greater number in Paraguay of a certain fly, which lays its eggs in the navels of these animals when first born. The increase of these flies, numerous as they are, must be habitually checked by some means, probably by other parasitic insects. Hence, if certain insectivorous birds were to decrease in Paraguay, the parasitic insects would probably increase; and this would lessen the number of the navel-frequenting flies—then cattle and horses would become feral, and this would certainly greatly alter (as indeed I have observed in parts of South America) the vegetation: this again would largely affect the insects; and this, as we have just seen in Staffordshire, the insectivorous birds, and so onwards in ever-increasing circles of complexity. Not that under nature the relations will ever be as simple as this. Battle within battle must be continually recurring with varying success; and yet in the long-run the forces are so nicely balanced, that the face of nature remains for long periods of time uniform, though assuredly the merest trifle would give the victory to one organic being over another. Nevertheless, so profound is our ignorance, and so high our presumption, that we marvel when we hear of the extinction of an organic being; and as we do not see the cause, we invoke cataclysms to desolate the world, or invent laws on the duration of the forms of life!

I am tempted to give one more instance showing how plants and animals, remote in the scale of nature, are bound together by a web of complex relations. I shall hereafter have occasion to show that the exotic Lobelia fulgens is never visited in my garden by insects, and consequently, from its peculiar structure, never sets a seed. Nearly all our orchidaceous plants absolutely require the visits of insects to remove their pollen-masses and thus to fertilise them. I find from experiments that humble-bees are almost indispensable to the fertilisation of the heartsease (Violo tricolor), for other bees do not visit this flower. I have also found that the visits of bees are necessary for the fertilisation of some kinds of clover; for instance, 20 heads of Dutch clover (Trifolium repens) yielded 2,290 seeds, but 20 other heads protected from bees produced not one. Again, 100 heads of red clover (T. pratense) produced 2,700 seeds, but the same number of protected heads produced not a single seed. Humble-bees alone visit red clover, as other bees cannot reach the nectar. It has been suggested that moths may fertilise the clovers; but I doubt whether they could do so in the case of the red clover, from their weight not being sufficient to depress the wing petals. Hence we may infer as highly probable that, if the whole genus of humble-bees became extinct or very rare in England, the heartsease and red clover would become very rare, or wholly disappear. The number of humble-bees in any district depends in a great measure upon the number of field-mice, which destroy their combs and nests; and Col. Newman, who has long attended to the habits of humble-bees, believes that "more than two-thirds of them are thus destroyed all over England." Now the number of mice is largely dependent, as every one knows, on the number of cats; and Col. Newman says, " Near villages and small towns I have found the nests of humble-bees more numerous than elsewhere, which I attribute to the number of cats that destroy the mice." Hence it is quite credible that the presence of a feline animal in large numbers in a district might determine, through the intervention first of mice and then of bees, the frequency of certain flowers in that district !

In the case of every species, many different checks, acting at different periods of life, and during different seasons or years, probably come into play; some one check or some few being generally the most potent; but all will concur in determining the average number or even the existence of the species. In some cases it can be shown that widely-different checks act on the same species in different districts. When we look at the plants and bushes clothing an entangled bank, we are tempted to attribute their proportional numbers and kinds to what we call chance. But how false a view is this ! Every one has heard that when an American forest is cut down, a very different vegetation springs up; but it has been observed that ancient Indian ruins in the Southern United States, which must formerly have been cleared of trees, now display the same beautiful diversity and proportion of kinds as in the surrounding virgin forest. What a struggle must have gone on during long centuries between the several kinds of trees, each annually scattering its seeds by the thousand; what war between insect and insect— between insects, snails, and other animals with birds and beasts of prey—all striving to increase, all feeding

on each other, or on the trees, their seeds and seedlings,
or on the other plants which first clothed the ground
and thus checked the growth of the trees! Throw up a
handful of feathers, and all fall to the ground according
to definite laws; but how simple is the problem where
each shall fall compared to that of the action and
reaction of the innumerable plants and animals which
have determined, in the course of centuries, the proportional numbers and kinds of trees now growing on
the old Indian ruins!

The dependency of one organic being on another, as
of a parasite on its prey, lies generally between beings
remote in the scale of nature. This is likewise sometimes the case with those which may be strictly said to
struggle with each other for existence, as in the case
of locusts and grass-feeding quadrupeds. But the
struggle will almost invariably be most severe between
the individuals of the same species, for they frequent
the same districts, require the same food, and are
exposed to the same dangers. In the case of varieties
of the same species, the struggle will generally be
almost equally severe, and we sometimes see the
contest soon decided: for instance, if several varieties of
wheat be sown together, and the mixed seed be resown,
some of the varieties which best suit the soil or climate,
or are naturally the most fertile, will beat the others
and so yield more seed, and will consequently in a few
years supplant the other varieties. To keep up a
mixed stock of even such extremely close varieties as
the variously-coloured sweet peas, they must be each
year harvested separately, and the seed then mixed in
due proportion, otherwise the weaker kinds will
steadily decrease in number and disappear. So again

with the varieties of sheep; it has been asserted that certain mountain-varieties will starve out other mountain-varieties, so that they cannot be kept together. The same result has followed from keeping together different varieties of the medicinal leech. It may even be doubted whether the varieties of any of our domestic plants or animals have so exactly the same strength, habits, and constitution, that the original proportions of a mixed stock (crossing being prevented) could be kept up for half-a-dozen generations, if they were allowed to struggle together, in the same manner as beings in a state of nature, and if the seed or young were not annually preserved in due proportion.

Struggle for Life most severe between Individuals and Varieties of the same Species.

As the species of the same genus usually have, though by no means invariably, much similarity in habits and constitution, and always in structure, the struggle will generally be more severe between them, if they come into competition with each other, than between the species of distinct genera. We see this in the recent extension over parts of the United States of one species of swallow having caused the decrease of another species. The recent increase of the misselthrush in parts of Scotland has caused the decrease of the song-thrush. How frequently we hear of one species of rat taking the place of another species under the most different climates! In Russia the small Asiatic cockroach has everywhere driven before it its great congener. In Australia the imported hive-bee is rapidly exterminating the small, stingless native bee.

One species of charlock has been known to supplant another species; and so in other cases. We can dimly see why the competition should be most severe between allied forms, which fill nearly the same place in the economy of nature; but probably in no one case could we precisely say why one species has been victorious over another in the great battle of life.

A corollary of the highest importance may be deduced from the foregoing remarks, namely, that the structure of every organic being is related, in the most essential yet often hidden manner, to that of all the other organic beings, with which it comes into competition for food or residence, or from which it has to escape, or on which it preys. This is obvious in the structure of the teeth and talons of the tiger; and in that of the legs and claws of the parasite which clings to the hair on the tiger's body. But in the beautifully plumed seed of the dandelion, and in the flattened and fringed legs of the water-beetle, the relation seems at first confined to the elements of air and water. Yet the advantage of plumed seeds no doubt stands in the closest relation to the land being already thickly clothed with other plants; so that the seeds may be widely distributed and fall on unoccupied ground. In the water-beetle, the structure of its legs, so well adapted for diving, allows it to compete with other aquatic insects, to hunt for its own prey, and to escape serving as prey to other animals.

The store of nutriment laid up within the seeds of many plants seems at first sight to have no sort of relation to other plants. But from the strong growth of young plants produced from such seeds, as peas and beans, when sown in the midst of long grass, it may be suspected that the chief use of the nutriment in the seed is to favour the growth of the seedlings, whilst struggling with other plants growing vigorously all around.

Look at a plant in the midst of its range, why does it not double or quadruple its numbers? We know that it can perfectly well withstand a little more heat or cold, dampness or dryness, for elsewhere it ranges into slightly hotter or colder, damper or drier districts. In this case we can clearly see that if we wish in imagination to give the plant the power of increasing in number, we should have to give it some advantage over its competitors, or over the animals which prey on it. On the confines of its geographical range, a change of constitution with respect to climate would clearly be an advantage to our plant; but we have reason to believe that only a few plants or animals range so far, that they are destroyed exclusively by the rigour of the climate. Not until we reach the extreme confines of life, in the Arctic regions or on the borders of an utter desert, will competition cease. The land may be extremely cold or dry, yet there will be competition between some few species, or between the individuals of the same species, for the warmest or dampest spots.

Hence we can see that when a plant or animal is placed in a new country amongst new competitors, the conditions of its life will generally be changed in an essential manner, although the climate may be exactly the same as in its former home. If its average numbers are to increase in its new home, we should have to modify it in a different way to what we should have had to do in its native country; for we should have to give it some advantage over a different set of competitors or enemies.

It is good thus to try in imagination to give to any one species an advantage over another. Probably in no single instance should we know what to do. This ought to convince us of our ignorance on the mutual relations of all organic beings; a conviction as necessary, as it is difficult to acquire. All that we can do, is to keep steadily in mind that each organic being is striving to increase in a geometrical ratio; that each at some period of its life, during some season of the year, during each generation or at intervals, has to struggle for life and to suffer great destruction. When we reflect on this struggle, we may console ourselves with the full belief, that the war of nature is not incessant, that no fear is felt, that death is generally prompt, and that the vigorous, the healthy, and the happy survive and multiply.

CHAPTER IV.

Natural Selection; or the Survival of the Fittest.

Natural Selection—its power compared with man's selection—its power on characters of trifling importance—its power at all ages and on both sexes—Sexual Selection—On the generality of intercrosses between individuals of the same species—Circumstances favourable and unfavourable to the results of Natural Selection, namely, intercrossing, isolation, number of individuals —Slow action—Extinction caused by Natural Selection—Divergence of Character, related to the diversity of inhabitants of any small area, and to naturalisation—Action of Natural Selection, through Divergence of Character, and Extinction, on the descendants from a common parent—Explains the grouping of all organic beings—Advance in organisation—Low forms preserved —Convergence of character—Indefinite multiplication of species —Summary.

How will the struggle for existence, briefly discussed in the last chapter, act in regard to variation? Can the principle of selection, which we have seen is so potent in the hands of man, apply under nature? I think we shall see that it can act most efficiently. Let the endless number of slight variations and individual differences occurring in our domestic productions, and, in a lesser degree, in those under nature, be borne in mind; as well as the strength of the hereditary tendency. Under domestication, it may be truly said that the whole organisation becomes in some degree plastic. But the variability, which we almost universally meet with in

H

our domestic productions, is not directly produced, as Hooker and Asa Gray have well remarked, by man; he can neither originate varieties, nor prevent their occurrence; he can only preserve and accumulate such as do occur. Unintentionally he exposes organic beings to new and changing conditions of life, and variability ensues; but similar changes of conditions might and do occur under nature. Let it also be borne in mind how infinitely complex and close-fitting are the mutual relations of all organic beings to each other and to their physical conditions of life; and consequently what infinitely varied diversities of structure might be of use to each being under changing conditions of life. Can it, then, be thought improbable, seeing that variations useful to man have undoubtedly occurred, that other variations useful in some way to each being in the great and complex battle of life, should occur in the course of many successive generations? If such do occur, can we doubt (remembering that many more individuals are born than can possibly survive) that individuals having any advantage, however slight, over others, would have the best chance of surviving and of procreating their kind? On the other hand, we may feel sure that any variation in the least degree injurious would be rigidly destroyed. This preservation of favourable individual differences and variations, and the destruction of those which are injurious, I have called Natural Selection, or the Survival of the Fittest. Variations neither useful nor injurious would not be affected by natural selection, and would be left either a fluctuating element, as perhaps we see in certain polymorphic species, or would ultimately become fixed, owing to the nature of the organism and the nature of the conditions.

Several writers have misapprehended or objected to the term Natural Selection. Some have even imagined that natural selection induces variability, whereas it implies only the preservation of such variations as arise and are beneficial to the being under its conditions of life. No one objects to agriculturists speaking of the potent effects of man's selection; and in this case the individual differences given by nature, which man for some object selects, must of necessity first occur. Others have objected that the term selection implies conscious choice in the animals which become modified; and it has even been urged that, as plants have no volition, natural selection is not applicable to them! In the literal sense of the word, no doubt, natural selection is a false term; but who ever objected to chemists speaking of the elective affinities of the various elements?—and yet an acid cannot strictly be said to elect the base with which it in preference combines. It has been said that I speak of natural selection as an active power or Deity; but who objects to an author speaking of the attraction of gravity as ruling the movements of the planets? Every one knows what is meant and is implied by such metaphorical expressions; and they are almost necessary for brevity. So again it is difficult to avoid personifying the word Nature; but I mean by Nature, only the aggregate action and product of many natural laws, and by laws the sequence of events as ascertained by us. With a little familiarity such superficial objections will be forgotten.

We shall best understand the probable course of natural selection by taking the case of a country undergoing some slight physical change, for instance, of climate. The proportional numbers of its inhabitants will almost immediately undergo a change, and some species will probably become extinct. We may conclude, from what we have seen of the intimate and complex manner in which the inhabitants of each country are bound together, that any change in the numerical proportions of the inhabitants, independently of the change of climate itself, would seriously affect the others. If the country were open on its borders, new forms would certainly immigrate, and this would likewise seriously disturb the relations of some of the former inhabitants. Let it be remembered how powerful the influence of a single introduced tree or mammal has been shown to be. But in the case of an island, or of a country partly surrounded by barriers, into which new and better adapted forms could not freely enter, we should then have places in the economy of nature which would assuredly be better filled up, if some of the original inhabitants were in some manner modified; for, had the area been open to immigration, these same places would have been seized on by intruders. In such cases, slight modifications, which in any way favoured the individuals of any species, by better adapting them to their altered conditions, would tend to be preserved; and natural selection would have free scope for the work of improvement.

We have good reason to believe, as shown in the first chapter, that changes in the conditions of life give a tendency to increased variability; and in the foregoing cases the conditions have changed, and this would manifestly be favourable to natural selection, by affording a better chance of the occurrence of profitable variations. Unless such occur, natural selection can do nothing. Under the term of "variations," it must never be forgotten that mere individual differences are included. As man

can produce a great result with his domestic animals and
plants by adding up in any given direction individual
differences, so could natural selection, but far more easily
from having incomparably longer time for action.   Nor
do I believe that any great physical change, as of climate,
or any unusual degree of isolation to check immigration,
is necessary in order that new and unoccupied places
should be left, for natural selection to fill up by improving some of the varying inhabitants.   For as all
the inhabitants of each country are struggling together
with nicely balanced forces, extremely slight modifications in the structure or habits of one species would
often give it an advantage over others; and still further
modifications of the same kind would often still further
increase the advantage, as long as the species continued
under the same conditions of life and profited by similar
means of subsistence and defence.   No country can be
named in which all the native inhabitants are now so
perfectly adapted to each other and to the physical conditions under which they live, that none of them could
be still better adapted or improved; for in all countries,
the natives have been so far conquered by naturalised
productions, that they have allowed some foreigners to
take firm possession of the land.   And as foreigners
have thus in every country beaten some of the natives,
we may safely conclude that the natives might have
been modified with advantage, so as to have better resisted the intruders.

As man can produce, and certainly has produced, a
great result by his methodical and unconscious means of
selection, what may not natural selection effect?   Man
can act only on external and visible characters: Nature,
if I may be allowed to personify the natural preservation

or survival of the fittest, cares nothing for appearances, except in so far as they are useful to any being. She can act on every internal organ, on every shade of constitutional difference, on the whole machinery of life. Man selects only for his own good: Nature only for that of the being which she tends. Every selected character is fully exercised by her, as is implied by the fact of their selection. Man keeps the natives of many climates in the same country; he seldom exercises each selected character in some peculiar and fitting manner; he feeds a long and a short beaked pigeon on the same food; he does not exercise a long-backed or long-legged quadruped in any peculiar manner; he exposes sheep with long and short wool to the same climate. He does not allow the most vigorous males to struggle for the females. He does not rigidly destroy all inferior animals, but protects during each varying season, as far as lies in his power, all his productions. He often begins his selection by some half-monstrous form; or at least by some modification prominent enough to catch the eye or to be plainly useful to him. Under nature, the slightest differences of structure or constitution may well turn the nicely-balanced scale in the struggle for life, and so be preserved. How fleeting are the wishes and efforts of man! how short his time! and consequently how poor will be his results, compared with those accumulated by Nature during whole geological periods! Can we wonder, then, that Nature's productions should be far " truer " in character than man's productions; that they should be infinitely better adapted to the most complex conditions of life, and should plainly bear the stamp of far higher workmanship?

It may metaphorically be said that natural selection

is daily and hourly scrutinising, throughout the world, the slightest variations; rejecting those that are bad, preserving and adding up all that are good; silently and insensibly working, *whenever and wherever opportunity offers,* at the improvement of each organic being in relation to its organic and inorganic conditions of life. We see nothing of these slow changes in progress, until the hand of time has marked the lapse of ages, and then so imperfect is our view into long-past geological ages, that we see only that the forms of life are now different from what they formerly were.

In order that any great amount of modification should be effected in a species, a variety when once formed must again, perhaps after a long interval of time, vary or present individual differences of the same favourable nature as before; and these must be again preserved, and so onwards step by step. Seeing that individual differences of the same kind perpetually recur, this can hardly be considered as an unwarrantable assumption. But whether it is true, we can judge only by seeing how far the hypothesis accords with and explains the general phenomena of nature. On the other hand, the ordinary belief that the amount of possible variation is a strictly limited quantity is likewise a simple assumption.

Although natural selection can act only through and for the good of each being, yet characters and structures, which we are apt to consider as of very trifling importance, may thus be acted on. When we see leaf-eating insects green, and bark-feeders mottled-grey; the alpine ptarmigan white in winter, the red-grouse the colour of heather, we must believe that these tints are of service to these birds and insects in preserving them from danger. Grouse, if not destroyed at some period of their lives, would increase in countless numbers ; they are known to suffer largely from birds of prey; and hawks are guided by eyesight to their prey—so much so, that on parts of the Continent persons are warned not to keep white pigeons, as being the most liable to destruction.   Hence natural selection might be effective in giving the proper colour to each kind of grouse, and in keeping that colour, when once acquired, true and constant.   Nor ought we to think that the occasional destruction of an animal of any particular colour would produce little effect: we should remember how essential it is in a flock of white sheep to destroy a lamb with the faintest trace of black.   We have seen how the colour of the hogs, which feed on the " paint-root " in Virginia, determines whether they shall live or die.   In plants, the down on the fruit and the colour of the flesh are considered by botanists as characters of the most trifling importance: yet we hear from an excellent horti-culturist, Downing, that in the United States smooth-skinned fruits suffer far more from a beetle, a Curculio, than those with down; that purple plums suffer far more from a certain disease than yellow plums ; whereas another disease attacks yellow-fleshed peaches far more than those with other coloured flesh.   If, with all the aids of art, these slight differences make a great difference in cultivating the several varieties, assuredly, in a state of nature, where the trees would have to struggle with other trees and with a host of enemies, such differences would effectually settle which variety, whether a smooth or downy, a yellow or purple fleshed fruit, should succeed.

In looking at many small points of difference between species, which, as far as our ignorance permits us to judge, seem quite unimportant, we must not forget that climate, food, &c., have no doubt produced some direct effect. It is also necessary to bear in mind that, owing to the law of correlation, when one part varies, and the variations are accumulated through natural selection, other modifications, often of the most unexpected nature, will ensue.

As we see that those variations which, under domestication, appear at any particular period of life, tend to reappear in the offspring at the same period;—for instance, in the shape, size, and flavour of the seeds of the many varieties of our culinary and agricultural plants; in the caterpillar and cocoon stages of the varieties of the silk-worm; in the eggs of poultry, and in the colour of the down of their chickens; in the horns of our sheep and cattle when nearly adult;—so in a state of nature natural selection will be enabled to act on and modify organic beings at any age, by the accumulation of variations profitable at that age, and by their inheritance at a corresponding age. If it profit a plant to have its seeds more and more widely disseminated by the wind, I can see no greater difficulty in this being effected through natural selection, than in the cotton-planter increasing and improving by selection the down in the pods on his cotton-trees. Natural selection may modify and adapt the larva of an insect to a score of contingencies, wholly different from those which concern the mature insect; and these modifications may effect, through correlation, the structure of the adult. So, conversely, modifications in the adult may affect the structure of the larva; but in all cases natural selection will ensure that they shall not be injurious: for if they were so, the species would become extinct.

Natural selection will modify the structure of the

young in relation to the parent, and of the parent in relation to the young. In social animals it will adapt the structure of each individual for the benefit of the whole community; if the community profits by the selected change. What natural selection cannot do, is to modify the structure of one species, without giving it any advantage, for the good of another species; and though statements to this effect may be found in works of natural history, I cannot find one case which will bear investigation. A structure used only once in an animal's life, if of high importance to it, might be modified to any extent by natural selection; for instance, the great jaws possessed by certain insects, used exclusively for opening the cocoon—or the hard tip to the beak of unhatched birds, used for breaking the egg. It has been asserted, that of the best short-beaked tumbler-pigeons a greater number perish in the egg than are able to get out of it; so that fanciers assist in the act of hatching. Now if nature had to make the beak of a full-grown pigeon very short for the bird's own advantage, the process of modification would be very slow, and there would be simultaneously the most rigorous selection of all the young birds within the egg, which had the most powerful and hardest beaks, for all with weak beaks would inevitably perish; or, more delicate and more easily broken shells might be selected, the thickness of the shell being known to vary like every other structure.

It may be well here to remark that with all beings there must be much fortuitous destruction, which can have little or no influence on the course of natural selection. For instance a vast number of eggs or seeds are annually devoured, and these could be modified through natural selection only if they varied in some manner which protected them from their enemies. Yet many of these eggs or seeds would perhaps, if not destroyed, have yielded individuals better adapted to their conditions of life than any of those which happened to survive. So again a vast number of mature animals and plants, whether or not they be the best adapted to their conditions, must be annually destroyed by accidental causes, which would not be in the least degree mitigated by certain changes of structure or constitution which would in other ways be beneficial to the species. But let the destruction of the adults be ever so heavy, if the number which can exist in any district be not wholly kept down by such causes,—or again let the destruction of eggs or seeds be so great that only a hundredth or a thousandth part are developed,—yet of those which do survive, the best adapted individuals, supposing that there is any variability in a favourable direction, will tend to propagate their kind in larger numbers than the less well adapted. If the numbers be wholly kept down by the causes just indicated, as will often have been the case, natural selection will be powerless in certain beneficial directions; but this is no valid objection to its efficiency at other times and in other ways; for we are far from having any reason to suppose that many species ever undergo modification and improvement at the same time in the same area.

Sexual Selection.

Inasmuch as peculiarities often appear under domestication in one sex and become hereditarily attached to that sex, so no doubt it will be under nature. Thus it is rendered possible for the two sexes to be modified

through natural selection in relation to different habits
of life, as is sometimes the case; or for one sex to be
modified in relation to the other sex, as commonly occurs.
This leads me to say a few words on what I have called
Sexual Selection.   This form of selection depends, not
on a struggle for existence in relation to other organic
beings or to external conditions, but on a struggle
between the individuals of one sex, generally the males,
for the possession of the other sex.   The result is not
death to the unsuccessful competitor, but few or no offspring.   Sexual selection is, therefore, less rigorous than
natural selection.   Generally, the most vigorous males,
those which are best fitted for their places in nature, will
leave most progeny.   But in many cases, victory depends
not so much on general vigour, as on having special
weapons, confined to the male sex.   A hornless stag or
spurless cock would have a poor chance of leaving
numerous offspring.   Sexual selection, by always allowing the victor to breed, might surely give indomitable
courage, length to the spur, and strength to the wing to
strike in the spurred leg, in nearly the same manner as
does the brutal cockfighter by the careful selection of
his best cocks.   How low in the scale of nature the law of
battle descends, I know not; male alligators have been
described as fighting, bellowing, and whirling round, like
Indians in a war-dance, for the possession of the females;
male salmons have been observed fighting all day long;
male stag-beetles sometimes bear wounds from the huge
mandibles of other males; the males of certain hymenopterous insects have been frequently seen by that inimitable observer M. Fabre, fighting for a particular female
who sits by, an apparently unconcerned beholder of the
struggle, and then retires with the conqueror.   The war

is, perhaps, severest between the males of polygamous animals, and these seem oftenest provided with special weapons. The males of carnivorous animals are already well armed; though to them and to others, special means of defence may be given through means of sexual selection, as the mane of the lion, and the hooked jaw to the male salmon; for the shield may be as important for victory, as the sword or spear.

Amongst birds, the contest is often of a more peaceful character. All those who have attended to the subject, believe that there is the severest rivalry between the males of many species to attract, by singing, the females. The rock-thrush of Guiana, birds of paradise, and some others, congregate; and successive males display with the most elaborate care, and show off in the best manner, their gorgeous plumage; they likewise perform strange antics before the females, which, standing by as spectators, at last choose the most attractive partner. Those who have closely attended to birds in confinement well know that they often take individual preferences and dislikes: thus Sir R. Heron has described how a pied peacock was eminently attractive to all his hen birds. I cannot here enter on the necessary details; but if man can in a short time give beauty and an elegant carriage to his bantams, according to his standard of beauty, I can see no good reason to doubt that female birds, by selecting, during thousands of generations, the most melodious or beautiful males, according to their standard of beauty, might produce a marked effect. Some well-known laws, with respect to the plumage of male and female birds, in comparison with the plumage of the young, can partly be explained through the action of sexual selection on variations occurring at different ages, and transmitted to the males alone or to both sexes at corresponding ages; but I have not space here to enter on this subject.

Thus it is, as I believe, that when the males and females of any animal have the same general habits of life, but differ in structure, colour, or ornament, such differences have been mainly caused by sexual selection: that is, by individual males having had, in successive generations, some slight advantage over other males, in their weapons, means of defence, or charms, which they have transmitted to their male offspring alone.   Yet, I would not wish to attribute all sexual differences to this agency: for we see in our domestic animals peculiarities arising and becoming attached to the male sex, which apparently have not been augmented through selection by man. The tuft of hair on the breast of the wild turkey-cock cannot be of any use, and it is doubtful whether it can be ornamental in the eyes of the female bird;—indeed, had the tuft appeared under domestication, it would have been called a monstrosity.

Illustrations of the Action of Natural Selection, or the Survival of the Fittest

In order to make it clear how, as I believe, natural selection acts, I must beg permission to give one or two imaginary illustrations.   Let us take the case of a wolf, which preys on various animals, securing some by craft, some by strength, and some by fleetness; and let us suppose that the fleetest prey, a deer for instance, had from any change in the country increased in numbers, or that other prey had decreased in numbers, during that season of the year when the wolf was hardest pressed for food. Under such circumstances the swiftest and slimmest wolves would have the best chance of surviving and so be preserved or selected,—provided always that they retained strength to master their prey at this or some other period of the year, when they were compelled to prey on other animals. I can see no more reason to doubt that this would be the result, than that man should be able to improve the fleetness of his greyhounds by careful and methodical selection, or by that kind of unconscious selection which follows from each man trying to keep the best dogs without any thought of modifying the breed.   I may add, that, according to Mr. Pierce, there are two varieties of the wolf inhabiting the Catskill Mountains, in the United States, one with a light greyhound-like form, which pursues deer, and the other more bulky, with shorter legs, which more frequently attacks the shepherd's flocks.

It should be observed that, in the above illustration, I speak of the slimmest individual wolves, and not of any single strongly-marked variation having been preserved. In former editions of this work I sometimes spoke as if this latter alternative had frequently occurred. I saw the great importance of individual differences, and this led me fully to discuss the results of unconscious selection by man, which depends on the preservation of all the more or less valuable individuals, and on the destruction of the worst. I saw, also, that the preservation in a state of nature of any occasional deviation of structure, such as a monstrosity, would be a rare event; and that, if at first preserved, it would generally be lost by subsequent intercrossing with ordinary individuals. Nevertheless, until reading an able and valuable article in the 'North British Review' (1867), I did not appreciate how rarely single variations, whether slight or strongly-marked, could be perpetuated. The author takes the case of a pair of animals, producing during their lifetime two hundred offspring, of which, from various causes of destruction, only two on an average survive to pro-create their kind. This is rather an extreme estimate for most of the higher animals, but by no means so for many of the lower organisms. He then shows that if a single individual were born, which varied in some manner, giving it twice as good a chance of life as that of the other individuals, yet the chances would be strongly against its survival. Supposing it to survive and to breed, and that half its young inherited the favourable variation; still, as the Reviewer goes on to show, the young would have only a slightly better chance of surviving and breeding; and this chance would go on decreasing in the succeeding generations. The justice of these remarks cannot, I think, be disputed. If, for instance, a bird of some kind could procure its food more easily by having its beak curved, and if one were born with its beak strongly curved, and which consequently flourished, nevertheless there would be a very poor chance of this one individual perpetuating its kind to the exclusion of the common form; but there can hardly be a doubt, judging by what we see taking place under domestication, that this result would follow from the preservation during many generations of a large number of individuals with more or less strongly curved beaks, and from the destruction of a still larger number with the straightest beaks.

It should not, however, be overlooked that certain rather strongly marked variations, which no one would rank as mere individual differences, frequently recur

owing to a similar organisation being similarly acted on
—of which fact numerous instances could be given with
our domestic productions.   In such cases, if the varying
individual did not actually transmit to its offspring its
newly-acquired character, it would undoubtedly transmit
to them, as long as the existing conditions remained the
same, a still stronger tendency to vary in the same
manner. There can also be little doubt that the tendency
to vary in the same manner has often been so strong
that all the individuals of the same species have been
similarly modified without the aid of any form of selection.   Or only a third, fifth, or tenth part of the individuals may have been thus affected, of which fact
several instances could be given.   Thus Graba estimates
that about one-fifth of the guillemots in the Faroe
Islands consist of a variety so well marked, that it was
formerly ranked as a distinct species under the name of
Uria lacrymans.   In cases of this kind, if the variation
were of a beneficial nature, the original form would soon
be supplanted by the modified form, through the survival
of the fittest.

To the effects of intercrossing in eliminating variations
of all kinds, I shall have to recur; but it may be here
remarked that most animals and plants keep to their
proper homes, and do not needlessly wander about; we
see this even with migratory birds, which almost always
return to the same spot.   Consequently each newly-
formed variety would generally be at first local, as seems
to be the common rule with varieties in a state of nature ;
so that similarly modified individuals would soon exist in
a small body together, and would often breed together.
If the new variety were successful in its battle for life,
it would slowly spread from a central district, competing

with and conquering the unchanged individuals on the margins of an ever-increasing circle.

It may be worth while to give another and more complex illustration of the action of natural selection. Certain plants excrete sweet juice, apparently for the sake of eliminating something injurious from the sap: this is effected, for instance, by glands at the base of the stipules in some Leguminosæ, and at the backs of the leaves of the common laurel. This juice, though small in quantity, is greedily sought by insects; but their visits do not in any way benefit the plant. Now, let us suppose that the juice or nectar was excreted from the inside of the flowers of a certain number of plants of any species. Insects in seeking the nectar would get dusted with pollen, and would often transport it from one flower to another. The flowers of two distinct individuals of the same species would thus get crossed; and the act of crossing, as can be fully proved, gives rise to vigorous seedlings, which consequently would have the best chance of flourishing and surviving. The plants which produced flowers with the largest glands or nectaries, excreting most nectar, would oftenest be visited by insects, and would oftenest be crossed; and so in the long-run would gain the upper hand and form a local variety. The flowers, also, which had their stamens and pistils placed, in relation to the size and habits of the particular insect which visited them, so as to favour in any degree the transportal of the pollen, would likewise be favoured. We might have taken the case of insects visiting flowers for the sake of collecting pollen instead of nectar; and as pollen is formed for the sole purpose of fertilisation, its destruction appears to be a simple loss to the plant; yet if a little pollen were carried, at first occasionally and then habitually, by the pollen-devouring insects from flower to flower, and a cross thus effected, although nine-tenths of the pollen were destroyed it might still be a great gain to the plant to be thus robbed; and the individuals which produced more and more pollen, and had larger anthers, would be selected.

When our plant, by the above process long continued, had been rendered highly attractive to insects, they would, unintentionally on their part, regularly carry pollen from flower to flower; and that they do this effectually, I could easily show by many striking facts. I will give only one, as likewise illustrating one step in the separation of the sexes of plants. Some holly-trees bear only male flowers, which have four stamens producing a rather small quantity of pollen, and a rudimentary pistil; other holly-trees bear only female flowers; these have a full-sized pistil, and four stamens with shrivelled anthers, in which not a grain of pollen can be detected. Having found a female tree exactly sixty yards from a male tree, I put the stigmas of twenty flowers, taken from different branches, under the microscope, and on all, without exception, there were a few pollen-grains, and on some a profusion. As the wind had set for several days from the female to the male tree, the pollen could not thus have been carried. The weather had been cold and boisterous, and therefore not favourable to bees, nevertheless every female flower which I examined had been effectually fertilised by the bees, which had flown from tree to tree in search of nectar. But to return to our imaginary case: as soon as the plant had been rendered so highly attractive to insects that pollen was regularly carried from flower to flower, another process might commence. No naturalist doubts the advantage of what has been called the "physiological division of labour;" hence we may believe that it would be advantageous to a plant to produce stamens alone in one flower or on one whole plant, and pistils alone in another flower or on another plant.   In plants under culture and placed under new conditions of life, sometimes the male organs and sometimes the female organs become more or less impotent; now if we suppose this to occur in ever so slight a degree under nature, then, as pollen is already carried regularly from flower to flower, and as a more complete separation of the sexes of our plant would be advantageous on the principle of the division of labour, individuals with this tendency more and more increased, would be continually favoured or selected, until at last a complete separation of the sexes might he effected.   It would take up too much space to show the various steps, through dimorphism and other means, by which the separation of the sexes in plants of various kinds is apparently now in progress; but I may add that some of the species of holly in North America, are, according to Asa Gray, in an exactly intermediate condition, or, as he expresses it, are more or less diœciously polygamous.

Let us now turn to the nectar-feeding insects; we may suppose the plant, of which we have been slowly increasing the nectar by continued selection, to be a common plant; and that certain insects depended in main part on its nectar for food.   I could give many facts showing how anxious bees are to save time: for instance, their habit of cutting holes and sucking the nectar at the bases of certain flowers, which with a very little more trouble, they can enter by the mouth. Bearing such facts in mind, it may be believed that under certain circumstances individual differences in the curvature or

length of the proboscis, &c., too slight to be appreciated
by us, might profit a bee or other insect, so that certain
individuals would be able to obtain their food more
quickly than others; and thus the communities to which
they belonged would flourish and throw off many swarms
inheriting the same peculiarities. The tubes of the
corolla of the common red and incarnate clovers (Trifolium pratense and incarnatum) do not on a hasty glance
appear to differ in length; yet the hive-bee can easily
suck the nectar out of the incarnate clover, but not out
of the common red clover, which is visited by humble-
bees alone; so that whole fields of the red clover offer
in vain an abundant supply of precious nectar to the
hive-bee. That this nectar is much liked by the hive-
bee is certain; for I have repeatedly seen, but only in the
autumn, many hive-bees sucking the flowers through
holes bitten in the base of the tube by humble-bees.
The difference in the length of the corolla in the two
kinds of clover, which determines the visits of the hive-
bee, must be very trifling; for I have been assured that
when red clover has been mown, the flowers of the second
crop are somewhat smaller, and that these are visited by
many hive-bees. I do not know whether this statement
is accurate; nor whether another published statement
can be trusted, namely, that the Ligurian bee, which is
generally considered a mere variety of the common hive-
bee, and which freely crosses with it, is able to reach and
suck the nectar of the red clover. Thus, in a country
where this kind of clover abounded, it might be a great
advantage to the hive-bee to have a slightly longer or
differently constructed proboscis. On the other hand, as
the fertility of this clover absolutely depends on bees
visiting the flowers, if humble-bees were to become rare in

any country, it might be a great advantage to the plant to have a shorter or more deeply divided corolla, so that the hive-bees should be enabled to suck its flowers. Thus I can understand how a flower and a bee might slowly become, either simultaneously or one after the other, modified and adapted to each other in the most perfect manner, by the continued preservation of all the individuals which presented slight deviations of structure mutually favourable to each other.

I am well aware that this doctrine of natural selection, exemplified in the above imaginary instances, is open to the same objections which were first urged against Sir Charles Lyell's noble views on "the modern changes of the earth, as illustrative of geology ;" but we now seldom hear the agencies which we see still at work, spoken of as trifling or insignificant, when used in explaining the excavation of the deepest valleys or the formation of long lines of inland cliffs. Natural selection acts only by the preservation and accumulation of small inherited modifications, each profitable to the preserved being; and as modern geology has almost banished such views as the excavation of a great valley by a single diluvial wave, so will natural selection banish the belief of the continued creation of new organic beings, or of any great and sudden modification in their structure.

On the Intercrossing of Individuals.

I must here introduce a short digression. In the case of animals and plants with separated sexes, it is of course obvious that two individuals must always (with the exception of the curious and not well understood cases of parthenogenesis) unite for each birth; but in the case of hermaphrodites this is far from obvious. Nevertheless there is reason to believe that with all hermaphrodites two individuals, either occasionally or habitually, concur for the reproduction of their kind. This view was long ago doubtfully suggested by Sprengel, Knight and Kölreuter. We shall presently see its importance; but I must here treat the subject with extreme brevity, though I have the materials prepared for an ample discussion. All vertebrate animals, all insects, and some other large groups of animals, pair for each birth. Modern research has much diminished the number of supposed hermaphrodites, and of real hermaphrodites a large number pair; that is, two individuals regularly unite for reproduction, which is all that concerns us. But still there are many hermaphrodite animals which certainly do not habitually pair, and a vast majority of plants are hermaphrodites. What reason, it may be asked, is there for supposing in these cases that two individuals ever concur in reproduction? As it is impossible here to enter on details, I must trust to some general considerations alone.

In the first place, I have collected so large a body of facts, and made so many experiments, showing, in accordance with the almost universal belief of breeders, that with animals and plants a cross between different varieties, or between individuals of the same variety but of another strain, gives vigour and fertility to the offspring; and on the other hand, that *close* interbreeding diminishes vigour and fertility; that these facts alone incline me to believe that it is a general law of nature that no organic being fertilises itself for a perpetuity of generations; but that a cross with another individual is occasionally—perhaps at long intervals of time—indispensable.

On the belief that this is a law of nature, we can, I
think, understand several large classes of facts, such as
the following, which on any other view are inexplicable.
Every hybridizer knows how unfavourable exposure to
wet is to the fertilisation of a flower, yet what a multitude
of flowers have their anthers and stigmas fully exposed
to the weather! If an occasional cross be indispensable,
notwithstanding that the plant's own anthers and pistil
stand so near each other as almost to insure self-fertilisation, the fullest freedom for the entrance of pollen from
another individual will explain the above state of exposure of the organs. Many flowers, on the other hand,
have their organs of fructification closely enclosed, as in
the great papilionaceous or pea-family; but these almost
invariably present beautiful and curious adaptations in
relation to the visits of insects. So necessary are the
visits of bees to many papilionaceous flowers, that their
fertility is greatly diminished if these visits be prevented.
Now, it is scarcely possible for insects to fly from flower
to flower, and not to carry pollen from one to the other,
to the great good of the plant. Insects act like a camel-
hair pencil, and it is sufficient, to ensure fertilisation, just
to touch with the same brush the anthers of one flower and
then the stigma of another; but it must not be supposed
that bees would thus produce a multitude of hybrids
between distinct species; for if a plant's own pollen and
that from another species are placed on the same stigma,
the former is so prepotent that it invariably and completely destroys, as has been shown by Gärtner, the
influence of the foreign pollen.

When the stamens of a flower suddenly spring towards
the pistil, or slowly move one after the other towards
it, the contrivance seems adapted solely to ensure self-

fertilisation; and no doubt it is useful for this end: but the agency of insects is often required to cause the stamens to spring forward, as Kölreuter has shown to be the case with the barberry; and in this very genus, which seems to have a special contrivance for self-fertilisation, it is well known that, if closely-allied forms or varieties are planted near each other, it is hardly possible to raise pure seedlings, so largely do they naturally cross. In numerous other cases, far from self-fertilisation being favoured, there are special contrivances which effectually prevent the stigma receiving pollen from its own flower, as I could show from the works of Sprengel and others, as well as from my own observations: for instance, in Lobelia fulgens, there is a really beautiful and elaborate contrivance by which all the infinitely numerous pollen-granules are swept out of the conjoined anthers of each flower, before the stigma of that individual flower is ready to receive them; and as this flower is never visited, at least in my garden, by insects, it never sets a seed, though by placing pollen from one flower on the stigma of another, I raise plenty of seedlings. Another species of Lobelia, which is visited by bees, seeds freely in my garden. In very many other cases, though there is no special mechanical contrivance to prevent the stigma receiving pollen from the same flower, yet, as Sprengel, and more recently Hildebrand, and others, have shown, and as I can confirm, either the anthers burst before the stigma is ready for fertilisation, or the stigma is ready before the pollen of that flower is ready, so that these so-named dichogamous plants have in fact separated sexes, and must habitually be crossed. So it is with the reciprocally dimorphic and trimorphic plants previously alluded to. How strange are these facts! How strange that the pollen and stigmatic surface of the same flower, though placed so close together, as if for the very purpose of self-fertilisation, should be in so many cases mutually useless to each other? How simply are these facts explained on the view of an occasional cross with a distinct individual being advantageous or indispensable!

If several varieties of the cabbage, radish, onion, and of some other plants, be allowed to seed near each other, a large majority of the seedlings thus raised turn out, as I have found, mongrels: for instance, I raised 233 seedling cabbages from some plants of different varieties growing near each other, and of these only 78 were true to their kind, and some even of these were not perfectly true. Yet the pistil of each cabbage-flower is surrounded not only by its own six stamens but by those of the many other flowers on the same plant; and the pollen of each flower readily gets on its own stigma without insect agency; for I have found that plants carefully protected from insects produce the full number of pods. How, then, comes it that such a vast number of the seedlings are mongrelized? It must arise from the pollen of a distinct *variety* having a prepotent effect over the flower's own pollen; and that this is part of the general law of good being derived from the intercrossing of distinct individuals of the same species. When distinct *species* are crossed the case is reversed, for a plant's own pollen is almost always prepotent over foreign pollen; but to this subject we shall return in a future chapter.

In the case of a large tree covered with innumerable flowers, it may be objected that pollen could seldom be carried from tree to tree, and at most only from flower to flower on the same tree; and flowers on the same tree can be considered as distinct individuals only in a limited sense. I believe this objection to be valid, but that nature has largely provided against it by giving to trees a strong tendency to bear flowers with separated sexes. When the sexes are separated, although the male and female flowers may be produced on the same tree, pollen must be regularly carried from flower to flower; and this will give a better chance of pollen being occasionally carried from tree to tree. That trees belonging to all Orders have their sexes more often separated than other plants, I find to be the case in this country; and at my request Dr. Hooker tabulated the trees of New Zealand, and Dr. Asa Gray those of the United States, and the result was as I anticipated. On the other hand, Dr. Hooker informs me that the rule does not hold good in Australia: but if most of the Australian trees are dichogamous, the same result would follow as if they bore flowers with separated sexes. I have made these few remarks on trees simply to call attention to the subject.

Turning for a brief space to animals: various terrestrial species are hermaphrodites, such as the land-mollusca and earth-worms; but these all pair. As yet I have not found a single terrestrial animal which can fertilise itself. This remarkable fact, which offers so strong a contrast with terrestrial plants, is intelligible on the view of an occasional cross being indispensable ; for owing to the nature of the fertilising element there are no means, analogous to the action of insects and of the wind with plants, by which an occasional cross could be effected with terrestrial animals without the concurrence of two individuals. Of aquatic animals, there are many self-fertilising hermaphrodites; but here the currents of water offer an obvious means for an occasional cross. As

in the case of flowers, I have as yet failed, after consulta-
tion with one of the highest authorities, namely, Professor
Huxley, to discover a single hermaphrodite animal with
the organs of reproduction so perfectly enclosed that
access from without, and the occasional influence of a
distinct individual, can be shown to be physically impossible. Cirripedes long appeared to me to present, under
this point of view, a case of great difficulty; but I
have been enabled, by a fortunate chance, to prove that
two individuals, though both are self-fertilising hermaphrodites, do sometimes cross.

It must have struck most naturalists as a strange
anomaly that, both with animals and plants, some species
of the same family and even of the same genus, though
agreeing closely with each other in their whole organisation, are hermaphrodites, and some unisexual. But if,
in fact, all hermaphrodites do occasionally intercross, the
difference between them and unisexual species is, as far
as function is concerned, very small.

From these several considerations and from the many
special facts which I have collected, but which I am
unable here to give, it appears that with animals and
plants an occasional intercross between distinct individuals is a very general, if not universal, law of nature.

*Circumstances favourable for the production of new forms
through Natural Selection.*

This is an extremely intricate subject. A great amount
of variability, under which term individual differences
are always included, will evidently be favourable. A
large number of individuals, by giving a better chance
within any given period for the appearance of profitable

variations, will compensate for a lesser amount of variability in each individual, and is, I believe, a highly important element of success. Though Nature grants long periods of time for the work of natural selection, she does not grant an indefinite period; for as all organic beings are striving to seize on each place in the economy of nature, if any one species does not become modified and improved in a corresponding degree with its competitors, it will be exterminated. Unless favourable variations be inherited by some at least of the offspring, nothing can be effected by natural selection. The tendency to reversion may often check or prevent the work; but as this tendency has not prevented man from forming by selection numerous domestic races, why should it prevail against natural selection ?

In the case of methodical selection, a breeder selects for some definite object, and if the individuals be allowed freely to intercross, his work will completely fail. But when many men, without intending to alter the breed, have a nearly common standard of perfection, and all try to procure and breed from the best animals, improvement surely but slowly follows from this unconscious process of selection, notwithstanding that there is no separation of selected individuals. Thus it will be under nature; for within a confined area, with some place in the natural polity not perfectly occupied, all the individuals varying in the right direction, though in different degrees, will tend to be preserved. But if the area be large, its several districts will almost certainly present different conditions of life; and then, if the same species undergoes modification in different districts, the newly-formed varieties will intercross on the confines of each. But we shall see in the sixth chapter that intermediate varieties, inhabiting intermediate districts, will in the long run generally be supplanted by one of the adjoining varieties. Intercrossing will chiefly affect those animals which unite for each birth and wander much, and which do not breed at a very quick rate. Hence with animals of this nature, for instance, birds, varieties will generally be confined to separated countries; and this I find to be the case. With hermaphrodite organisms which cross only occasionally, and likewise with animals which unite for each birth, but which wander little and can increase at a rapid rate, a new and improved variety might be quickly formed on any one spot, and might there maintain itself in a body and afterwards spread, so that the individuals of the new variety would chiefly cross together. On this principle, nurserymen always prefer saving seed from a large body of plants, as the chance of intercrossing is thus lessened.

Even with animals which unite for each birth, and which do not propagate rapidly, we must not assume that free intercrossing would always eliminate the effects of natural selection ; for I can bring forward a considerable body of facts showing that within the same area, two varieties of the same animal may long remain distinct, from haunting different stations, from breeding at slightly different seasons, or from the individuals of each variety preferring to pair together.

Intercrossing plays a very important part in nature by keeping the individuals of the same species, or of the same variety, true and uniform in character. It will obviously thus act far more efficiently with those animals which unite for each birth; but, as already stated, we have reason to believe that occasional intercrosses take place with all animals and plants. Even if these take place only at long intervals of time, the young thus produced will gain so much in vigour and fertility over the offspring from long-continued self-fertilisation, that they will have a better chance of surviving and propagating their kind; and thus in the long run the influence of crosses, even at rare intervals, will be great. With respect to organic beings extremely low in the scale, which do not propagate sexually, nor conjugate, and which cannot possibly intercross, uniformity of character can be retained by them under the same conditions of life, only through the principle of inheritance, and through natural selection which will destroy any individuals departing from the proper type. If the conditions of life change and the form undergoes modification, uniformity of character can be given to the modified offspring, solely by natural selection preserving similar favourable variations.

Isolation, also, is an important element in the modification of species through natural selection. In a confined or isolated area, if not very large, the organic and inorganic conditions of life will generally be almost uniform; so that natural selection will tend to modify all the varying individuals of the same species in the same manner. Intercrossing with the inhabitants of the surrounding districts will, also, be thus prevented. Moritz Wagner has lately published an interesting essay on this subject, and has shown that the service rendered by isolation in preventing crosses between newly-formed varieties is probably greater even than I supposed. But from reasons already assigned I can by no means agree with this naturalist, that migration and isolation are necessary elements for the formation of new species. The importance of isolation is likewise great in preventing, after any physical change in the conditions, such as of climate, elevation of the land, &c., the immigration of better adapted organisms; and thus new places in the natural economy of the district will be left open to be filled up by the modificaticn of the old inhabitants. Lastly, isolation will give time for a new variety to be improved at a slow rate; and this may sometimes be of much importance. If, however, an isolated area be very small, either from being surrounded by barriers, or from having very peculiar physical conditions, the total number of the inhabitants will be small; and this will retard the production of new species through natural selection, by decreasing the chances of favourable variations arising.

The mere lapse of time by itself does nothing, either for or against natural selection. I state this because it has been erroneously asserted that the element of time has been assumed by me to play an all-important part in modifying species, as if all the forms of life were necessarily undergoing change through some innate law. Lapse of time is only so far important, and its importance in this respect is great, that it gives a better chance of beneficial variations arising and of their being selected, accumulated, and fixed. It likewise tends to increase the direct action of the physical conditions of life, in relation to the constitution of each organism.

If we turn to nature to test the truth of these remarks, and look at any small isolated area, such as an oceanic island, although the number of species inhabiting it is small, as we shall see in our chapter on Geographical Distribution; yet of these species a very large proportion are endemic,—that is, have been produced there and nowhere else in the world. Hence an oceanic island at first sight seems to have been highly favourable for the production of new species. But we may thus deceive ourselves, for to ascertain whether a small isolated area, or a large open area like a continent, has been most favourable for the production of new organic forms, we ought to make the comparison within equal times; and this we are incapable of doing.

Although isolation is of great importance in the production of new species, on the whole I am inclined to believe that largeness of area is still more important, especially for the production of species which shall prove capable of enduring for a long period, and of spreading widely. Throughout a great and open area, not only will there be a better chance of favourable variations, arising from the large number of individuals of the same species there supported, but the conditions of life are much more complex from the large number of already existing species; and if some of these many species become modified and improved, others will have to be improved in a corresponding degree, or they will be exterminated. Each new form, also, as soon as it has been much improved, will be able to spread over the open and continuous area, and will thus come into competition with many other forms. Moreover, great areas, though now continuous, will often, owing to former oscillations of level, have existed in a broken condition; so that the good effects of isolation will generally, to a certain extent, have concurred. Finally, I conclude that, although small isolated areas have been in some respects highly favourable for the production of new species, yet that the course of modification will generally have been more rapid on large areas; and what is more important, that the new forms produced on large areas, which already have been victorious over

K

many competitors, will be those that will spread most
widely, and will give rise to the greatest number of new
varieties and species.   They will thus play a more important part in the changing history of the organic world.

In accordance with this view, we can, perhaps, understand some facts which will be again alluded to in our
chapter on Geographical Distribution; for instance, the
fact of the productions of the smaller continent of
Australia now yielding before those of the larger Europæo-
Asiatic area.   Thus, also, it is that continental productions have everywhere become so largely naturalised on
islands.   On a small island, the race for life will have
been less severe, and there will have been less modification and less extermination.   Hence, we can understand
how it is that the flora of Madeira, according to Oswald
Heer, resembles to a certain extent the extinct tertiary
flora of Europe.   All fresh-water basins, taken together,
make a small area compared with that of the sea or of the
land.   Consequently, the competition between fresh-water
productions will have been less severe than elsewhere;
new forms will have been then more slowly produced, and
old forms more slowly exterminated.   And it is in fresh-
water basins that we find seven genera of Ganoid fishes,
remnants of a once preponderant order: and in fresh
water we find some of the most anomalous forms now
known in the world as the Ornithorhynchus and Lepidosiren, which, like fossils, connect to a certain extent
orders at present widely sundered in the natural scale.
These anomalous forms may be called living fossils; they
have endured to the present day, from having inhabited
a confined area, and from having been exposed to less
varied, and therefore less severe, competition.

To sum up, as far as the extreme intricacy of the

subject permits, the circumstances favourable and unfavourable for the production of new species through
natural selection. I conclude that for terrestrial productions a large continental area, which has undergone many
oscillations of level, will have been the most favourable
for the production of many new forms of life, fitted to
endure for a long time and to spread widely. Whilst
the area existed as a continent, the inhabitants will
have been numerous in individuals and kinds, and will
have been subjected to severe competition. When converted by subsidence into large separate islands, there
will still have existed many individuals of the same
species on each island: intercrossing on the confines of
the range of each new species will have been checked :
after physical changes of any kind, immigration will
have been prevented, so that new places in the polity of
each island will have had to be filled up by the modification of the old inhabitants ; and time will have been
allowed for the varieties in each to become well modified
and perfected. When, by renewed elevation, the islands
were reconverted into a continental area, there will
again have been very severe competition : the most
favoured or improved varieties will have been enabled
to spread : there will have been much extinction of the
less improved forms, and the relative porportional numbers of the various inhabitants of the reunited continent
will again have been changed ; and again there will have
been a fair field for natural selection to improve still
further the inhabitants, and thus to produce new species.

That natural selection generally acts with extreme
slowness I fully admit. It can act only when there are
places in the natural polity of a district which can be
better occupied by the modification of some of its existing

inhabitants. The occurrence of such places will often depend on physical changes, which generally take place very slowly, and on the immigration of better adapted forms being prevented. As some few of the old inhabitants become modified, the mutual relations of others will often be disturbed; and this will create new places, ready to be filled up by better adapted forms; but all this will take place very slowly. Although all the individuals of the same species differ in some slight degree from each other, it would often be long before differences of the right nature in various parts of the organisation might occur. The result would often be greatly retarded by free intercrossing. Many will exclaim that these several causes are amply sufficient to neutralise the power of natural selection. I do not believe so. But I do believe that natural selection will generally act very slowly, only at long intervals of time, and only on a few of the inhabitants of the same region. I further believe that these slow, intermittent results accord well with what geology tells us of the rate and manner at which the inhabitants of the world have changed.

Slow though the process of selection may be, if feeble man can do much by artificial selection, I can see no limit to the amount of change, to the beauty and complexity of the coadaptations between all organic beings, one with another and with their physical conditions of life, which may have been affected in the long course of time through nature's power of selection, that is by the survival of the fittest.

Extinction caused by Natural Selection.

This subject will be more fully discussed in our chapter on Geology; but it must here be alluded to from being intimately connected with natural selection. Natural selection acts solely through the preservation of variations in some way advantageous, which consequently endure. Owing to the high geometrical rate of increase of all organic beings, each area is already fully stocked with inhabitants; and it follows from this, that as the favoured forms increase in number, so, generally, will the less favoured decrease and become rare. Rarity, as geology tells us, is the precursor to extinction. We can see that any form which is represented by few individuals will run a good chance of utter extinction, during great fluctuations in the nature of the seasons, or from a temporary increase in the number of its enemies. But we may go further than this; for, as new forms are produced, unless we admit that specific forms can go on indefinitely increasing in number, many old forms must become extinct. That the number of specific forms has not indefinitely increased, geology plainly tells us; and we shall presently attempt to show why it is that the number of species throughout the world has not become immeasurably great.

We have seen that the species which are most numerous in individuals have the best chance of producing favourable variations within any given period. We have evidence of this, in the facts stated in the second chapter, showing that it is the common and diffused or dominant species which offer the greatest number of recorded varieties. Hence, rare species will be less quickly modified or improved within any given period; they will consequently be beaten in the race for life by the modified and improved descendants of the commoner species.

From these several considerations I think it inevitably follows, that as new species in the course of time are formed through natural selection, others will become rarer and rarer, and finally extinct. The forms which stand in closest competition with those undergoing modification and improvement, will naturally suffer most. And we have seen in the chapter on the Struggle for Existence that it is the most closely-allied forms,—varieties of the same species, and species of the same genus or of related genera,—which, from having nearly the same structure, constitution, and habits, generally come into the severest competition with each other; consequently, each new variety or species, during the progress of its formation, will generally press hardest on its nearest kindred, and tend to exterminate them. We see the same process of extermination amongst our domesticated productions, through the selection of improved forms by man. Many curious instances could be given showing how quickly new breeds of cattle, sheep, and other animals, and varieties of flowers, take the place of older and inferior kinds. In Yorkshire, it is historically known that the ancient black cattle were displaced by the long-horns, and that these " were swept away by the short-horns " (I quote the words of an agricultural writer) " as if by some murderous pestilence."

Divergence of Character.

The principle, which I have designated by this term, is of high importance, and explains, as I believe, several important facts. In the first place, varieties, even strongly-marked ones, though having somewhat of the character of species—as is shown by the hopeless doubts in many cases how to rank them—yet certainly differ far less from each other than do good and distinct species. Nevertheless, according to my view, varieties are species in the process of formation, or are, as I have called them, incipient species. How, then, does the lesser difference between varieties become augmented into the greater difference between species? That this does habitually happen, we must infer from most of the innumerable species throughout nature presenting well-marked differences; whereas varieties, the supposed prototypes and parents of future well-marked species, present slight and ill-defined differences. Mere chance, as we may call it, might cause one variety to differ in some character from its parents, and the offspring of this variety again to differ from its parent in the very same character and in a greater degree; but this alone would never account for so habitual and large a degree of difference as that between the species of the same genus.

As has always been my practice, I have sought light on this head from our domestic productions. We shall here find something analogous. It will be admitted that the production of races so different as short-horn and Hereford cattle, race and cart horses, the several breeds of pigeons, &c., could never have been effected by the mere chance accumulation of similar variations during many successive generations. In practice, a fancier is, for instance, struck by a pigeon having a slightly shorter beak; another fancier is struck by a pigeon having a rather longer beak; and on the acknowledged principle that "fanciers do not and will not admire a medium standard, but like extremes," they both go on (as has actually occurred with the sub-breeds of the tumbler-pigeon) choosing and breeding from birds with longer and longer beaks, or with shorter and shorter beaks. Again, we may suppose that at an early period of history, the men of one nation or district required swifter horses, whilst those of another required stronger and bulkier horses. The early differences would be very slight; but, in the course of time, from the continued selection of swifter horses in the one case, and of stronger ones in the other, the differences would become greater, and would be noted as forming two sub-breeds. Ultimately, after the lapse of centuries, these sub-breeds would become converted into two well-established and distinct breeds. As the differences became greater, the inferior animals with intermediate characters, being neither very swift nor very strong, would not have been used for breeding, and will thus have tended to disappear. Here, then, we see in man's productions the action of what may be called the principle of divergence, causing differences, at first barely appreciable, steadily to increase, and the breeds to diverge in character, both from each other and from their common parent.

But how, it may be asked, can any analogous principle apply in nature? I believe it can and does apply most efficiently (though it was a long time before I saw how), from the simple circumstance that the more diversified the descendants from any one species become in structure, constitution, and habits, by so much will they be better enabled to seize on many and widely diversified places in the polity of nature, and so be enabled to increase in numbers.

We can clearly discern this in the case of animals with simple habits. Take the case of a carnivorous quadruped, of which the number that can be supported in any country has long ago arrived at its full average. If its natural power of increase be allowed to act, it can succeed in increasing (the country not undergoing any change in conditions) only by its varying descendants seizing on places at present occupied by other animals : some of them, for instance, being enabled to feed on new kinds of prey, either dead or alive; some inhabiting new stations, climbing trees, frequenting water, and some perhaps becoming less carnivorous. The more diversified in habits and structure the descendants of our carnivorous animals become, the more places they will be enabled to occupy. What applies to one animal will apply throughout all time to all animals—that is, if they vary—for otherwise natural selection can effect nothing. So it will be with plants. It has been experimentally proved, that if a plot of ground be sown with one species of grass, and a similar plot be sown with several distinct genera of grasses, a greater number of plants and a greater weight of dry herbage can be raised in the latter than in the former case. The same has been found to hold good when one variety and several mixed varieties of wheat have been sown on equal spaces of ground. Hence, if any one species of grass were to go on varying, and the varieties were continually selected which differed from each other in the same manner, though in a very slight degree, as do the distinct species and genera of grasses, a greater number of individual plants of this species, including its modified descendants, would succeed in living on the same piece of ground. And we know that each species and each variety of grass is annually sowing almost countless seeds; and is thus striving, as it may be said, to the utmost to increase in number. Consequently, in the course of many thousand generations, the most distinct varieties of any one species of grass would have the best chance of succeeding and of increasing in numbers, and thus of supplanting the less distinct varieties; and varieties, when rendered very distinct from each other, take the rank of species.

The truth of the principle that the greatest amount of life can be supported by great diversification of structure, is seen under many natural circumstances. In an extremely small area, especially if freely open to immigration, and where the contest between individual and individual must be very severe, we always find great diversity in its inhabitants. For instance, I found that a piece of turf, three feet by four in size, which had been exposed for many years to exactly the same conditions, supported twenty species of plants, and these belonged to eighteen genera and to eight orders, which shows how much these plants differed from each other. So it is with the plants and insects on small and uniform islets : also in small ponds of fresh water. Farmers find that they can raise most food by a rotation of plants belonging to the most different orders : nature follows what may be called a simultaneous rotation. Most of the animals and plants which live close round any small piece of ground, could live on it (supposing its nature not to be in any way peculiar), and may be said to be striving to the utmost to live there; but, it is seen, that where they come into the closest competition, the advantages of diversification of structure, with the accompanying differences of habit and constitution, determine that the inhabitants, which thus jostle each other most closely, shall, as a general rule, belong to what we call different genera and orders.

The same principle is seen in the naturalisation of plants through man's agency in foreign lands. It might have been expected that the plants which would succeed in becoming naturalised in any land would generally have been closely allied to the indigenes; for these are commonly looked at as specially created and adapted for their own country. It might also, perhaps, have been expected that naturalised plants would have belonged to a few groups more especially adapted to certain stations in their new homes. But the case is very different; and Alph. de Candolle has well remarked, in his great and admirable work, that floras gain by naturalisation, proportionally with the number of the native genera and species, far more in new genera than in new species. To give a single instance: in the last edition of Dr. Asa Gray's 'Manual of the Flora of the Northern United States,' 260 naturalised plants are enumerated, and these belong to 162 genera. We thus see that these naturalised plants are of a highly diversified nature. They differ, moreover, to a large extent, from the indigenes, for out of the 162 naturalised genera, no less than 100 genera are not there indigenous, and thus a large proportional addition is made to the genera now living in the United States.

By considering the nature of the plants or animals which have in any country struggled successfully with the indigenes, and have there become naturalised, we may gain some crude idea in what manner some of the natives would have to be modified, in order to gain an advantage over their compatriots; and we may at least infer that diversification of structure, amounting to new generic differences, would be profitable to them.

The advantage of diversification of structure in the inhabitants of the same region is, in fact, the same as that of the physiological division of labour in the organs of the same individual body—a subject so well elucidated by Milne Edwards. No physiologist doubts that a stomach adapted to digest vegetable matter alone, or flesh alone, draws most nutriment from these substances. So in the general economy of any land, the more widely and perfectly the animals and plants are diversified for different habits of life, so will a greater number of individuals be capable of there supporting themselves. A set of animals, with their organisation but little diversified, could hardly compete with a set more perfectly diversified in structure. It may be doubted, for instance, whether the Australian marsupials, which are divided into groups differing but little from each other, and feebly representing, as Mr. Waterhouse and others have remarked, our carnivorous, ruminant, and rodent mammals, could successfully compete with these well-developed orders. In the Australian mammals, we see the process of diversification in an early and incomplete stage of development.

The Probable Effects of the Action of Natural Selection through Divergence of Character and Extinction, on the Descendants of a Common Ancestor.

After the foregoing discussion, which has been much compressed, we may assume that the modified descendants of any one species will succeed so much the better as they become more diversified in structure, and are thus enabled to encroach on places occupied by other beings. Now let us see how this principle of benefit being derived from divergence of character, combined with the principles of natural selection and of extinction, tends to act.

The accompanying diagram will aid us in understanding this rather perplexing subject. Let A to L represent the species of a genus large in its own country; these species are suppossd to resemble each other in unequal degrees, as is so generally the case in nature, and as is

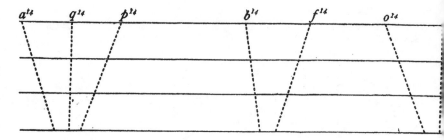

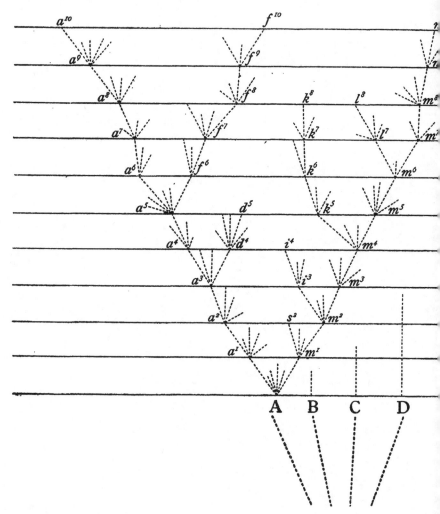

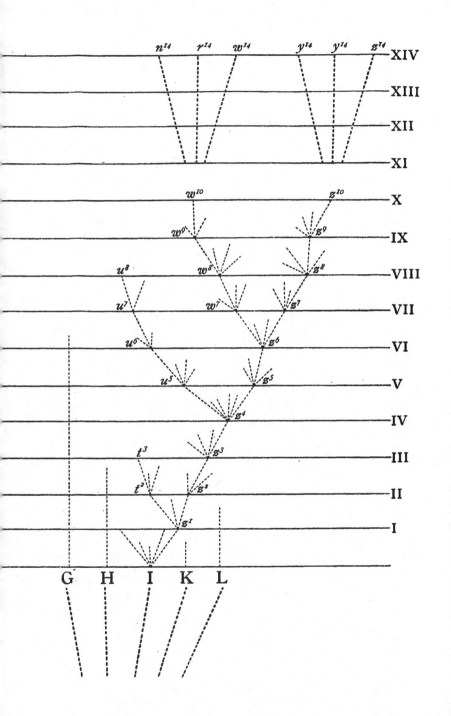

represented in the diagram by the letters standing at unequal distances. I have said a large genus, because as we saw in the second chapter, on an average more species vary in large genera than in small genera; and the varying species of the large genera present a greater number of varieties. We have, also, seen that the species, which are the commonest and the most widely diffused, vary more than do the rare and restricted species. Let (A) be a common, widely-diffused, and varying species, belonging to a genus large in its own country. The branching and diverging dotted lines of unequal lengths proceeding from (A), may represent its varying offspring. The variations are supposed to be extremely slight, but of the most diversified nature; they are not supposed all to appear simultaneously, but often after long intervals of time; nor are they all supposed to endure for equal periods. Only those variations which are in some way profitable will be preserved or naturally selected. And here the importance of the principle of benefit derived from divergence of character comes in; for this will generally lead to the most different or divergent variations (represented by the outer dotted lines) being preserved and accumulated by natural selection. When a dotted line reaches one of the horizontal lines, and is there marked by a small numbered letter, a sufficient amount of variation is supposed to have been accumulated to form it into a fairly well-marked variety, such as would be thought worthy of record in a systematic work.

The intervals between the horizontal lines in the diagram, may represent each a thousand or more generations. After a thousand generations, species (A) is supposed to have produced two fairly well-marked varieties, namely a^1 and m^1. These two varieties will generally

still be exposed to the same conditions which made
their parents variable, and the tendency to variability is
in itself hereditary; consequently they will likewise
tend to vary, and commonly in nearly the same manner
as did their parents.   Moreover, these two varieties,
being only slightly modified forms, will tend to inherit
those advantages which made their parent (A) more
numerous than most of the other inhabitants of the same
country; they will also partake of those more general
advantages which made the genus to which the parent-
species belonged, a large genus in its own country.   And
all these circumstances are favourable to the production
of new varieties.

If, then, these two varieties be variable, the most
divergent of their variations will generally be preserved
during the next thousand generations.   And after this
interval, variety a^1 is supposed in the diagram to have
produced variety a^2, which will, owing to the principle
of divergence, differ more from (A) than did variety a^1.
Variety m^1 is supposed to have produced two varieties,
namely m^2 and s^2, differing from each other, and more
considerably from their common parent (A).   We may
continue the process by similar steps for any length of
time; some of the varieties, after each thousand generations, producing only a single variety, but in a more and
more modified condition, some producing two or three
varieties, and some failing to produce any.   Thus the
varieties or modified descendants of the common parent
(A), will generally go on increasing in number and
diverging in character.   In the diagram the process is
represented up to the ten-thousandth generation, and
under a condensed and simplified form up to the fourteen-
thousandth generation.

But I must here remark that I do not suppose that the process ever goes on so regularly as is represented in the diagram, though in itself made somewhat irregular, nor that it goes on continuously; it is far more probable that each form remains for long periods unaltered, and then again undergoes modification. Nor do I suppose that the most divergent varieties are invariably preserved: a medium form may often long endure, and may or may not produce more than one modified descendant; for natural selection will always act according to the nature of the places which are either unoccupied or not perfectly occupied by other beings; and this will depend on infinitely complex relations. But as a general rule, the more diversified in structure the descendants from any one species can be rendered, the more places they will be enabled to seize on, and the more their modified progeny will increase. In our diagram the line of succession is broken at regular intervals by small numbered letters marking the successive forms which have become sufficiently distinct to be recorded as varieties. But these breaks are imaginary, and might have been inserted anywhere, after intervals long enough to allow the accumulation of a considerable amount of divergent variation.

As all the modified descendants from a common and widely-diffused species, belonging to a large genus, will tend to partake of the same advantages which made their parent successful in life, they will generally go on multiplying in number as well as diverging in character: this is represented in the diagram by the several divergent branches proceeding from (A). The modified offspring from the later and more highly improved branches in the lines of descent, will, it is probable, often take the place of, and so destroy, the earlier and less improved branches: this is represented in the diagram by some of the lower branches not reaching to the upper horizontal lines. In some cases no doubt the process of modification will be confined to a single line of descent, and the number of modified descendants will not be increased; although the amount of divergent modification may have been augmented. This case would be represented in the diagram, if all the lines proceeding from (A) were removed, excepting that from a^1 to a^{10}. In the same way the English race-horse and English pointer have apparently both gone on slowly diverging in character from their original stocks, without either having given off any fresh branches or races.

After ten thousand generations, species (A) is supposed to have produced three forms, a^{10}, f^{10}, and m^{10}, which, from having diverged in character during the successive generations, will have come to differ largely, but perhaps unequally, from each other and from their common parent. If we suppose the amount of change between each horizontal line in our diagram to be excessively small, these three forms may still be only well-marked varieties; but we have only to suppose the steps in the process of modification to be more numerous or greater in amount, to convert these three forms into doubtful or at least into well-defined species. Thus the diagram illustrates the steps by which the small differences distinguishing varieties are increased into the larger differences distinguishing species. By continuing the same process for a greater number of generations (as shown in the diagram in a condensed and simplified manner), we get eight species, marked by the letters between a^{14} and m^{14}, all descended from (A). Thus, as I believe, species are multiplied and genera are formed.

In a large genus it is probable that more than one species would vary. In the diagram I have assumed that a second species (I) has produced, by analogous steps, after ten thousand generations, either two well-marked varieties (w^{10} and z^{10}) or two species, according to the amount of change supposed to be represented between the horizontal lines. After fourteen thousand generations, six new species, marked by the letters n^{14} to z^{14}, are supposed to have been produced. In any genus, the species which are already very different in character from each other, will generally tend to produce the greatest number of modified descendants; for these will have the best chance of seizing on new and widely different places in the polity of nature: hence in the diagram I have chosen the extreme species (A), and the nearly extreme species (I), as those which have largely varied, and have given rise to new varieties and species. The other nine species (marked by capital letters) of our original genus, may for long but unequal periods continue to transmit unaltered descendants; and this is shown in the diagram by the dotted lines unequally prolonged upwards.

But during the process of modification, represented in the diagram, another of our principles, namely that of extinction, will have played an important part. As in each fully stocked country natural selection necessarily acts by the selected form having some advantage in the struggle for life over other forms, there will be a constant tendency in the improved descendants of any one species to supplant and exterminate in each stage of descent their predecessors and their original progenitor. For it should be remembered that the competition will generally be most severe between those forms which are most

L

nearly related to each other in habits, constitution, and structure. Hence all the intermediate forms between the earlier and later states, that is between the less and more improved states of the same species, as well as the original parent-species itself, will generally tend to become extinct. So it probably will be with many whole collateral lines of descent, which will be conquered by later and improved lines. If, however, the modified offspring of a species get into some distinct country, or become quickly adapted to some quite new station, in which offspring and progenitor do not come into competition, both may continue to exist.

If, then, our diagram be assumed to represent a considerable amount of modification, species (A) and all the earlier varieties will have become extinct, being replaced by eight new species (a^{14} to m^{14}); and species (I) will be replaced by six (n^{14} to z^{14}) new species.

But we may go further than this. The original species of our genus were supposed to resemble each other in unequal degrees, as is so generally the case in nature; species (A) being more nearly related to B, C, and D, than to the other species; and species (I) more to G, H, K, L, than to the others. These two species (A) and (I) were also supposed to be very common and widely diffused species, so that they must originally have had some advantage over most of the other species of the genus. Their modified descendants, fourteen in number at the fourteen-thousandth generation, will probably have inherited some of the same advantages: they have also been modified and improved in a diversified manner at each stage of descent, so as to have become adapted to many related places in the natural economy of their country. It seems, therefore, extremely probable that they will have taken the places of, and thus exterminated, not only their parents (A) and (I), but likewise some of the original species which were most nearly related to their parents. Hence very few of the original species will have transmitted offspring to the fourteen-thousandth generation. We may suppose that only one, (F), of the two species (E and F) which were least closely related to the other nine original species, has transmitted descendants to this late stage of descent.

The new species in our diagram descended from the original eleven species, will now be fifteen in number. Owing to the divergent tendency of natural selection, the extreme amount of difference in character between species a^{14} and z^{14} will be much greater than that between the most distinct of the original eleven species. The new species, moreover, will be allied to each other in a widely different manner. Of the eight descendants from (A) the three marked a^{14}, q^{14}, p^{14}, will be nearly related from having recently branched off from a^{10}; b^{14}, and f^{14}, from having diverged at an earlier period from a^5, will be in some degree distinct from the three first-named species; and lastly, o^{14}, e^{14}, and m^{14}, will be nearly related one to the other, but, from having diverged at the first commencement of the process of modification, will be widely different from the other five species, and may constitute a sub-genus or a distinct genus.

The six descendants from (I) will form two sub-genera or genera. But as the original species (I) differed largely from (A), standing nearly at the extreme end of the original genus, the six descendants from (I) will, owing to inheritance alone, differ considerably from the eight descendants from (A); the two groups, moreover, are supposed to have gone on diverging in different directions.

The intermediate species, also (and this is a very important consideration), which connected the original species (A) and (I), have all become, excepting (F), extinct, and have left no descendants. Hence the six new species descended from (I), and the eight descendants from (A), will have to be ranked as very distinct genera, or even as distinct sub-families.

Thus it is, as I believe, that two or more genera are produced by descent with modification, from two or more species of the same genus. And the two or more parent-species are supposed to be descended from some one species of an earlier genus. In our diagram, this is indicated by the broken lines, beneath the capital letters, converging in sub-branches downwards towards a single point; this point represents a species, the supposed progenitor of our several new sub-genera and genera.

It is worth while to reflect for a moment on the character of the new species F^{14}, which is supposed not to have diverged much in character, but to have retained the form of (F), either unaltered or altered only in a slight degree. In this case, its affinities to the other fourteen new species will be of a curious and circuitous nature. Being descended from a form which stood between the parent-species (A) and (I), now supposed to be extinct and unknown, it will be in some degree intermediate in character between the two groups descended from these two species. But as these two groups have gone on diverging in character from the type of their parents, the new species (F^{14}) will not be directly intermediate between them, but rather between types of the two groups; and every naturalist will be able to call such cases before his mind.

In the diagram, each horizontal line has hitherto been

supposed to represent a thousand generations, but each
may represent a million or more generations; it máy
also represent a section of the successive strata of the
earth's crust including extinct remains.   We shall, when
we come to our chapter on Geology, have to refer again
to this subject, and I think we shall then see that the
diagram throws light on the affinities of extinct beings,
which, though generally belonging to the same orders,
families, or genera, with those now living, yet are often,
in some degree, intermediate in character between existing groups; and we can understand this fact, for the
extinct species lived at various remote epochs when the
branching lines of descent had diverged less.

I see no reason to limit the process of modification,
as now explained, to the formation of genera alone.   If,
in the diagram, we suppose the amount of change represented by each successive group of diverging dotted lines
to be great, the forms marked a^{14} to p^{14}, those marked
b^{14} and f^{14}, and those marked o^{14} to m^{14}, will form three
very distinct genera.   We shall also have two very dis-
tinct genera descended from (I), differing widely from
the descendants of (A).   These two groups of genera
will thus form two distinct families, or orders, according
to the amount of divergent modification supposed to be
represented in the diagram.   And the two new families,
or orders, are descended from two species of the original
genus, and these are supposed to be descended from some
still more ancient and unknown form.

We have seen that in each country it is the species
belonging to the larger genera which oftenest present
varieties or incipient species.   This, indeed, might have
been expected; for, as natural selection acts through
one form having some advantage over other forms in the

struggle for existence, it will chiefly act on those which already have some advantage; and the largeness of any group shows that its species have inherited from a common ancestor some advantage in common. Hence, the struggle for the production of new and modified descendants will mainly lie between the larger groups which are all trying to increase in number. One large group will slowly conquer another large group, reduce its numbers, and thus lessen its chance of further variation and improvement. Within the same large group, the later and more highly perfected sub-groups, from branching out and seizing on many new places in the polity of Nature, will constantly tend to supplant and destroy the earlier and less improved sub-groups. Small and broken groups and sub-groups will finally disappear. Looking to the future, we can predict that the groups of organic beings which are now large and triumphant, and which are least broken up, that is, which have as yet suffered least extinction, will, for a long period, continue to increase. But which groups will ultimately prevail, no man can predict; for we know that many groups, formerly most extensively developed, have now become extinct. Looking still more remotely to the future, we may predict that, owing to the continued and steady increase of the larger groups, a multitude of smaller groups will become utterly extinct, and leave no modified descendants; and consequently that, of the species living at any one period, extremely few will transmit descendants to a remote futurity. I shall have to return to this subject in the chapter on Classification, but I may add that as, according to this view, extremely few of the more ancient species have transmitted descendants to the present day, and, as all the descendants of the

same species form a class, we can understand how it is
that there exists so few classes in each main division of
the animal and vegetable kingdoms. Although few of
the most ancient species have left modified descendants,
yet, at remote geological periods, the earth may have
been almost as well peopled with species of many genera
families, orders, and classes, as at the present time.

On the Degree to which Organisation tends to advance.

Natural Selection acts exclusively by the preservation
and accumulation of variations, which are beneficial
under the organic and inorganic conditions to which
each creature is exposed at all periods of life. The ultimate result is that each creature tends to become more
and more improved in relation to its conditions. This
improvement inevitably leads to the gradual advancement of the organisation of the greater number of living
beings throughout the world. But here we enter on a
very intricate subject, for naturalists have not defined to
each other's satisfaction what is meant by an advance in
organisation. Amongst the vertebrata the degree of intellect and an approach in structure to man clearly come
into play. It might be thought that the amount of
change which the various parts and organs pass through
in their development from the embryo to maturity would
suffice as a standard of comparison; but there are cases,
as with certain parasitic crustaceans, in which several
parts of the structure become less perfect, so that the
mature animal cannot be called higher than its larva.
Von Baer's standard seems the most widely applicable
and the best, namely, the amount of differentiation of the
parts of the same organic being, in the adult state as I

should be inclined to add, and their specialisation for different functions; or, as Milne Edwards would express it, the completeness of the division of physiological labour. But we shall see how obscure this subject is if we look, for instance, to fishes, amongst which some naturalists rank those as highest which, like the sharks, approach nearest to amphibians; whilst other naturalists rank the common bony or teleostean fishes as the highest, inasmuch as they are most strictly fish-like, and differ most from the other vertebrate classes. We see still more plainly the obscurity of the subject by turning to plants, amongst which the standard of intellect is of course quite excluded; and here some botanists rank those plants as highest which have every organ, as sepals, petals, stamens, and pistils, fully developed in each flower; whereas other botanists, probably with more truth, look at the plants which have their several organs much modified and reduced in number as the highest.

If we take as the standard of high organisation, the amount of differentiation and specialisation of the several organs in each being when adult (and this will include the advancement of the brain for intellectual purposes), natural selection clearly leads towards this standard: for all physiologists admit that the specialisation of organs, inasmuch as in this state they perform their functions better, is an advantage to each being; and hence the accumulation of variations tending towards specialisation is within the scope of natural selection. On the other hand, we can see, bearing in mind that all organic beings are striving to increase at a high ratio and to seize on every unoccupied or less well occupied place in the economy of nature, that it is quite possible for natural selection gradually to fit a being to a situation in which several organs would be superfluous or useless : in such cases there would be retrogression in the scale of organisation. Whether organisation on the whole has actually advanced from the remotest geological periods to the present day will be more conveniently discussed in our chapter on Geological Succession.

But it may be objected that if all organic beings thus tend to rise in the scale, how is it that throughout the world a multitude of the lowest forms still exist; and how is it that in each great class some forms are far more highly developed than others ? Why have not the more highly developed forms everywhere supplanted and exterminated the lower ? Lamarck, who believed in an innate and inevitable tendency towards perfection in all organic beings, seems to have felt this difficulty so strongly, that he was led to suppose that new and simple forms are continually being produced by spontaneous generation. Science has not as yet proved the truth of this belief, whatever the future may reveal. On our theory the continued existence of lowly organisms offers no difficulty ; for natural selection, or the survival of the fittest, does not necessarily include progressive development—it only takes advantage of such variations as arise and are beneficial to each creature under its complex relations of life. And it may be asked what advantage, as far as we can see, would it be to an infusorian animalcule—to an intestinal worm—or even to an earthworm, to be highly organised. If it were no advantage, these forms would be left, by natural selection, unimproved or but little improved, and might remain for indefinite ages in their present lowly condition. And geology tells us that some of the lowest forms, as the infusoria and rhizopods, have remained for an enormous period in nearly their present state. But to suppose that most of the many now existing low forms have not in the least advanced since the first dawn of life would be extremely rash; for every naturalist who has dissected some of the beings now ranked as very low in the scale, must have been struck with their really wondrous and beautiful organisation.

Nearly the same remarks are applicable if we look to the different grades of organisation within the same great group; for instance, in the vertebrata, to the co-existence of mammals and fish—amongst mammalia, to the co-existence of man and the ornithorhynchus—amongst fishes, to the co-existence of the shark and the lancelet (Amphioxus), which latter fish in the extreme simplicity of its structure approaches the invertebrate classes.   But mammals and fish hardly come into competition with each other; the advancement of the whole class of mammals, or or certain members in this class, to the highest grade would not lead to their taking the place of fishes. Physiologists believe that the brain must be bathed by warm blood to be highly active, and this requires aërial respiration; so that warm - blooded mammals when inhabiting the water lie under a disadvantage in having to come continually to the surface to breathe.   With fishes, members of the shark family would not tend to supplant the lancelet; for the lancelet, as I hear from Fritz Müller, has as sole companion and competitor on the barren sandy shore of South Brazil, an anomalous annelid.   The three lowest orders of mammals, namely, marsupials, edentata, and rodents, co-exist in South America in the same region with numerous monkeys, and probably interfere little with each other. Although organisation, on the whole, may have advanced and be still advancing throughout the world, yet the scale will always present many degrees of perfection ; for the high advancement of certain whole classes, or of certain members of each class, does not at all necessarily lead to the extinction of those groups with which they do not enter into close competition. In some cases, as we shall hereafter see, lowly organised forms appear to have been preserved to the present day, from inhabiting confined or peculiar stations, where they have been subjected to less severe competition, and where their scanty numbers have retarded the chance of favourable variations arising.

Finally, I believe that many lowly organised forms now exist throughout the world, from various causes. In some cases variations or individual differences of a favourable nature may never have arisen for natural selection to act on and accumulate. In no case, probably, has time sufficed for the utmost possible amount of development. In some few cases there has been what we must call retrogression of organisation. But the main cause lies in the fact that under very simple conditions of life a high organisation would be of no service, —possibly would be of actual disservice, as being of a more delicate nature, and more liable to be put out of order and injured.

Looking to the first dawn of life, when all organic beings, as we may believe, presented the simplest structure, how, it has been asked, could the first steps in the advancement or differentiation of parts have arisen ? Mr. Herbert Spencer would probably answer that, as soon as simple unicellular organism came by growth or division to be compounded of several cells, or became attached to any supporting surface, his law "that homologous units of any order become differentiated in proportion as their relations to incident forces become different" would come into action. But as we have no facts to guide us, speculation on the subject is almost useless. It is, however, an error to suppose that there would be no struggle for existence, and, consequently, no natural selection, until many forms had been produced: variations in a single species inhabiting an isolated station might be beneficial, and thus the whole mass of individuals might be modified, or two distinct forms might arise. But, as I remarked towards the close of the Introduction, no one ought to feel surprise at much remaining as yet unexplained on the origin of species, if we make due allowance for our profound ignorance on the mutual relations of the inhabitants of the world at the present time, and still more so during past ages.

Convergence of Character.

Mr. H. C. Watson thinks that I have overrated the importance of divergence of character (in which, however, he apparently believes), and that convergence, as it may be called, has likewise played a part. If two species, belonging to two distinct though allied genera, had both produced a large number of new and divergent forms, it is conceivable that these might approach each other so closely that they would have all to be classed under the same genus; and thus the descendants of two distinct genera would converge into one. But it would in most cases be extremely rash to attribute to convergence a close and general similarity of structure in the modified descendants of widely distinct forms. The shape of a crystal is determined solely by the molecular forces, and it is not surprising that dissimilar substances should sometimes assume the same form; but with organic beings we should bear in mind that the form of each depends on an infinitude of complex relations, namely on the variations which have arisen, these being due to causes far too intricate to be followed out,—on the nature of the variations which have been preserved or selected, and this depends on the surrounding physical conditions, and in a still higher degree on the surrounding organisms with which each being has come into competition,—and lastly, on inheritance (in itself a fluctuating element) from innumerable progenitors, all of which have had their forms determined through equally complex relations. It is incredible that the descendants of two organisms, which had originally differed in a marked manner, should ever afterwards converge so closely as to lead to a near approach to identity throughout their whole organisation. If this had occurred, we should meet with the same form, independently of genetic connection, recurring in widely separated geological formations; and the balance of evidence is opposed to any such an admission.

Mr. Watson has also objected that the continued action of natural selection, together with divergence of character, would tend to make an indefinite number of specific forms. As far as mere inorganic conditions are concerned ,it seems probable that a sufficient number of species would soon become adapted to all considerable diversities of heat, moisture, &c. ; but I fully admit that the mutual relations of organic beings are more important; and as the number of species in any country goes on increasing, the organic conditions of life must become more and more complex. Consequently there

seems at first sight no limit to the amount of profitable
diversification of structure, and therefore no limit to
the number of species which might be produced.   We
do not know that even the most prolific area is fully
stocked with specific forms : at the Cape of Good Hope
and in Australia, which support such an astonishing
number of species, many European plants have become
naturalised.   But geology shows us, that from an early
part of the tertiary period the number of species of
shells, and that from the middle part of this same period
the number of mammals, has not greatly or at all
increased.   What then checks an indefinite increase in
the number of species ?   The amount of life (I do not
mean the number of specific forms) supported on an area
must have a limit, depending so largely as it does on
physical conditions; therefore, if an area be inhabited
by very many species, each or nearly each species will
be represented by few individuals; and such species
will be liable to extermination from accidental fluctuations in the nature of the seasons or in the number of
their enemies.   The process of extermination in such
cases would be rapid, whereas the production of new
species must always be slow.   Imagine the extreme case
of as many species as individuals in England, and the
first severe winter or very dry summer would exterminate thousands on thousands of species.   Rare species,
and each species will become rare if the number of
species in any country becomes indefinitely increased,
will, on the principle often explained, present within a
given period few favourable variations; consequently,
the process of giving birth to new specific forms would
thus be retarded.   When any species becomes very rare,
close interbreeding will help to exterminate it; authors

have thought that this comes into play in accounting for the deterioration of the Aurochs in Lithuania, of Red Deer in Scotland, and of Bears in Norway, &c. Lastly, and this I am inclined to think is the most important element, a dominant species, which has already beaten many competitors in its own home, will tend to spread and supplant many others. Alph. de Candolle has shown that those species which spread widely, tend generally to spread *very* widely; consequently, they will tend to supplant and exterminate several species in several areas, and thus check the inordinate increase of specific forms throughout the world. Dr. Hooker has recently shown that in the S.E. corner of Australia, where, apparently, there are many invaders from different quarters of the globe, the endemic Australian species have been greatly reduced in number. How much weight to attribute to these several considerations I will not pretend to say; but conjointly they must limit in each country the tendency to an indefinite augmentation of specific forms.

Summary of Chapter.

If under changing conditions of life organic beings present individual differences in almost every part of their structure, and this cannot be disputed; if there be, owing to their geometrical rate of increase, a severe struggle for life at some age, season, or year, and this certainly cannot be disputed; then, considering the infinite complexity of the relations of all organic beings to each other and to their conditions of life, causing an infinite diversity in structure, constitution, and habits, to be advantageous to them, it would be a most extraordinary fact if no variations had ever occurred useful to each being's own welfare, in the same manner as so many variations have occurred useful to man. But if variations useful to any organic being ever do occur, assuredly individuals thus characterised will have the best chance of being preserved in the struggle for life; and from the strong principle of inheritance, these will tend to produce offspring similarly characterised. This principle of preservation, or the survival of the fittest, I have called Natural Selection. It leads to the improvement of each creature in relation to its organic and inorganic conditions of life; and consequently, in most cases, to what must be regarded as an advance in organisation. Nevertheless, low and simple forms will long endure if well fitted for their simple conditions of life.

Natural selection, on the principle of qualities being inherited at corresponding ages, can modify the egg, seed, or young, as easily as the adult. Amongst many animals, sexual selection will have given its aid to ordinary selection, by assuring to the most vigorous and best adapted males the greatest number of offspring. Sexual selection will also give characters useful to the males alone, in their struggles or rivalry with other males; and these characters will be transmitted to one sex or to both sexes, according to the form of inheritance which prevails.

Whether natural selection has really thus acted in adapting the various forms of life to their several conditions and stations, must be judged by the general tenor and balance of evidence given in the following chapters. But we have already seen how it entails extinction; and how largely extinction has acted in the world's history, geology plainly declares. Natural selection, also, leads to divergence of character; for the more organic beings diverge in structure, habits, and constitution, by so much the more can a large number be supported on the area,—of which we see proof by looking to the inhabitants of any small spot, and to the productions naturalised in foreign lands. Therefore, during the modification of the descendants of any one species, and during the incessant struggle of all species to increase in numbers, the more diversified the descendants become, the better will be their chance of success in the battle for life. Thus the small differences distinguishing varieties of the same species, steadily tend to increase, till they equal the greater differences between species of the same genus, or even of distinct genera.

We have seen that it is the common, the widely-diffused and widely-ranging species, belonging to the larger genera within each class, which vary most; and these tend to transmit to their modified offspring that superiority which now makes them dominant in their own countries. Natural selection, as has just been remarked, leads to divergence of character and to much extinction of the less improved and intermediate forms of life. On these principles, the nature of the affinities, and the generally well-defined distinctions between the innumerable organic beings in each class throughout the world, may be explained. It is a truly wonderful fact—the wonder of which we are apt to overlook from familiarity —that all animals and all plants throughout all time and space should be related to each other in groups, subordinate to groups, in the manner which we everywhere behold—namely, varieties of the same species most closely related, species of the same genus less closely and unequally related, forming sections and sub-genera,

M

species of distinct genera much less closely related, and genera related in different degrees, forming sub-families, families, orders, sub-classes and classes. The several subordinate groups in any class cannot be ranked in a single file, but seem clustered round points, and these round other points, and so on in almost endless cycles. If species had been independently created, no explanation would have been possible of this kind of classification; but it is explained through inheritance and the complex action of natural selection, entailing extinction and divergence of character, as we have seen illustrated in the diagram.

The affinities of all the beings of the same class have sometimes been represented by a great tree. I believe this simile largely speaks the truth. The green and budding twigs may represent existing species; and those produced during former years may represent the long succession of extinct species. At each period of growth all the growing twigs have tried to branch out on all sides, and to overtop and kill the surrounding twigs and branches, in the same manner as species and groups of species have at all times overmastered other species in the great battle for life. The limbs divided into great branches, and these into lesser and lesser branches, were themselves once, when the tree was young, budding twigs; and this connection of the former and present buds by ramifying branches may well represent the classification of all extinct and living species in groups subordinate to groups. Of the many twigs which flourished when the tree was a mere bush, only two or three, now grown into great branches, yet survive and bear the other branches; so with the species which lived during long-past geological periods, very few have left

living and modified descendants.   From the first growth
of the tree, many a limb and branch has decayed and
dropped off; and these fallen branches of various sizes
may represent those whole orders, families, and genera
which have now no living representatives, and which
are known to us only in a fossil state.   As we here and
there see a thin straggling branch springing from a fork
low down in a tree, and which by some chance has been
favoured and is still alive on its summit, so we occasionally
see an animal like the Ornithorhynchus or Lepidosiren,
which in some small degree connects by its affinities two
large branches of life, and which has apparently been
saved from fatal competition by having inhabited a
protected station.   As buds give rise by growth to fresh
buds, and these, if vigorous, branch out and overtop on
all sides many a feebler branch, so by generation I
believe it has been with the great Tree of Life, which fills
with its dead and broken branches the crust of the
earth, and covers the surface with its ever-branching
and beautiful ramifications.

CHAPTER V.

Laws of Variation.

Effects of changed conditions—Use and disuse, combined with natural selection; organs of flight and of vision—Acclimatisation —Correlated variation—Compensation and economy of growth— False correlations—Multiple, rudimentary, and lowly organised structures variable—Parts developed in an unusual manner are highly variable: specific characters more variable than generic: secondary sexual characters variable—Species of the same genus vary in an analogous manner—Reversions to long-lost characters —Summary.

I HAVE hitherto sometimes spoken as if the variations —so common and multiform with organic beings under domestication, and in a lesser degree with those under nature—were due to chance. This, of course, is a wholly incorrect expression, but it serves to acknowledge plainly our ignorance of the cause of each particular variation. Some authors believe it to be as much the function of the reproductive system to produce individual differences, or slight deviations of structure, as to make the child like its parents. But the fact of variations and monstrosities occurring much more frequently under domestication than under nature, and the greater variability of species having wide ranges than of those with restricted ranges, lead to the conclusion that variability is generally related to the conditions of life to which each species has been exposed during several successive generations. In the first chapter I attempted to show that changed conditions act in two ways, directly on the whole organisation or on certain parts alone, and indirectly through the reproductive system. In all cases there are two factors, the nature of the organism, which is much the most important of the two, and the nature of the conditions. The direct action of changed conditions leads to definite or indefinite results. In the latter case the organisation seems to become plastic, and we have much fluctuating variability. In the former case the nature of the organism is such that it yields readily, when subjected to certain conditions, and all, or nearly all the individuals become modified in the same way.

It is very difficult to decide how far changed conditions, such as of climate, food, &c., have acted in a definite manner. There is reason to believe that in the course of time the effects have been greater than can be proved by clear evidence. But we may safely conclude that the innumerable complex co-adaptations of structure, which we see throughout nature between various organic beings, cannot be attributed simply to such action. In the following cases the conditions seem to have produced some slight definite effect: E. Forbes asserts that shells at their southern limit, and when living in shallow water, are more brightly coloured than those of the same species from further north or from a greater depth; but this certainly does not always hold good. Mr. Gould believes that birds of the same species are more brightly coloured under a clear atmosphere, than when living near the coast or on islands; and Wollaston is convinced that residence near the sea affects the colours of insects. Moquin-

Tandon gives a list of plants which, when growing near the sea-shore, have their leaves in some degree fleshy, though not elsewhere fleshy. These slightly varying organisms are interesting in as far as they present characters analogous to those possessed by the species which are confined to similiar conditions.

When a variation is of the slightest use to any being, we cannot tell how much to attribute to the accumulative action of natural selection, and how much to the definite action of the conditions of life. Thus, it is well known to furriers that animals of the same species have thicker and better fur the further north they live; but who can tell how much of this difference may be due to the warmest-clad individuals having been favoured and preserved during many generations, and how much to the action of the severe climate? for it would appear that climate has some direct action on the hair of our domestic quadrupeds.

Instances could be given of similar varieties being produced from the same species under external conditions of life as different as can well be conceived; and, on the other hand, of dissimilar varieties being produced under apparently the same external conditions. Again, innumerable instances are known to every naturalist, of species keeping true, or not varying at all, although living under the most opposite climates. Such considerations as these incline me to lay less weight on the direct action of the surrounding conditions, than on a tendency to vary, due to causes of which we are quite ignorant.

In one sense the conditions of life may be said, not only to cause variability, either directly or indirectly, but likewise to include natural selection, for the conditions determine whether this or that variety shall survive. But when man is the selecting agent, we clearly see that the two elements of change are distinct; variability is in some manner excited, but it is the will of man which accumulates the variations in certain directions; and it is this latter agency which answers to the survival of the fittest under nature.

Effects of the increased Use and Disuse of Parts, as controlled by Natural Selection.

From the facts alluded to in the first chapter, I think there can be no doubt that use in our domestic animals has strengthened and enlarged certain parts, and disuse diminished them; and that such modifications are inherited. Under free nature, we have no standard of comparison, by which to judge of the effects of long-continued use or disuse, for we know not the parent-forms; but many animals possess structures which can be best explained by the effects of disuse. As Professor Owen has remarked, there is no greater anomaly in nature than a bird that cannot fly; yet there are several in this state. The logger-headed duck of South America can only flap along the surface of the water, and has its wings in nearly the same condition as the domestic Aylesbury duck: it is a remarkable fact that the young birds, according to Mr. Cunningham, can fly, while the adults have lost this power. As the larger ground-feeding birds seldom take flight except to escape danger, it is probable that the nearly wingless condition of several birds, now inhabiting or which lately inhabited several oceanic islands, tenanted by no beast of prey, has been caused by disuse. The ostrich indeed inhabits continents, and is exposed to danger from which it cannot escape by flight, but it can defend itself by kicking its enemies, as efficiently as many quadrupeds. We may believe that the progenitor of the ostrich genus had habits like those of the bustard, and that, as the size and weight of its body were increased during successive generations, its legs were used more, and its wings less, until they became incapable of flight.

Kirby has remarked (and I have observed the same fact) that the anterior tarsi, or feet, of many male dung-feeding beetles are often broken off; he examined seventeen specimens in his own collection, and not one had even a relic left. In the Onites apelles the tarsi are so habitually lost, that the insect has been described as not having them. In some other genera they are present, but in a rudimentary condition. In the Ateuchus or sacred beetle of the Egyptians, they are totally deficient. The evidence that accidental mutilations can be inherited is at present not decisive; but the remarkable cases observed by Brown-Séquard in guinea-pigs, of the inherited effects of operations, should make us cautious in denying this tendency. Hence it will perhaps be safest to look at the entire absence of the anterior tarsi in Ateuchus, and their rudimentary condition in some other genera, not as cases of inherited mutilations, but as due to the effects of long-continued disuse; for as many dung-feeding beetles are generally found with their tarsi lost, this must happen early in life; therefore the tarsi cannot be of much importance or be much used by these insects.

In some cases we might easily put down to disuse

modifications of structure which are wholly, or mainly, due to natural selection. Mr. Wollaston has discovered the remarkable fact that 200 beetles, out of the 550 species (but more are now known) inhabiting Madeira, are so far deficient in wings that they cannot fly; and that, of the twenty-nine endemic genera, no less than twenty-three have all their species in this condition! Several facts,—namely, that beetles in many parts of the world are frequently blown to sea and perish; that the beetles in Madeira, as observed by Mr. Wollaston, lie much concealed, until the wind lulls and the sun shines; that the proportion of wingless beetles is larger on the exposed Desertas than in Madeira itself; and especially the extraordinary fact, so strongly insisted on by Mr. Wollaston, that certain large groups of beetles, elsewhere excessively numerous, which absolutely require the use of their wings, are here almost entirely absent;—these several considerations make me believe that the wingless condition of so many Madeira beetles is mainly due to the action of natural selection, combined probably with disuse. For during many successive generations each individual beetle which flew least, either from its wings having been ever so little less perfectly developed or from indolent habit, will have had the best chance of surviving from not being blown out to sea; and, on the other hand, those beetles which most readily took to flight would oftenest have been blown to sea, and thus destroyed.

The insects in Madeira which are not ground-feeders, and which, as certain flower-feeding coleoptera and lepidoptera, must habitually use their wings to gain their subsistence, have, as Mr. Wollaston suspects, their wings not at all reduced, but even enlarged. This is quite compatible with the action of natural selection. For when a new insect first arrived on the island, the tendency of natural selection to enlarge or to reduce the wings, would depend on whether a greater number of individuals were saved by successfully battling with the winds, or by giving up the attempt and rarely or never flying. As with mariners shipwrecked near a coast, it would have been better for the good swimmers if they had been able to swim still further, whereas it would have been better for the bad swimmers if they had not been able to swim at all and had stuck to the wreck.

The eyes of moles and of some burrowing rodents are rudimentary in size , and in some cases are quite covered by skin and fur. This state of the eyes is probably due to gradual reduction from disuse, but aided perhaps by natural selection. In South America, a burrowing rodent, the tuco-tuco, or Ctenomys, is even more subterranean in its habits than the mole; and I was assured by a Spaniard, who had often caught them, that they were frequently blind. One which I kept alive was certainly in this condition, the cause, as appeared on dissection, having been inflammation of the nictitating membrane. As frequent inflammation of the eyes must be injurious to any animal, and as eyes are certainly not necessary to animals having subterranean habits, a reduction in their size, with the adhesion of the eyelids and growth of fur over them, might in such case be an advantage; and if so, natural selection would aid the effects of disuse.

It is well known that several animals, belonging to the most different classes, which inhabit the caves of Carniola and of Kentucky, are blind. In some of the crabs the foot-stalk for the eye remains, though the eye

is gone;—the stand for the telescope is there, though
the telescope with its glasses has been lost. As it is
difficult to imagine that eyes, though useless, could be
in any way injurious to animals living in darkness,
their loss may be attributed to disuse. In one of the
blind animals, namely, the cave-rat (Neotoma), two of
which were captured by Professor Silliman at above
half a mile distance from the mouth of the cave, and
therefore not in the profoundest depths, the eyes were
lustrous and of large size; and these animals, as I am
informed by Professor Silliman, after having been
exposed for about a month to a graduated light,
acquired a dim perception of objects.

It is difficult to imagine conditions of life more
similar than deep limestone caverns under a nearly
similar climate; so that, in accordance with the old
view of the blind animals having been separately
created for the American and European caverns, very
close similarity in their organisation and affinities
might have been expected. This is certainly not the
case if we look at the two whole faunas; and with
respect to the insects alone, Schiödte has remarked,
"We are accordingly prevented from considering the
entire phenomenon in any other light than something
purely local, and the similarity which is exhibited in a
few forms between the Mammoth cave (in Kentucky)
and the caves in Carniola, otherwise than as a very
plain expression of that analogy which subsists
generally between the fauna of Europe and of North
America." On my view we must suppose that
American animals, having in most cases ordinary
powers of vision, slowly migrated by successive
generations from the outer world into the deeper and

deeper recesses of the Kentucky caves, as did European animals into the caves of Europe. We have some evidence of this gradation of habit; for, as Schiödte remarks, "We accordingly look upon the subterranean faunas as small ramifications which have penetrated into the earth from the geographically limited faunas of the adjacent tracts, and which, as they extended themselves into darkness, have been accommodated to surrounding circumstances. Animals not far remote from ordinary forms, prepare the transition from light to darkness. Next follow those that are constructed for twilight; and, last of all, those destined for total darkness, and whose formation is quite peculiar." These remarks of Schiödte's, it should be understood, apply not to the same, but to distinct species. By the time that an animal had reached, after numberless generations, the deepest recesses, disuse will on this view have more or less perfectly obliterated its eyes, and natural selection will often have effected other changes, such as an increase in the length of the antennæ or palpi, as a compensation for blindness. Notwithstanding such modifications, we might expect still to see in the cave-animals of America, affinities to the other inhabitants of that continent, and in those of Europe to the inhabitants of the European continent. And this is the case with some of the American cave-animals, as I hear from Professor Dana; and some of the European cave-insects are very closely allied to those of the surrounding country. It would be difficult to give any rational explanation of the affinities of the blind cave-animals to the other inhabitants of the two continents on the ordinary view of their independent creation. That several of the inhabitants of the caves of the Old and New Worlds should be closely related, we might expect from the well-known relationship of most of their other productions. As a blind species of Bathyscia is found in abundance on shady rocks far from caves, the loss of vision in the cave-species of this one genus has probably had no relation to its dark habitation; for it is natural that an insect already deprived of vision should readily become adapted to dark caverns. Another blind genus (Anophthalmus) offers this remarkable peculiarity, that the species, as Mr. Murray observes, have not as yet been found anywhere except in caves; yet those which inhabit the several caves of Europe and America are distinct; but it is possible that the progenitors of these several species, whilst they were furnished with eyes, may formerly have ranged over both continents, and then have become extinct, excepting in their present secluded abodes. Far from feeling surprise that some of the cave-animals should be very anomalous, as Agassiz has remarked in regard to the blind fish, the Amblyopsis, and as is the case with the blind Proteus with reference to the reptiles of Europe, I am only surprised that more wrecks of ancient life have not been preserved, owing to the less severe competition to which the scanty inhabitants of these dark abodes will have been exposed.

Acclimatisation.

Habit is hereditary with plants, as in the period of flowering, in the time of sleep, in the amount of rain requisite for seeds to germinate, &c., and this leads me to say a few words on acclimatisation. As it is extremely common for distinct species belonging to the

same genus to inhabit hot and cold countries, if it
be true that all the species of the same genus are
descended from a single parent-form, acclimatisation
must be readily effected during a long course of descent.
It is notorious that each species is adapted to the
climate of its own home: species from an arctic or even
from a temperate region cannot endure a tropical
climate, or conversely. So again, many succulent
plants cannot endure a damp climate. But the degree
of adaptation of species to the climates under which
they live is often overrated. We may infer this from
our frequent inability to predict whether or not an
imported plant will endure our climate, and from the
number of plants and animals brought from different
countries which are here perfectly healthy. We have
reason to believe that species in a state of nature are
closely limited in their ranges by the competition of
other organic beings quite as much as, or more than, by
adaptation to particular climates. But whether or not
this adaptation is in most cases very close, we have
evidence with some few plants, of their becoming, to a
certain extent, naturally habituated to different temperatures; that is, they become acclimatised: thus the pines
and rhododendrons, raised from seed collected by Dr.
Hooker from the same species growing at different
heights on the Himalaya, were found to possess in this
country different constitutional powers of resisting cold.
Mr. Thwaites informs me that he has observed similar
facts in Ceylon; analogous observations have been made
by Mr. H. C. Watson on European species of plants
brought from the Azores to England; and I could give
other cases. In regard to animals, several authentic
instances could be adduced of species having largely

extended, within historical times, their range from warmer to cooler latitudes, and conversely; but we do not positively know that these animals were strictly adapted to their native climate, though in all ordinary cases we assume such to be the case; nor do we know that they have subsequently become specially acclimatised to their new homes, so as to be better fitted for them than they were at first.

As we may infer that our domestic animals were originally chosen by uncivilised man because they were useful and because they bred readily under confinement, and not because they were subsequently found capable of far-extended transportation, the common and extraordinary capacity in our domestic animals of not only withstanding the most different climates, but of being perfectly fertile (a far severer test) under them, may be used as an argument that a large proportion of other animals now in a state of nature could easily be brought to bear widely different climates. We must not, however, push the foregoing argument too far, on account of the probable origin of some of our domestic animals from several wild stocks; the blood, for instance, of a tropical and arctic wolf may perhaps be mingled in our domestic breeds. The rat and mouse cannot be considered as domestic animals, but they have been transported by man to many parts of the world, and now have a far wider range than any other rodent; for they live under the cold climate of Faroe in the north and of the Falklands in the south, and on many an island in the torrid zones. Hence adaptation to any special climate may be looked at as a quality readily grafted on an innate wide flexibility of constitution, common to most animals. On this view, the capacity of enduring the most different climates by man himself and by his domestic animals, and the fact of the extinct elephant and rhinoceros having formerly endured a glacial climate, whereas the living species are now all tropical or sub-tropical in their habits, ought not to be looked at as anomalies, but as examples of a very common flexibility of constitution, brought, under peculiar circumstances, into action.

How much of the acclimatisation of species to any peculiar climate is due to mere habit, and how much to the natural selection of varieties having different innate constitutions, and how much to both means combined, is an obscure question. That habit or custom has some influence, I must believe, both from analogy and from the incessant advice given in agricultural works, even in the ancient Encyclopædias of China, to be very cautious in transporting animals from one district to another. And as it is not likely that man should have succeeded in selecting so many breeds and sub-breeds with constitutions specially fitted for their own districts, the result must, I think, be due to habit. On the other hand, natural selection would inevitably tend to preserve those individuals which were born with constitutions best adapted to any country which they inhabited. In treatises on many kinds of cultivated plants, certain varieties are said to withstand certain climates better than others; this is strikingly shown in works on fruit-trees published in the United States, in which certain varieties are habitually recommended for the northern and others for the southern States; and as most of these varieties are of recent origin, they cannot owe their constitutional differences to habit. The case of the Jerusalem artichoke, which is never propagated in England by seed, and of which consequently new varieties have not been produced, has even been advanced, as proving that acclimatisation cannot be effected, for it is now as tender as ever it was! The case, also, of the kidney-bean has been often cited for a similar purpose, and with much greater weight; but until someone will sow, during a score of generations, his kidney-beans so early that a very large proportion are destroyed by frost, and then collect seed from the few survivors, with care to prevent accidental crosses, and then again get seed from these seedlings, with the same precautions, the experiment cannot be said to have been tried. Nor let it be supposed that differences in the constitution of seedling kidney-beans never appear, for an account has been published how much more hardy some seedlings are than others; and of this fact I have myself observed striking instances.

On the whole, we may conclude that habit, or use and disuse, have, in some cases, played a considerable part in the modification of the constitution and structure; but that the effects have often been largely combined with, and sometimes overmastered by, the natural selection of innate variations.

Correlated Variation.

I mean by this expression that the whole organisation is so tied together during its growth and development, that when slight variations in any one part occur, and are accumulated through natural selection, other parts become modified. This is a very important subject, most imperfectly understood, and no doubt wholly different classes of facts may be here easily confounded

N

together. We shall presently see that simple inheritance often gives the false appearance of correlation. One of the most obvious real cases is, that variations of structure arising in the young or larvæ naturally tend to affect the structure of the mature animal. The several parts of the body which are homologous, and which, at an early embryonic period, are identical in structure, and which are necessarily exposed to similar conditions, seem eminently liable to vary in a like manner: we see this in the right and left sides of the body varying in the same manner; in the front and hind legs, and even in the jaws and limbs, varying together, for the lower jaw is believed by some anatomists to be homologous with the limbs. These tendencies, I do not doubt, may be mastered more or less completely by natural selection; thus a family of stags once existed with an antler only on one side; and if this had been of any great use to the breed, it might probably have been rendered permanent by selection.

Homologous parts, as has been remarked by some authors, tend to cohere; this is often seen in monstrous plants: and nothing is more common than the union of homologous parts in normal structures, as in the union of the petals into a tube. Hard parts seem to affect the form of adjoining soft parts; it is believed by some authors that with birds the diversity in the shape of the pelvis causes the remarkable diversity in the shape of their kidneys. Others believe that the shape of the pelvis in the human mother influences by pressure the shape of the head of the child. In snakes, according to Schlegel, the form of the body and the manner of swallowing determine the position and form of several of the most important viscera.

The nature of the bond is frequently quite obscure. M. Is. Geoffroy St. Hilaire has forcibly remarked, that certain malconformations frequently, and that others rarely, co-exist, without our being able to assign any reason. What can be more singular than the relation in cats between complete whiteness and blue eyes with deafness, or between the tortoise-shell colour and the female sex; or in pigeons between their feathered feet and skin betwixt the outer toes, or between the presence of more or less down on the young pigeon when first hatched, with the future colour of its plumage; or, again, the relation between the hair and teeth in the naked Turkish dog, though here no doubt homology comes into play? With respect to this latter case of correlation, I think it can hardly be accidental, that the two orders of mammals which are most abnormal in their dermal covering, viz., Cetacea (whales) and Edentata (armadilloes, scaly ant-eaters, &c.), are likewise on the whole the most abnormal in their teeth; but there are so many exceptions to this rule, as Mr. Mivart has remarked, that it has little value.

I know of no case better adapted to show the importance of the laws of correlation and variation, independently of utility and therefore of natural selection, than that of the difference between the outer and inner flowers in some Compositous and Umbelliferous plants. Every one is familiar with the difference between the ray and central florets of, for instance, the daisy, and this difference is often accompanied with the partial or complete abortion of the reproductive organs. But in some of these plants, the seeds also differ in shape and sculpture. These differences have sometimes been

attributed to the pressure of the involucra on the
florets, or to their mutual pressure, and the shape of the
seeds in the ray-florets of some Compositæ countenances
this idea; but with the Umbelliferæ, it is by no means,
as Dr. Hooker informs me, the species with the densest
heads which most frequently differ in their inner and
outer flowers. It might have been thought that the
development of the ray-petals by drawing nourishment
from the reproductive organs causes their abortion; but
this can hardly be the sole cause, for in some Compositæ
the seeds of the outer and inner florets differ, without
any difference in the corolla. Possibly these several
differences may be connected with the different flow of
nutriment towards the central and external flowers : we
know, at least, that with irregular flowers, those nearest
to the axis are most subject to peloria, that is to
become abnormally symmetrical. I may add, as an
instance of this fact, and as a striking case of correlation,
that in many pelargoniums, the two upper petals in the
central flower of the truss often lose their patches of
darker colour; and when this occurs, the adherent
nectary is quite aborted; the central flower thus
becoming peloric or regular. When the colour is
absent from only one of the two upper petals, the
nectary is not quite aborted but is much shortened.

With respect to the development of the corolla,
Sprengel's idea that the ray-florets serve to attract
insects, whose agency is highly advantageous or necessary for the fertilisation of these ₜlants, is highly
probable ; and if so, natural selection may have come
into play. But with respect to the seeds, it seems
impossible that their differences in shape, which are not
always correlated with any difference in the corolla,

can be in any way beneficial : yet in the Umbelliferæ these differences are of such apparent importance—the seeds being sometimes orthospermous in the exterior flowers and cœlospermous in the central flowers,—that the elder De Candolle founded his main divisions in the order on such characters. Hence modifications of structure, viewed by systematists as of high value, may be wholly due to the laws of variation and correlation, without being, as far as we can judge, of the slightest service to the species.

We may often falsely attribute to correlated variation structures which are common to whole groups of species, and which in truth are simply due to inheritance ; for an ancient progenitor may have acquired through natural selection some one modification in structure, and, after thousands of generations, some other and independent modification ; and these two modifications, having been transmitted to a whole group of descendants with diverse habits, would naturally be thought to be in some necessary manner correlated. Some other correlations are apparently due to the manner in which natural selection can alone act. For instance, Alph. de Candolle has remarked that winged seeds are never found in fruits which do not open ; I should explain this rule by the impossibility of seeds gradually becoming winged through natural selection, unless the capsules were open ; for in this case alone could the seeds, which were a little better adapted to be wafted by the wind, gain an advantage over others less well fitted for wide dispersal.

Compensation and Economy of Growth.

The elder Geoffroy and Goethe propounded, at about the same time, their law of compensation or balancement of growth; or, as Goethe expressed it, "in order to spend on one side, nature is forced to economise on the other side." I think this holds true to a certain extent with our domestic productions: if nourishment flows to one part or organ in excess, it rarely flows, at least in excess, to another part; thus it is difficult to get a cow to give much milk and to fatten readily. The same varieties of the cabbage do not yield abundant and nutritious foliage and a copious supply of oil-bearing seeds. When the seeds in our fruits become atrophied, the fruit itself gains largely in size and quality. In our poultry, a large tuft of feathers on the head is generally accompanied by a diminished comb, and a large beard by diminished wattles. With species in a state of nature it can hardly be maintained that the law is of universal application; but many good observers, more especially botanists, believe in its truth. I will not, however, here give any instances, for I see hardly any way of distinguishing between the effects, on the one hand, of a part being largely developed through natural selection and another and adjoining part being reduced by this same process or by disuse, and, on the other hand, the actual withdrawal of nutriment from one part owing to the excess of growth in another and adjoining part.

I suspect, also, that some of the cases of compensation which have been advanced, and likewise some other facts, may be merged under a more general principle, namely, that natural selection is continually trying to economise every part of the organisation. If under changed conditions of life a structure, before useful, becomes less useful, its diminution will be favoured, for it will profit the individual not to have its nutriment wasted in building up an useless structure. I can thus only understand a fact with which I was much struck when examining cirripedes, and of which many analogous instances could be given: namely, that when a cirripede is parasitic within another cirripede and is thus protected, it loses more or less completely its own shell or carapace. This is the case with the male Ibla, and in a truly extraordinary manner with the Proteolepas: for the carapace in all other cirripedes consists of the three highly-important anterior segments of the head enormously developed, and furnished with great nerves and muscles; but in the parasitic and protected Proteolepas, the whole anterior part of the head is reduced to the merest rudiment attached to the bases of the prehensile antennæ. Now the saving of a large and complex structure, when rendered superfluous, would be a decided advantage to each successive individual of the species; for in the struggle for life to which every animal is exposed, each would have a better chance of supporting itself, by less nutriment being wasted.

Thus, as I believe, natural selection will tend in the long run to reduce any part of the organisation, as soon as it becomes, through changed habits, superfluous, without by any means causing some other part to be largely developed in a corresponding degree. And, conversely, that natural selection may perfectly well succeed in largely developing an organ without requiring as a necessary compensation the reduction of some adjoining part.

Multiple, Rudimentary, and Lowly-organised Structures
are Variable.

It seems to be a rule, as remarked by Is. Geoffroy St. Hilaire, both with varieties and species, that when any part or organ is repeated many times in the same individual (as the vertebræ in snakes, and the stamens in polyandrous flowers) the number is variable; whereas the same part or organ, when it occurs in lesser numbers, is constant. The same author as well as some botanists have further remarked that multiple parts are extremely liable to vary in structure. As " vegetative repetition," to use Prof. Owen's expression, is a sign of low organisation, the foregoing statements accord with the common opinion of naturalists, that beings which stand low in the scale of nature are more variable than those which are higher. I presume that lowness here means that the several parts of the organisation have been but little specialised for particular functions; and as long as the same part has to perform diversified work, we can perhaps see why it should remain variable, that is, why natural selection should not have preserved or rejected each little deviation of form so carefully as when the part has to serve for some one special purpose. In the same way that a knife which has to cut all sorts of things may be of almost any shape; whilst a tool for some particular purpose must be of some particular shape. Natural selection, it should never be forgotten, can act solely through and for the advantage of each being.

Rudimentary parts, as it is generally admitted, are apt to be highly variable. We shall have to recur to this subject; and I will here only add that their variability seems to result from their uselessness, and consequently from natural selection having had no power to check deviations in their structure.

A Part developed in any Species in an extraordinary degree or manner, in comparison with the same Part in allied Species, tends to be highly variable.

Several years ago I was much struck by a remark, to the above effect, made by Mr. Waterhouse. Professor Owen, also, seems to have come to a nearly similar conclusion. It is hopeless to attempt to convince any one of the truth of the above proposition without giving the long array of facts which I have collected, and which cannot possibly be here introduced. I can only state my conviction that it is a rule of high generality. I am aware of several causes of error, but I hope that I have made due allowance for them. It should be understood that the rule by no means applies to any part, however unusually developed, unless it be unusually developed in one species or in a few species in comparison with the same part in many closely allied species. Thus, the wing of a bat is a most abnormal structure in the class of mammals, but the rule would not apply here, because the whole group of bats possesses wings; it would apply only if some one species had wings developed in a remarkable manner in comparison with the other species of the same genus. The rule applies very strongly in the case of secondary sexual characters, when displayed in any unusual manner. The term, secondary sexual characters, used by Hunter, relates to characters which are attached to one sex, but are not directly connected with the act of

reproduction. The rule applies to males and females;
but more rarely to the females, as they seldom offer
remarkable secondary sexual characters. The rule
being so plainly applicable in the case of secondary
sexual characters, may be due to the great variability
of these characters, whether or not displayed in any
unusual manner—of which fact I think there can be
little doubt. But that our rule is not confined to
secondary sexual characters is clearly shown in the case
of hermaphrodite cirripedes; I particularly attended to
Mr. Waterhouse's remark, whilst investigating this
Order, and I am fully convinced that the rule almost
always holds good. I shall, in a future work, give a
list of all the more remarkable cases; I will here give
only one, as it illustrates the rule in its largest application. The opercular valves of sessile cirripedes (rock
barnacles) are, in every sense of the word, very
important structures, and they differ extremely little
even in distinct genera; but in the several species of
one genus, Pyrgoma, these valves present a marvellous
amount of diversification; the homologous valves in the
different species being sometimes wholly unlike in
shape; and the amount of variation in the individuals
of the same species is so great, that it is no exaggeration
to state that the varieties of the same species differ
more from each other in the characters derived from
these important organs, than do the species belonging
to other distinct genera.

As with birds the individuals of the same species,
inhabiting the same country, vary extremely little, I
have particularly attended to them; and the rule
certainly seems to hold good in this class. I cannot
make out that it applies to plants, and this would have

seriously shaken my belief in its truth, had not the great variability in plants made it particularly difficult to compare their relative degrees of variability.

When we see any part or organ developed in a remarkable degree or manner in a species, the fair presumption is that it is of high importance to that species: nevertheless it is in this case eminently liable to variation. Why should this be so? On the view that each species has been independently created, with all its parts as we now see them, I can see no explanation. But on the view that groups of species are descended from some other species, and have been modified through natural selection, I think we can obtain some light. First let me make some preliminary remarks. If, in our domestic animals, any part or the whole animal be neglected, and no selection be applied, that part (for instance, the comb in the Dorking fowl) or the whole breed will cease to have a uniform character: and the breed may be said to be degenerating. In rudimentary organs, and in those which have been but little specialised for any particular purpose, and perhaps in polymorphic groups, we see a nearly parallel case; for in such cases natural selection either has not or cannot have come into full play, and thus the organisation is left in a fluctuating condition. But what here more particularly concerns us is, that those points in our domestic animals, which at the present time are undergoing rapid change by continued selection, are also eminently liable to variation. Look at the individuals of the same breed of the pigeon, and see what a prodigious amount of difference there is in the beaks of tumblers, in the beaks and wattle of carriers, in the carriage and tail of fantails, &c., these

being the points now mainly attended to by English
fanciers. Even in the same sub-breed, as in that of the
short-faced tumbler, it is notoriously difficult to breed
nearly perfect birds, many departing widely from the
standard. There may truly be said to be a constant
struggle going on between, on the one hand, the
tendency to reversion to a less perfect state, as well as
an innate tendency to new variations, and, on the other
hand, the power of steady selection to keep the breed
true. In the long run selection gains the day, and we
do not expect to fail so completely as to breed a bird as
coarse as a common tumbler pigeon from a good short-
faced strain. But as long as selection is rapidly going
on, much variability in the parts undergoing modification may always be expected.

Now let us turn to nature. When a part has been
developed in an extraordinary manner in any one
species, compared with the other species of the same
genus, we may conclude that this part has undergone
an extraordinary amount of modification since the
period when the several species branched off from the
common progenitor of the genus. This period will
seldom be remote in any extreme degree, as species
rarely endure for more than one geological period. An
extraordinary amount of modification implies an unusually large and long-continued amount of variability,
which has continually been accumulated by natural
selection for the benefit of the species. But as the
variability of the extraordinarily developed part or organ
has been so great and long-continued within a period
not excessively remote, we might, as a general rule, still
expect to find more variability in such parts than in
other parts of the organisation which have remained for

a much longer period nearly constant. And this, I am convinced, is the case. That the struggle between natural selection on the one hand, and the tendency to reversion and variability on the other hand, will in the course of time cease; and that the most abnormally developed organs may be made constant, I see no reason to doubt. Hence, when an organ, however abnormal it may be, has been transmitted in approximately the same condition to many modified descendants, as in the case of the wing of the bat, it must have existed, according to our theory, for an immense period in nearly the same state; and thus it has come not to be more variable than any other structure. It is only in those cases in which the modification has been comparatively recent and extraordinarily great that we ought to find the *generative variability*, as it may be called, still present in a high degree. For in this case the variability will seldom as yet have been fixed by the continued selection of the individuals varying in the required manner and degree, and by the continued rejection of those tending to revert to a former and less-modified condition.

Specific Characters more Variable than Generic Characters.

The principle discussed under the last heading may be applied to our present subject. It is notorious that specific characters are more variable than generic. To explain by a simple example what is meant: if in a large genus of plants some species had blue flowers and some had red, the colour would be only a specific character, and no one would be surprised at one of the blue species varying into red, or conversely; but if all the species had blue flowers, the colour would become a generic character, and its variation would be a more unusual circumstance. I have chosen this example because the explanation which most naturalists would advance is not here applicable, namely, that specific characters are more variable than generic, because they are taken from parts of less physiological importance than those commonly used for classing genera. I believe this explanation is partly, yet only indirectly, true; I shall, however, have to return to this point in the chapter on Classification. It would be almost superfluous to adduce evidence in support of the statement, that ordinary specific characters are more variable than generic; but with respect to important characters, I have repeatedly noticed in works on natural history, that when an author remarks with surprise that some important organ or part, which is generally very constant throughout a large group of species, *differs* considerably in closely-allied species, it is often *variable* in the individuals of the same species. And this fact shows that a character, which is generally of generic value, when it sinks in value and becomes only of specific value, often becomes variable, though its physiological importance may remain the same. Something of the same kind applies to monstrosities: at least Is. Geoffroy St. Hilaire apparently entertains no doubt, that the more an organ normally differs in the different species of the same group, the more subject it is to anomalies in the individuals.

On the ordinary view of each species having been independently created, why should that part of the structure, which differs from the same part in other

independently-created species of the same genus, be
more variable than those parts which are closely alike
in the several species ?   I do not see that any explanation can be given.   But on the view that species are
only strongly marked and fixed varieties, we might
expect often to find them still continuing to vary in
those parts of their structure which have varied within
a moderately recent period, and which have thus come
to differ.   Or to state the case in another manner :—
the points in which all the species of a genus resemble
each other, and in which they differ from allied genera,
are called generic characters; and these characters may
be attributed to inheritance from a common progenitor,
for it can rarely have happened that natural selection
will have modified several distinct species, fitted to
more or less widely-different habits, in exactly the
same manner : and as these so-called generic characters
have been inherited from before the period when the
several species first branched off from their common
progenitor, and subsequently have not varied or come
to differ in any degree, or only in a slight degree, it is
not probable that they should vary at the present day.
On the other hand, the points in which species differ
from other species of the same genus are called specific
characters; and as these specific characters have varied
and come to differ since the period when the species
branched off from a common progenitor, it is probable
that they should still often be in some degree variable,
—at least more variable than those parts of the
organisation which have for a very long period remained
constant.

Secondary Sexual Characters Variable.—I think it
will be admitted by naturalists, without my entering on

details, that secondary sexual characters are highly variable. It will also be admitted that species of the same group differ from each other more widely in their secondary sexual characters, than in other parts of their organisation: compare, for instance, the amount of difference between the males of gallinaceous birds, in which secondary sexual characters are strongly displayed, with the amount of difference between the females. The cause of the original variability of these characters is not manifest; but we can see why they should not have been rendered as constant and uniform as others, for they are accumulated by sexual selection, which is less rigid in its action than ordinary selection, as it does not entail death, but only gives fewer offspring to the less favoured males. Whatever the cause may be of the variability of secondary sexual characters, as they are highly variable, sexual selection will have had a wide scope for action, and may thus have succeeded in giving to the species of the same group a greater amount of difference in these than in other respects.

It is a remarkable fact, that the secondary differences between the two sexes of the same species are generally displayed in the very same parts of the organisation in which the species of the same genus differ from each other. Of this fact I will give in illustration the two first instances which happen to stand on my list; and as the differences in these cases are of a very unusual nature, the relation can hardly be accidental. The same number of joints in the tarsi is a character common to very large groups of beetles, but in the Engidæ, as Westwood has remarked, the number varies greatly; and the number likewise differs in the two sexes of the same species. Again in the fossorial hymenoptera, the neuration of the wings is a character of the highest importance, because common to large groups; but in certain genera the neuration differs in the different species, and likewise in the two sexes of the same species. Sir J. Lubbock has recently remarked, that several minute crustaceans offer excellent illustrations of this law. "In Pontella, for instance, the sexual characters are afforded mainly by the anterior antennæ and by the fifth pair of legs: the specific differences also are principally given by these organs." This relation has a clear meaning on my view: I look at all the species of the same genus as having as certainly descended from a common progenitor, as have the two sexes of any one species. Consequently, whatever part of the structure of the common progenitor, or of its early descendants, became variable, variations of this part would, it is highly probable, be taken advantage of by natural and sexual selection, in order to fit the several species to their several places in the economy or nature, and likewise to fit the two sexes of the same species to each other, or to fit the males to struggle with other males for the possession of the females.

Finally, then, I conclude that the greater variability of specific characters, or those which distinguish species from species, than of generic characters, or those which are possessed by all the species;—that the frequent extreme variability of any part which is developed in a species in an extraordinary manner in comparison with the same part in its congeners; and the slight degree of variability in a part, however extraordinarily it may be developed, if it be common to a whole group of

o

species;—that the great variability of secondary sexual characters, and their great difference in closely allied species;—that secondary sexual and ordinary specific differences are generally displayed in the same parts of the organisation,—are all principles closely connected together. All being mainly due to the species of the same group being the descendants of a common progenitor, from whom they have inherited much in common,—to parts which have recently and largely varied being more likely still to go on varying than parts which have long been inherited and have not varied —to natural selection having more or less completely, according to the lapse of time, overmastered the tendency to reversion and to further variability,—to sexual selection being less rigid than ordinary selection,—and to variations in the same parts having been accumulated by natural and sexual selection, and having been thus adapted for secondary sexual, and for ordinary purposes.

Distinct Species present analogous Variations, so that a Variety of one Species often assumes a Character proper to an allied Species, or reverts to some of the Characters of an early Progenitor.—These propositions will be most readily understood by looking to our domestic races. The most distinct breeds of the pigeon, in countries widely apart, present sub-varieties with reversed feathers on the head, and with feathers on the feet,— characters not possessed by the aboriginal rock-pigeon; these then are analogous variations in two or more distinct races. The frequent presence of fourteen or even sixteen tail-feathers in the pouter may be con- sidered as a variation representing the normal structure of another race, the fantail. I presume that no one will doubt that all such analogous variations are due to the several races of the pigeon having inherited from a common parent the same constitution and tendency to variation, when acted on by similar unknown influences.   In the vegetable kingdom we have a case of analogous variation, in the enlarged stems, or as commonly called roots, of the Swedish turnip and Ruta baga, plants which several botanists rank as varieties produced by cultivation from a common parent: if this be not so, the case will then be one of analogous variation in two so-called distinct species; and to these a third may be added, namely, the common turnip.   According to the ordinary view of each species having been independently created, we should have to attribute this similarity in the enlarged stems of these three plants, not to the *vera causa* of community of descent, and a consequent tendency to vary in a like manner, but to three separate yet closely related acts of creation.   Many similar cases of analogous variation have been observed by Naudin in the great gourd-family, and by various authors in our cereals.   Similar cases occurring with insects under natural conditions have lately been discussed with much ability by Mr. Walsh, who has grouped them under his law of Equable Variability.

With pigeons, however, we have another case, namely, the occasional appearance in all the breeds, of slaty-blue birds with two black bars on the wings, white loins, a bar at the end of the tail, with the outer feathers externally edged near their basis with white. As all these marks are characteristic of the parent rock-pigeon, I presume that no one will doubt that this is a case of reversion, and not of a new yet analogous variation appearing in the several breeds.   We may, I think, confidently come to this conclusion, because, as we have seen, these coloured marks are eminently liable to appear in the crossed offspring of two distinct and differently coloured breeds; and in this case there is nothing in the external conditions of life to cause the reappearance of the slaty-blue, with the several marks, beyond the influence of the mere act of crossing on the laws of inheritance.

No doubt it is a very surprising fact that characters should reappear after having been lost for many, probably for hundreds of generations. But when a breed has been crossed only once by some other breed, the offspring occasionally show for many generations a tendency to revert in character to the foreign breed— some say, for a dozen or even a score of generations. After twelve generations, the proportion of blood, to use a common expression, from one ancestor, is only 1 in 2048 ; and yet, as we see, it is generally believed that a tendency to reversion is retained by this remnant of foreign blood. In a breed which has not been crossed, but in which *both* parents have lost some character which their progenitor possessed, the tendency, whether strong or weak, to reproduce the lost character might, as was formerly remarked, for all that we can see to the contrary, be transmitted for almost any number of generations. When a character which has been lost in a breed, reappears after a great number of generations, the most probable hypothesis is, not that one individual suddenly takes after an ancestor removed by some hundred generations, but that in each successive generation the character in question has been lying latent, and at last, under unknown favourable conditions, is developed. With the barb-pigeon, for instance, which very rarely produces a blue bird, it is probable that there is a latent tendency in each generation to produce blue plumage. The abstract improbability of such a tendency being transmitted through a vast number of generations, is not greater than that of quite useless or rudimentary organs being similarly transmitted. A mere tendency to produce a rudiment is indeed sometimes thus inherited.

As all the species of the same genus are supposed to be descended from a common progenitor, it might be expected that they would occasionally vary in an analogous manner; so that the varieties of two or more species would resemble each other, or that a variety of one species would resemble in certain characters another and distinct species,—this other species being, according to our view, only a well-marked and permanent variety. But characters exclusively due to analogous variation would probably be of an unimportant nature, for the preservation of all functionally important characters will have been determined through natural selection, in accordance with the different habits of the species. It might further be expected that the species of the same genus would occasionally exhibit reversions to long lost characters. As, however, we do not know the common ancestor of any natural group, we cannot distinguish between reversionary and analogous characters. If, for instance, we did not know that the parent rock-pigeon was not feather-footed or turn-crowned, we could not have told, whether such characters in our domestic breeds were reversions or only analogous variations; but we might have inferred that the blue colour was a case of reversion from the number of the markings, which are correlated with this tint, and which would not probably have all appeared together from simple

variation.   More especially we might have inferred
this, from the blue colour and the several marks so often
appearing when differently coloured breeds are crossed.
Hence, although under nature it must generally be left
doubtful, what cases are reversions to formerly existing
characters, and what are new but analogous variations,
yet we ought, on our theory, sometimes to find the
varying offspring of a species assuming characters which
are already present in other members of the same group.
And this undoubtedly is the case.

The difficulty in distinguishing variable species is
largely due to the varieties mocking, as it were, other
species of the same genus.   A considerable catalogue,
also, could be given of forms intermediate between two
other forms, which themselves can only doubtfully be
ranked as species; and this shows, unless all these
closely allied forms be considered as independently
created species, that they have in varying assumed some
of the characters of the others.   But the best evidence
of analogous variations is afforded by parts or organs
which are generally constant in character, but which
occasionally vary so as to resemble, in some degree, the
same part or organ in an allied species.   I have collected a long list of such cases; but here, as before, I lie
under the great disadvantage of not being able to give
them.   I can only repeat that such cases certainly
occur, and seem to me very remarkable.

I will, however, give one curious and complex case,
not indeed as affecting any important character, but
from occurring in several species of the same genus,
partly under domestication and partly under nature.
It is a case almost certainly of reversion.   The ass
sometimes has very distinct transverse bars on its legs,

like those on the legs of the zebra: it has been asserted that these are plainest in the foal, and, from inquiries which I have made, I believe this to be true. The stripe on the shoulder is sometimes double, and is very variable in length and outline. A white ass, but *not* an albino, has been described without either spinal or shoulder stripe : and these stripes are sometimes very obscure, or actually quite lost, in dark-coloured asses. The koulan of Pallas is said to have been seen with a double shoulder-stripe. Mr. Blyth has seen a specimen of the hemionus with a distinct shoulder-stripe, though it properly has none ; and I have been informed by Colonel Poole that the foals of this species are generally striped on the legs, and faintly on the shoulder. The quagga, though so plainly barred like a zebra over the body, is without bars on the legs ; but Dr. Gray has figured one specimen with very distinct zebra-like bars on the hocks.

With respect to the horse, I have collected cases in England of the spinal stripe in horses of the most distinct breeds, and of *all* colours : transverse bars on the legs are not rare in duns, mouse-duns, and in one instance in a chestnut: a faint shoulder-stripe may sometimes be seen in duns, and I have seen a trace in a bay horse. My son made a careful examination and sketch for me of a dun Belgian cart-horse with a double stripe on each shoulder and with leg-stripes ; I have myself seen a dun Devonshire pony, and a small dun Welsh pony has been carefully described to me, both with *three* parallel stripes on each shoulder.

In the north-west part of India the Kattywar breed of horses is so generally striped, that, as I hear from Colonel Poole, who examined this breed for the Indian Government, a horse without stripes is not considered as purely-bred. The spine is always striped; the legs are generally barred; and the shoulder-stripe, which is sometimes double and sometimes treble, is common; the side of the face, moreover, is sometimes striped. The stripes are often plainest in the foal; and sometimes quite disappear in old horses. Colonel Poole has seen both gray and bay Kattywar horses striped when first foaled. I have also reason to suspect, from information given me by Mr. W. W. Edwards, that with the English race-horse the spinal stripe is much commoner in the foal than in the full-grown animal. I have myself recently bred a foal from a bay mare (offspring of a Turkoman horse and a Flemish mare) by a bay English race-horse; this foal when a week old was marked on its hinder quarters and on its forehead with numerous, very narrow, dark, zebra-like bars, and its legs were feebly striped: all the stripes soon disappeared completely. Without here entering on further details, I may state that I have collected cases of leg and shoulder stripes in horses of very different breeds in various countries from Britain to Eastern China; and from Norway in the north to the Malay Archipelago in the south. In all parts of the world these stripes occur far oftenest in duns and mouse-duns; by the term dun a large range of colour is included, from one between brown and black to a close approach to cream-colour.

I am aware that Colonel Hamilton Smith, who has written on this subject, believes that the several breeds of the horse are descended from several aboriginal species—one of which, the dun, was striped; and that the above-described appearances are all due to ancient crosses with the dun stock. But this view may be safely rejected; for it is highly improbable that the heavy Belgian cart-horse, Welsh ponies, Norwegian cobs, the lanky Kattywar race, &c., inhabiting the most distant parts of the world, should all have been crossed with one supposed aboriginal stock.

Now let us turn to the effects of crossing the several species of the horse-genus. Rollin asserts, that the common mule from the ass and horse is particularly apt to have bars on its legs; according to Mr. Gosse, in certain parts of the United States about nine out of ten mules have striped legs. I once saw a mule with its legs so much striped that any one might have thought that it was a hybrid-zebra; and Mr. W. C. Martin, in his excellent treatise on the horse, has given a figure of a similar mule. In four coloured drawings, which I have seen, of hybrids between the ass and zebra, the legs were much more plainly barred than the rest of the body; and in one of them there was a double shoulder-stripe. In Lord Morton's famous hybrid, from a chestnut mare and male quagga, the hybrid, and even the pure offspring subsequently produced from the same mare by a black Arabian sire, were much more plainly barred across the legs than is even the pure quagga. Lastly, and this is another most remarkable case, a hybrid has been figured by Dr. Gray (and he informs me that he knows of a second case) from the ass and the hemionus; and this hybrid, though the ass only occasionally has stripes on his legs and the hemionus has none and has not even a shoulder-stripe, nevertheless had all four legs barred, and had three short shoulder-stripes, like those on the dun Devonshire and Welsh ponies, and even had some zebra-like stripes on the sides of its face. With respect to this last fact, I was so convinced that not even a stripe of colour appears from what is commonly called chance, that I was led solely from the occurrence of the face-stripes on this hybrid from the ass and hemionus to ask Colonel Poole whether such face-stripes ever occurred in the eminently striped Kattywar breed of horses, and was, as we have seen, answered in the affirmative.

What now are we to say to these several facts? We see several distinct species of the horse-genus becoming, by simple variation, striped on the legs like a zebra, or striped on the shoulders like an ass. In the horse we see this tendency strong whenever a dun tint appears— a tint which approaches to that of the general colouring of the other species of the genus. The appearance of the stripes is not accompanied by any change of form or by any other new character. We see this tendency to become striped most strongly displayed in hybrids from between several of the most distinct species. Now observe the case of the several breeds of pigeons: they are descended from a pigeon (including two or three sub-species or geographical races) of a bluish colour, with certain bars and other marks; and when any breed assumes by simple variation a bluish tint, these bars and other marks invariably reappear; but without any other change of form or character. When the oldest and truest breeds of various colours are crossed, we see a strong tendency for the blue tint and bars and marks to reappear in the mongrels. I have stated that the most probable hypothesis to account for the reappearance of very ancient characters, is—that there is a *tendency* in the young of each successive generation to produce the long-lost character, and that this tendency, from unknown causes, sometimes prevails.

And we have just seen that in several species of the horse-genus the stripes are either plainer or appear more commonly in the young than in the old. Call the breeds of pigeons, some of which have bred true for centuries, species; and how exactly parallel is the case with that of the species of the horse-genus! For myself, I venture confidently to look back thousands on thousands of generations, and I see an animal striped like a zebra, but perhaps otherwise very differently constructed, the common parent of our domestic horse (whether or not it be descended from one or more wild stocks) of the ass, the hemionus, quagga, and zebra.

He who believes that each equine species was independently created, will, I presume, assert that each species has been created with a tendency to vary, both under nature and under domestication, in this particular manner, so as often to become striped like the other species of the genus; and that each has been created with a strong tendency, when crossed with species inhabiting distant quarters of the world, to produce hybrids resembling in their stripes, not their own parents, but other species of the genus. To admit this view is, as it seems to me, to reject a real for an unreal, or at least for an unknown, cause. It makes the works of God a mere mockery and deception; I would almost as soon believe with the old and ignorant cosmogonists, that fossil shells had never lived, but had been created in stone so as to mock the shells living on the sea-shore.

Summary.—Our ignorance of the laws of variation is profound. Not in one case out of a hundred can we pretend to assign any reason why this or that part has varied. But whenever we have the means of instituting a comparison, the same laws appear to have acted in producing the lesser differences between varieties of the same species, and the greater differences between species of the same genus. Changed conditions generally induce mere fluctuating variability, but sometimes they cause direct and definite effects ; and these may become strongly marked in the course of time, though we have not sufficient evidence on this head. Habit in producing constitutional peculiarities and use in strengthening and disuse in weakening and diminishing organs, appear in many cases to have been potent in in their effects. Homologous parts tend to vary in the same manner, and homologous parts tend to cohere. Modifications in hard parts and in external parts sometimes affect softer and internal parts. When one part is largely developed, perhaps it tends to draw nourishment from the adjoining parts ; and every part of the structure which can be saved without detriment will be saved. Changes of structure at an early age may affect parts subsequently developed ; and many cases of correlated variation, the nature of which we are unable to understand, undoubtedly occur. Multiple parts are variable in number and in structure, perhaps arising from such parts not having been closely special- ised for any particular function, so that their modifica- tions have not been closely checked by natural selection. It follows probably from this same cause, that organic beings low in the scale are more variable than those standing higher in the scale, and which have their whole organisation more specialised. Rudimentary organs, from being useless, are not regulated by natural selection, and hence are variable. Specific characters— that is, the characters which have come to differ since the several species of the same genus branched off from

a common parent—are more variable than generic
characters, or those which have long been inherited,
and have not differed within this same period. In
these remarks we have referred to special parts or
organs being still variable, because they have recently
varied and thus come to differ; but we have also seen
in the second chapter that the same principle applies to
the whole individual; for in a district where many
species of a genus are found—that is, where there has
been much former variation and differentiation, or where
the manufactory of new specific forms has been actively
at work—in that district and amongst these species, we
now find, on an average, most varieties. Secondary
sexual characters are highly variable, and such characters differ much in the species of the same group.
Variability in the same parts of the organisation has
generally been taken advantage of in giving secondary
sexual differences to the two sexes of the same species,
and specific differences to the several species of the
same genus. Any part or organ developed to an
extraordinary size or in an extraordinary manner, in
comparison with the same part or organ in the allied
species, must have gone through an extraordinary
amount of modification since the genus arose; and thus
we can understand why it should often still be variable
in a much higher degree than other parts; for variation
is a long-continued and slow process, and natural
selection will in such cases not as yet have had time to
overcome the tendency to further variability and to
reversion to a less modified state. But when a species
with any extraordinarily-developed organ has become
the parent of many modified descendants—which on
our view must be a very slow process, requiring a long

lapse of time—in this case, natural selection has succeeded in giving a fixed character to the organ, in however extraordinary a manner it may have been developed. Species inheriting nearly the same constitution from a common parent, and exposed to similar influences, naturally tend to present analogous variations, or these same species may occasionally revert to some of the characters of their ancient progenitors. Although new and important modifications may not arise from reversion and analogous variation, such modifications will add to the beautiful and harmonious diversity of nature.

Whatever the cause may be of each slight difference between the offspring and their parents—and a cause for each must exist—we have reason to believe that it is the steady accumulation of beneficial differences which has given rise to all the more important modifications of structure in relation to the habits of each species.

CHAPTER VI.

DIFFICULTIES OF THE THEORY.

Difficulties of the theory of descent with modification—Absence or rarity of transitional varieties—Transitions in habits of life—Diversified habits in the same species—Species with habits widely different from those of their allies—Organs of extreme perfection—Modes of transition—Cases of difficulty—Natura non facit saltum—Organs of small importance—Organs not in all cases absolutely perfect—The law of Unity of Type and of the Conditions of Existence embraced by the theory of Natural Selection.

LONG before the reader has arrived at this part of my work, a crowd of difficulties will have occurred to him. Some of them are so serious that to this day I can hardly reflect on them without being in some degree staggered; but, to the best of my judgment, the greater number are only apparent, and those that are real are not, I think, fatal to the theory.

These difficulties and objections may be classed under the following heads:—First, why, if species have descended from other species by fine gradations, do we not everywhere see innumerable transitional forms? Why is not all nature in confusion, instead of the species being, as we see them, well defined?

Secondly, is it possible that an animal having, for instance, the structure and habits of a bat, could have been formed by the modification of some other animal with widely different habits and structure? Can we believe that natural selection could produce, on the one hand, an organ of trifling importance, such as the tail of a giraffe, which serves as a fly-flapper, and, on the other hand, an organ so wonderful as the eye ?

Thirdly, can instincts be acquired and modified through natural selection ? What shall we say to the instinct which leads the bee to make cells, and which has practically anticipated the discoveries of profound mathematicians ?

Fourthly, how can we account for species, when crossed, being sterile and producing sterile offspring, whereas, when varieties are crossed, their fertility is unimpaired ?

The two first heads will here be discussed; some miscellaneous objections in the following chapter; Instinct and Hybridism in the two succeeding chapters.

On the Absence or Rarity of Transitional Varieties.— As natural selection acts solely by the preservation of profitable modifications, each new form will tend in a fully-stocked country to take the place of, and finally to exterminate, its own less improved parent-form and other less-favoured forms with which it comes into competition. Thus extinction and natural selection go hand in hand. Hence, if we look at each species as descended from some unknown form, both the parent and all the transitional varieties will generally have been exterminated by the very process of the formation and perfection of the new form.

But, as by this theory innumerable transitional forms must have existed, why do we not find them embedded in countless numbers in the crust of the earth ? It will be more convenient to discuss this question in the chapter on the Imperfection of the Geological Record ;

and I will here only state that I believe the answer mainly lies in the record being incomparably less perfect than is generally supposed. The crust of the earth is a vast museum; but the natural collections have been imperfectly made, and only at long intervals of time.

But it may be urged that when several closely-allied species inhabit the same territory, we surely ought to find at the present time many transitional forms. Let us take a simple case: in travelling from north to south over a continent, we generally meet at successive intervals with closely allied or representative species, evidently filling nearly the same place in the natural economy of the land. These representative species often meet and interlock; and as the one becomes rarer and rarer, the other becomes more and more frequent, till the one replaces the other. But if we compare these species where they intermingle, they are generally as absolutely distinct from each other in every detail of structure as are specimens taken from the metropolis inhabited by each. By my theory these allied species are descended from a common parent; and during the process of modification, each has become adapted to the conditions of life of its own region, and has supplanted and exterminated its original parent-form and all the transitional varieties between its past and present states. Hence we ought not to expect at the present time to meet with numerous transitional varieties in each region, though they must have existed there, and may be embedded there in a fossil condition. But in the intermediate region, having intermediate conditions of life, why do we not now find closely-linking intermediate varieties? This difficulty for a long time quite

P

confounded me. But I think it can be in large part explained.

In the first place we should be extremely cautious in inferring, because an area is now continuous, that it has been continuous during a long period. Geology would lead us to believe that most continents have been broken up into islands even during the later tertiary periods; and in such islands distinct species might have been separately formed without the possibility of inter-mediate varieties existing in the intermediate zones. By changes in the form of the land and of climate, marine areas now continuous must often have existed within recent times in a far less continuous and uniform condition than at present. But I will pass over this way of escaping from the difficulty; for I believe that many perfectly defined species have been formed on strictly continuous areas; though I do not doubt that the formerly broken condition of areas now continuous, has played an important part in the formation of new species, more especially with freely-crossing and wander-ing animals.

In looking at species as they are now distributed over a wide area, we generally find them tolerably numerous over a large territory, then becoming somewhat abruptly rarer and rarer on the confines, and finally disappearing. Hence the neutral territory between two representative species is generally narrow in comparison with the territory proper to each. We see the same fact in ascending mountains, and sometimes it is quite remark-able how abruptly, as Alph. de Candolle has observed, a common alpine species disappears. The same fact has been noticed by E. Forbes in sounding the depths of the sea with the dredge. To those who look at climate and the physical conditions of life as the all-important elements of distribution, these facts ought to cause surprise, as climate and height or depth graduate away insensibly. But when we bear in mind that almost every species, even in its metropolis, would increase immensely in numbers, were it not for other competing species; that nearly all either prey on or serve as prey for others; in short, that each organic being is either directly or indirectly related in the most important manner to other organic beings,—we see that the range of the inhabitants of any country by no means exclusively depends on insensibly changing physical conditions, but in a large part on the presence of other species, on which it lives, or by which it is destroyed, or with which it comes into competition; and as these species are already defined objects, not blending one into another by insensible gradations, the range of any one species, depending as it does on the range of others, will tend to be sharply defined. Moreover, each species on the confines of its range, where it exists in lessened numbers, will, during fluctuations in the number of its enemies or of its prey, or in the nature of the seasons, be extremely liable to utter extermination; and thus its geographical range will come to be still more sharply defined.

As allied or representative species, when inhabiting a continuous area, are generally distributed in such a manner that each has a wide range, with a comparatively narrow neutral territory between them, in which they become rather suddenly rarer and rarer; then, as varieties do not essentially differ from species, the same rule will probably apply to both; and if we take a varying species inhabiting a very large area, we shall have

to adapt two varieties to two large areas, and a third
variety to a narrow intermediate zone. The intermediate variety, consequently, will exist in lesser
numbers from inhabiting a narrow and lesser area; and
practically, as far as I can make out, this rule holds
good with varieties in a state of nature. I have met
with striking instances of the rule in the case of
varieties intermediate between well-marked varieties in
the genus Balanus. And it would appear from information given me by Mr. Watson, Dr. Asa Gray, and
Mr. Wollaston, that generally, when varieties intermediate between two other forms occur, they are much
rarer numerically than the forms which they connect.
Now, if we may trust these facts and inferences, and conclude that varieties linking two other varieties together
generally have existed in lesser numbers than the
forms which they connect, then we can understand why
intermediate varieties should not endure for very long
periods :—why, as a general rule, they should be exterminated and disappear, sooner than the forms which
they originally linked together.

For any form existing in lesser numbers would, as
already remarked, run a greater chance of being exterminated than one existing in large numbers; and in
this particular case the intermediate form would be
eminently liable to the inroads of closely-allied forms
existing on both sides of it. But it is a far more
important consideration, that during the process of
further modification, by which two varieties are
supposed to be converted and perfected into two
distinct species, the two which exist in larger numbers,
from inhabiting larger areas, will have a great advantage
over the intermediate variety, which exists in smaller

numbers in a narrow and intermediate zone. For forms existing in larger numbers will have a better chance, within any given period, of presenting further favourable variations for natural selection to seize on, than will the rarer forms which exist in lesser numbers. Hence, the more common forms, in the race for life, will tend to beat and supplant the less common forms, for these will be more slowly modified and improved. It is the same principle which, as I believe, accounts for the common species in each country, as shown in the second chapter, presenting on an average a greater number of well-marked varieties than do the rarer species. I may illustrate what I mean by supposing three varieties of sheep to be kept, one adapted to an extensive mountainous region; a second to a comparatively narrow, hilly tract; and a third to the wide plains at the base; and that the inhabitants are all trying with equal steadiness and skill to improve their stocks by selection; the chances in this case will be strongly in favour of the great holders on the mountains or on the plains, improving their breeds more quickly than the small holders on the intermediate narrow, hilly tract; and consequently the improved mountain or plain breed will soon take the place of the less improved hill breed; and thus the two breeds, which originally existed in greater numbers, will come into close contact with each other, without the interposition of the supplanted, intermediate hill variety.

To sum up, I believe that species come to be tolerably well-defined objects, and do not at any one period present an inextricable chaos of varying and intermediate links: first, because new varieties are very slowly formed, for variation is a slow process, and natural selection can do nothing until favourable individual differences or variations occur, and until a place in the natural polity of the country can be better filled by some modification of some one or more of its inhabitants. And such new places will depend on slow changes of climate, or on the occasional immigration of new inhabitants, and, probably, in a still more important degree, on some of the old inhabitants becoming slowly modified, with the new forms thus produced and the old ones acting and reacting on each other. So that, in any one region and at any one time, we ought to see only a few species presenting slight modifications of structure in some degree permanent; and this assuredly we do see.

Secondly, areas now continuous must often have existed within the recent period as isolated portions, in which many forms, more especially amongst the classes which unite for each birth and wander much, may have separately been rendered sufficiently distinct to rank as representative species. In this case, intermediate varieties between the several representative species and their common parent, must formerly have existed within each isolated portion of the land, but these links during the process of natural selection will have been supplanted and exterminated, so that they will no longer be found in a living state.

Thirdly, when two or more varieties have been formed in different portions of a strictly continuous area, intermediate varieties will, it is probable, at first have been formed in the intermediate zones, but they will generally have had a short duration. For these intermediate varieties will, from reasons already assigned (namely from what we know of the actual distribution of closely allied or representative species, and likewise of acknowledged varieties), exist in the intermediate zones in lesser numbers than the varieties which they tend to connect. From this cause alone the intermediate varieties will be liable to accidental extermination; and during the process of further modification through natural selection, they will almost certainly be beaten and supplanted by the forms which they connect; for these from existing in greater numbers will, in the aggregate, present more varieties, and thus be further improved through natural selection and gain further advantages.

Lastly, looking not to any one time, but to all time, if my theory be true, numberless intermediate varieties, linking closely together all the species of the same group, must assuredly have existed; but the very process of natural selection constantly tends, as has been so often remarked, to exterminate the parent-forms and the intermediate links. Consequently evidence of their former existence could be found only amongst fossil remains, which are preserved, as we shall attempt to show in a future chapter, in an extremely imperfect and intermittent record.

On the Origin and Transitions of Organic Beings with peculiar Habits and Structure.—It has been asked by the opponents of such views as I hold, how, for instance, could a land carnivorous animal have been converted into one with aquatic habits; for how could the animal in its transitional state have subsisted? It would be easy to show that there now exist carnivorous animals presenting close intermediate grades from strictly terrestrial to aquatic habits; and as each exists by a struggle for life, it is clear that each must be well adapted to its place in nature. Look at the Mustela vison of North America, which has webbed feet, and which resembles an otter in its fur, short legs, and form of tail. During the summer this animal dives for and preys on fish, but during the long winter it leaves the frozen waters, and preys, like other pole-cats, on mice and land animals. If a different case had been taken, and it had been asked how an insectivorous quadruped could possibly have been converted into a flying bat, the question would have been far more difficult to answer. Yet I think such difficulties have little weight.

Here, as on other occasions, I lie under a heavy disadvantage, for, out of the many striking cases which I have collected, I can give only one or two instances of transitional habits and structures in allied species; and of diversified habits, either constant or occasional, in the same species. And it seems to me that nothing less than a long list of such cases is sufficient to lessen the difficulty in any particular case like that of the bat.

Look at the family of squirrels; here we have the finest gradation from animals with their tails only slightly flattened, and from others, as Sir J. Richardson has remarked, with the posterior part of their bodies rather wide and with the skin on their flanks rather full, to the so-called flying squirrels; and flying squirrels have their limbs and even the base of the tail united by a broad expanse of skin, which serves as a parachute and allows them to glide through the air to an astonishing distance from tree to tree. We cannot doubt that each structure is of use to each kind of squirrel in its own country, by enabling it to escape birds or beasts of prey, to collect food more quickly, or, as there is reason to believe, to lessen the danger from occasional falls. But it does not follow from this fact that the structure of each squirrel is the best that it is possible to conceive under all possible conditions. Let the climate and vegetation change, let other competing rodents or new beasts of prey immigrate, or old ones become modified, and all analogy would lead us to believe that some at least of the squirrels would decrease in numbers or become exterminated, unless they also became modified and improved in structure in a corresponding manner. Therefore, I can see no difficulty, more especially under changing conditions of life, in the continued preservation of individuals with fuller and fuller flank-membranes, each modification being useful, each being propagated, until, by the accumulated effects of this process of natural selection, a perfect so-called flying squirrel was produced.

Now look at the Galeopithecus or so-called flying lemur, which formerly was ranked amongst bats, but is now believed to belong to the Insectivora. An extremely wide flank-membrane stretches from the corners of the jaw to the tail, and includes the limbs with the elongated fingers. This flank-membrane is furnished with an extensor muscle. Although no graduated links of structure, fitted for gliding through the air, now connect the Galeopithecus with the other Insectivora, yet there is no difficulty in supposing that such links formerly existed, and that each was developed in the same manner as with the less perfectly gliding squirrels; each grade of structure having been useful to its possessor. Nor can I see any insuperable difficulty in further believing that the membrane connected fingers and fore-arm of the Galeopithecus might have been greatly lengthened by natural selection; and this, as far as the organs of flight are concerned, would have converted the animal into a bat. In certain bats in which the wing-membrane extends from the top of the shoulder to the tail and includes the hind-legs, we perhaps see traces of an apparatus originally fitted for gliding through the air rather than for flight.

If about a dozen genera of birds were to become extinct, who would have ventured to surmise that birds might have existed which used their wings solely as flappers, like the logger-headed duck (Micropterus of Eyton); as fins in the water and as front-legs on the land, like the penguin; as sails, like the ostrich; and functionally for no purpose, like the Apteryx? Yet the structure of each of these birds is good for it, under the conditions of life to which it is exposed, for each has to live by a struggle; but it is not necessarily the best possible under all possible conditions. It must not be inferred from these remarks that any of the grades of wing-structure here alluded to, which perhaps may all be the result of disuse, indicate the steps by which birds actually acquired their perfect power of flight; but they serve to show what diversified means of transition are at least possible.

Seeing that a few members of such water-breathing classes as the Crustacea and Mollusca are adapted to live on the land; and seeing that we have flying birds and mammals, flying insects of the most diversified types, and formerly had flying reptiles, it is conceivable that flying-fish, which now glide far through the air, slightly rising and turning by the aid of their fluttering fins, might have been modified into perfectly winged animals. If this had been effected, who would have ever imagined that in an early transitional state they had been the inhabitants of the open ocean, and had used their incipient organs of flight exclusively, as far as we know, to escape being devoured by other fish?

When we see any structure highly perfected for any particular habit, as the wings of a bird for flight, we should bear in mind that animals displaying early transitional grades of the structure will seldom have survived to the present day, for they will have been supplanted by their successors, which were gradually rendered more perfect through natural selection. Furthermore, we may conclude that transitional states between structures fitted for very different habits of life will rarely have been developed at an early period in great numbers and under many subordinate forms. Thus, to return to our imaginary illustration of the flying-fish, it does not seem probable that fishes capable of true flight would have been developed under many subordinate forms, for taking prey of many kinds in many ways, on the land and in the water, until their organs of flight had come to a high stage of perfection, so as to have given them a decided advantage over other animals in the battle for life. Hence the chance of discovering species with transitional grades of structure in a fossil condition will always be less, from their having existed in lesser numbers, than in the case of species with fully developed structures.

I will now give two or three instances both of diversified and of changed habits in the individuals of the same species. In either case it would be easy for natural selection to adapt the structure of the animal to its changed habits, or exclusively to one of its several habits. It is, however, difficult to decide, and im-

material for us, whether habits generally change first
and structure afterwards; or whether slight modifications of structure lead to changed habits; both
probably often occurring almost simultaneously. Of
cases of changed habits it will suffice merely to allude
to that of the many British insects which now feed on
exotic plants, or exclusively on artificial substances.
Of diversified habits innumerable instances could be
given : I have often watched a tyrant flycatcher (Sauro-
phagus sulphuratus) in South America, hovering over
one spot and then proceeding to another, like a kestrel,
and at other times standing stationary on the margin of
water, and then dashing into it like a kingfisher at a
fish. In our own country the larger titmouse (Parus
major) may be seen climbing branches, almost like a
creeper; it sometimes, like a shrike, kills small birds
by blows on the head; and I have many times seen
and heard it hammering the seeds of the yew on a
branch, and thus breaking them like a nuthatch. In
North America the black bear was seen by Hearne
swimming for hours with widely open mouth, thus
catching, almost like a whale, insects in the water.

As we sometimes see individuals following habits
different from those proper to their species and to the
other species of the same genus, we might expect that
such individuals would occasionally give rise to new
species, having anomalous habits, and with their
structure either slightly or considerably modified from
that of their type. And such instances occur in nature.
Can a more striking instance of adaptation be given
than that of a woodpecker for climbing trees and seizing
insects in the chinks of the bark ? Yet in North
America there are woodpeckers which feed largely on

fruit, and others with elongated wings which chase insects on the wing.   On the plains of La Plata, where hardly a tree grows, there is a woodpecker (Colaptes campestris) which has two toes before and two behind, a long pointed tongue, pointed tail-feathers, sufficiently stiff to support the bird in a vertical position on a post, but not so stiff as in the typical woodpeckers, and a straight strong beak.   The beak, however, is not so straight or so strong as in the typical woodpeckers, but it is strong enough to bore into wood.   Hence this Colaptes in all the essential parts of its structure is a woodpecker.   Even in such trifling characters as the colouring, the harsh tone of the voice, and undulatory flight, its close blood-relationship to our common woodpecker is plainly declared ; yet, as I can assert, not only from my own observations, but from those of the accurate Azara, in certain large districts it does not climb trees, and it makes its nest in holes in banks! In certain other districts, however, this same woodpecker, as Mr. Hudson states, frequents trees, and bores holes in the trunk for its nest.   I may mention as another illustration of the varied habits of this genus, that a Mexican Colaptes has been described by De Saussure as boring holes into hard wood in order to lay up a store of acorns.

Petrels are the most aërial and oceanic of birds, but in the quiet sounds of Tierra del Fuego, the Puffinuria berardi, in its general habits, in its astonishing power of diving, in its manner of swimming and of flying when made to take flight, would be mistaken by any one for an auk or a grebe ; nevertheless it is essentially a petrel, but with many parts of its organisation profoundly modified in relation to its new habits of life ;

whereas the woodpecker of La Plata has had its structure only slightly modified.   In the case of the water-ouzel, the acutest observer by examining its dead body would never have suspected its sub-aquatic habits; yet this bird, which is allied to the thrush family, subsists by diving—using its wings under water, and grasping stones with its feet.   All the members of the great order of Hymenopterous insects are terrestrial, excepting the genus Proctotrupes, which Sir John Lubbock has discovered to be aquatic in its habits; it often enters the water and dives about by the use not of its legs but of its wings, and remains as long as four hours beneath the surface; yet it exhibits no modification in structure in accordance with its abnormal habits.

He who believes that each being has been created as we now see it, must occasionally have felt surprise when he has met with an animal having habits and structure not in agreement.   What can be plainer than that the webbed feet of ducks and geese are formed for swimming?   Yet there are upland geese with webbed feet which rarely go near the water; and no one except Audubon has seen the frigate-bird, which has all its four toes webbed, alight on the surface of the ocean. On the other hand, grebes and coots are eminently aquatic, although their toes are only bordered by membrane.   What seems plainer than that the long toes, not furnished with membrane of the Grallatores are formed for walking over swamps and floating plants?—the water-hen and landrail are members of this order, yet the first is nearly as aquatic as the coot, and the second nearly as terrestrial as the quail or partridge.   In such cases, and many others could be given, habits have changed without a corresponding change of structure. The webbed feet of the upland goose may be said to have become almost rudimentary in function, though not in structure. In the frigate-bird, the deeply scooped membrane between the toes shows that structure has begun to change.

He who believes in separate and innumerable acts of creation may say, that in these cases it has pleased the Creator to cause a being of one type to take the place of one belonging to another type; but this seems to me only re-stating the fact in dignified language. He who believes in the struggle for existence and in the principle of natural selection, will acknowledge that every organic being is constantly endeavouring to increase in numbers; and that if any one being varies ever so little, either in habits or structure, and thus gains an advantage over some other inhabitant of the same country, it will seize on the place of that inhabitant, however different that may be from its own place. Hence it will cause him no surprise that there should be geese and frigate-birds with webbed feet, living on the dry land and rarely alighting on the water, that there should be long-toed corncrakes, living in meadows instead of in swamps; that there should be woodpeckers where hardly a tree grows; that there should be diving thrushes and diving Hymenoptera, and petrels with the habits of auks.

Organs of extreme Perfection and Complication.

To suppose that the eye with all its inimitable contrivances for adjusting the focus to different distances, for admitting different amounts of light, and for the correction of spherical and chromatic aberration, could have been formed by natural selection, seems, I freely confess, absurd in the highest degree. When it was first said that the sun stood still and the world turned round, the common sense of mankind declared the doctrine false; but the old saying of *Vox populi*, *vox Dei*, as every philosopher knows, cannot be trusted in science. Reason tells me, that if numerous gradations from a simple and imperfect eye to one complex and perfect can be shown to exist, each grade being useful to its possessor, as is certainly the case; if further, the eye ever varies and the variations be inherited, as is likewise certainly the case; and if such variations should be useful to any animal under changing conditions of life, then the difficulty of believing that a perfect and complex eye could be formed by natural selection, though insuperable by our imagination, should not be considered as subversive of the theory. How a nerve comes to be sensitive to light, hardly concerns us more than how life itself originated; but I may remark that, as some of the lowest organisms, in which nerves cannot be detected, are capable of perceiving light, it does not seem impossible that certain sensitive elements in their sarcode should become aggregated and developed into nerves, endowed with this special sensibility.

In searching for the gradations through which an organ in any species has been perfected, we ought to look exclusively to its lineal progenitors; but this is scarcely ever possible, and we are forced to look to other species and genera of the same group, that is to the collateral descendants from the same parent-form, in order to see what gradations are possible, and for the chance of some gradations having been transmitted in an unaltered or little altered condition. But the state of the same organ in distinct classes may incidentally throw light on the steps by which it has been perfected.

The simplest organ which can be called an eye consists of an optic nerve, surrounded by pigment-cells and covered by translucent skin, but without any lens or other refractive body. We may, however, according to M. Jourdain, descend even a step lower and find aggregates of pigment-cells, apparently serving as organs of vision, without any nerves, and resting merely on sarcodic tissue. Eyes of the above simple nature are not capable of distinct vision, and serve only to distinguish light from darkness. In certain star-fishes, small depressions in the layer of pigment which surrounds the nerve are filled, as described by the author just quoted, with transparent gelatinous matter, projecting with a convex surface, like the cornea in the higher animals. He suggests that this serves not to form an image, but only to concentrate the luminous rays and render their perception more easy. In this concentration of the rays we gain the first and by far the most important step towards the formation of a true, picture-forming eye; for we have only to place the naked extremity of the optic nerve, which in some of the lower animals lies deeply buried in the body, and in some near the surface, at the right distance from the concentrating apparatus, and an image will be formed on it.

In the great class of the Articulata, we may start from an optic nerve simply coated with pigment, the latter sometimes forming a sort of pupil, but destitute of a lens or other optical contrivance. With insects it is now known that the numerous facets on the cornea

q

of their great compound eyes form true lenses, and that the cones include curiously modified nervous filaments. But these organs in the Articulata are so much diversified that Müller formerly made three main classes with seven subdivisions, besides a fourth main class of aggregated simple eyes.

When we reflect on these facts, here given much too briefly, with respect to the wide, diversified, and graduated range of structure in the eyes of the lower animals; and when we bear in mind how small the number of all living forms must be in comparison with those which have become extinct, the difficulty ceases to be very great in believing that natural selection may have converted the simple apparatus of an optic nerve, coated with pigment and invested by transparent membrane, into an optical instrument as perfect as is possessed by any member of the Articulate Class.

He who will go thus far, ought not to hesitate to go one step further, if he finds on finishing this volume that large bodies of facts, otherwise inexplicable, can be explained by the theory of modification through natural selection; he ought to admit that a structure even as perfect as an eagle's eye might thus be formed, although in this case he does not know the transitional states. It has been objected that in order to modify the eye and still preserve it as a perfect instrument, many changes would have to be effected simultaneously, which, it is assumed, could not be done through natural selection; but as I have attempted to show in my work on the variation of domestic animals, it is not necessary to suppose that the modifications were all simultaneous, if they were extremely slight and gradual. Different kinds of modification would, also, serve for the

same general purpose: as Mr. Wallace has remarked,
"if a lens has too short or too long a focus, it may be
amended either by an alteration of curvature, or an
alteration of density; if the curvature be irregular, and
the rays do not converge to a point, then any increased
regularity of curvature will be an improvement. So
the contraction of the iris and the muscular movements
of the eye are neither of them essential to vision, but
only improvements which might have been added and
perfected at any stage of the construction of the
instrument." Within the highest division of the
animal kingdom, namely, the Vertebrata, we can start
from an eye so simple, that it consists, as in the
lancelet, of a little sack of transparent skin, furnished
with a nerve and lined with pigment, but destitute of
any other apparatus. In fishes and reptiles, as Owen
has remarked, "the range of gradations of dioptric
structures is very great." It is a significant fact that
even in man, according to the high authority of
Virchow, the beautiful crystalline lens is formed in the
embryo by an accumulation of epidermic cells, lying
in a sack-like fold of the skin; and the vitreous body
is formed from embryonic sub-cutaneous tissue. To
arrive, however, at a just conclusion regarding the
formation of the eye, with all its marvellous yet not
absolutely perfect characters, it is indispensable that
the reason should conquer the imagination; but I have
felt the difficulty far too keenly to be surprised at
others hesitating to extend the principle of natural
selection to so startling a length.

It is scarcely possible to avoid comparing the eye
with a telescope. We know that this instrument has
been perfected by the long-continued efforts of the

highest human intellects; and we naturally infer that
the eye has been formed by a somewhat analogous
process. But may not this inference be presumptuous?
Have we any right to assume that the Creator works by
intellectual powers like those of man? If we must
compare the eye to an optical instrument, we ought in
imagination to take a thick layer of transparent tissue,
with spaces filled with fluid, and with a nerve sensitive to light beneath, and then suppose every part
of this layer to be continually changing slowly in
density, so as to separate into layers of different densities
and thicknesses, placed at different distances from each
other, and with the surfaces of each layer slowly changing
in form. Further we must suppose that there is a
power, represented by natural selection or the survival
of the fittest, always intently watching each slight
alteration in the transparent layers; and carefully
preserving each which, under varied circumstances, in
any way or in any degree, tends to produce a distincter
image. We must suppose each new state of the
instrument to be multiplied by the million; each to be
preserved until a better one is produced, and then the
old ones to be all destroyed. In living bodies, variation
will cause the slight alterations, generation will
multiply them almost infinitely, and natural selection
will pick out with unerring skill each improvement.
Let this process go on for millions of years; and during
each year on millions of individuals of many kinds;
and may we not believe that a living optical instrument
might thus be formed as superior to one of glass, as the
works of the Creator are to those of man?

Modes of Transition.

If it could be demonstrated that any complex organ existed, which could not possibly have been formed by numerous, successive, slight modifications, my theory would absolutely break down. But I can find out no such case. No doubt many organs exist of which we do not know the transitional grades, more especially if we look to much-isolated species, round which, according to the theory, there has been much extinction. Or again, if we take an organ common to all the members of a class, for in this latter case the organ must have been originally formed at a remote period, since which all the many members of the class have been developed; and in order to discover the early transitional grades through which the organ has passed, we should have to look to very ancient ancestral forms, long since become extinct.

We should be extremely cautious in concluding that an organ could not have been formed by transitional gradations of some kind. Numerous cases could be given amongst the lower animals of the same organ performing at the same time wholly distinct functions; thus in the larva of the dragon-fly and in the fish Cobites the alimentary canal respires, digests, and excretes. In the Hydra, the animal may be turned inside out, and the exterior surface will then digest and the stomach respire. In such cases natural selection might specialise, if any advantage were thus gained, the whole or part of an organ, which had previously performed two functions, for one function alone, and thus by insensible steps greatly change its nature. Many plants are known which regularly produce at the same time differently constructed flowers; and if such plants were to produce one kind alone, a great change would be effected with comparative suddenness in the character of the species. It is, however, probable that the two sorts of flowers borne by the same plant were originally differentiated by finely graduated steps, which may still be followed in some few cases.

Again, two distinct organs, or the same organ under two very different forms, may simultaneously perform in the same individual the same function, and this is an extremely important means of transition: to give one instance,—there are fish with gills or branchiæ that breathe the air dissolved in the water, at the same time that they breathe free air in their swimbladders, this latter organ being divided by highly vascular partitions and having a ductus pneumaticus for the supply of air. To give another instance from the vegetable kingdom: plants climb by three distinct means, by spirally twining, by clasping a support with their sensitive tendrils, and by the emission of aërial rootlets; these three means are usually found in distinct groups, but some few species exhibit two of the means, or even all three, combined in the same individual. In all such cases one of the two organs might readily be modified and perfected so as to perform all the work, being aided during the progress of modification by the other organ; and then this other organ might be modified for some other and quite distinct purpose, or be wholly obliterated.

The illustration of the swimbladder in fishes is a good one, because it shows us clearly the highly important fact that an organ originally constructed for one purpose, namely, flotation, may be converted into one for a widely different purpose, namely, respiration. The swimbladder has, also, been worked in as an accessory to the auditory organs of certain fishes. All physiologists admit that the swimbladder is homologous, or "ideally similar" in position and structure with the lungs of the higher vertebrate animals: hence there is no reason to doubt that the swimbladder has actually been converted into lungs, or an organ used exclusively for respiration.

According to this view it may be inferred that all vertebrate animals with true lungs are descended by ordinary generation from an ancient and unknown prototype, which was furnished with a floating apparatus or swimbladder. We can thus, as I infer from Owen's interesting description of these parts, understand the strange fact that every particle of food and drink which we swallow has to pass over the orifice of the trachea, with some risk of falling into the lungs, notwithstanding the beautiful contrivance by which the glottis is closed. In the higher Vertebrata the branchiæ have wholly disappeared—but in the embryo the slits on the sides of the neck and the loop-like course of the arteries still mark their former position. But it is conceivable that the now utterly lost branchiæ might have been gradually worked in by natural selection for some distinct purpose: for instance, Landois has shown that the wings of insects are developed from the tracheæ; it is therefore highly probable that in this great class organs which once served for respiration have been actually converted into organs for flight.

In considering transitions of organs, it is so important to bear in mind the probability of conversion from one function to another, that I will give another instance.

Pedunculated cirripedes have two minute folds of skin, called by me the ovigerous frena, which serve, through the means of a sticky secretion, to retain the eggs until they are hatched within the sack. These cirripedes have no branchiæ, the whole surface of the body and of the sack, together with the small frena, serving for respiration. The Balanidæ or sessile cirripedes, on the other hand, have no ovigerous frena, the eggs lying loose at the bottom of the sack, within the well-enclosed shell; but they have, in the same relative position with the frena, large, much-folded membranes, which freely communicate with the circulatory lacunæ of the sack and body, and which have been considered by all naturalists to act as branchiæ. Now I think no one will dispute that the ovigerous frena in the one family are strictly homologous with the branchiæ of the other family; indeed, they graduate into each other. Therefore it need not be doubted that the two little folds of skin, which originally served as ovigerous frena, but which, likewise, very slightly aided in the act of respiration, have been gradually converted by natural selection into branchiæ, simply through an increase in their size and the obliteration of their adhesive glands. If all pedunculated cirripedes had become extinct, and they have suffered far more extinction than have sessile cirripedes, who would ever have imagined that the branchiæ in this latter family had originally existed as organs for preventing the ova from being washed out of the sack?

There is another possible mode of transition, namely, through the acceleration or retardation of the period of reproduction. This has lately been insisted on by Prof. Cope and others in the United States. It is now known that some animals are capable of reproduction at a very early age, before they have acquired their perfect characters ; and if this power became thoroughly well developed in a species, it seems probable that the adult stage of development would sooner or later be lost ; and in this case, especially if the larva differed much from the mature form, the character of the species would be greatly changed and degraded.   Again, not a few animals, after arriving at maturity, go on changing in character during nearly their whole lives.   With mammals, for instance, the form of the skull is often much altered with age, of which Dr. Murie has given some striking instances with seals ; every one knows how the horns of stags become more and more branched, and the plumes of some birds become more finely developed, as they grow older. Prof. Cope states that the teeth of certain lizards change much in shape with advancing years ; with crustaceans not only many trivial, but some important parts assume a new character, as recorded by Fritz Müller, after maturity.   In all such cases,—and many could be given,—if the age for reproduction were retarded, the character of the species, at least in its adult state, would be modified ; nor is it improbable that the previous and earlier stages of development would in some cases be hurried through and finally lost.   Whether species have often or ever been modified through this comparatively sudden mode of transition, I can form no opinion ; but if this has occurred, it is probable that the differences between the young and the mature, and between the mature and the old, were primordially acquired by graduated steps.

Special Difficulties of the Theory of Natural Selection.

Although we must be extremely cautious in concluding that any organ could not have been produced by successive, small, transitional gradations, yet undoubtedly serious cases of difficulty occur.

One of the most serious is that of neuter insects, which are often differently constructed from either the males or fertile females ; but this case will be treated of in the next chapter.   The electric organs of fishes offer another case of special difficulty ; for it is impossible to conceive by what steps these wondrous organs have been produced.   But this is not surprising, for we do not even know of what use they are.   In the Gymnotus and Torpedo they no doubt serve as powerful means of defence, and perhaps for securing prey ; yet in the Ray, as observed by Matteucci, an analogous organ in the tail manifests but little electricity, even when the animal is greatly irritated; so little, that it can hardly be of any use for the above purposes.   Moreover, in the Ray, besides the organ just referred to, there is, as Dr. R. M'Donnell has shown, another organ near the head, not known to be electrical, but which appears to be the real homologue of the electric battery in the Torpedo. It is generally admitted that there exists between these organs and ordinary muscle a close analogy, in intimate structure, in the distribution of the nerves, and in the manner in which they are acted on by various reagents. It should, also, be especially observed that muscular contraction is accompanied by an electrical discharge ; and, as Dr. Radcliffe insists, " in the electrical apparatus of the torpedo during rest, there would seem to be a charge in every respect like that which is met with in muscle and nerve during rest, and the discharge of the torpedo, instead of being peculiar, may be only another form of the discharge which attends upon the action of muscle and motor nerve." Beyond this we cannot at present go in the way of explanation; but as we know so little about the uses of these organs, and as we know nothing about the habits and structure of the progenitors of the existing electric fishes, it would be extremely bold to maintain that no serviceable transitions are possible by which these organs might have been gradually developed.

These organs appear at first to offer another and far more serious difficulty; for they occur in about a dozen kinds of fish, of which several are widely remote in their affinities. When the same organ is found in several members of the same class, especially if in members having very different habits of life, we may generally attribute its presence to inheritance from a common ancestor; and its absence in some of the members to loss through disuse or natural selection. So that, if the electric organs had been inherited from some one ancient progenitor, we might have expected that all electric fishes would have been specially related to each other; but this is far from the case. Nor does geology at all lead to the belief that most fishes formerly possessed electric organs, which their modified descendants have now lost. But when we look at the subject more closely, we find in the several fishes provided with electric organs, that these are situated in different parts of the body,—that they differ in construction, as in the arrangement of the plates, and, according to Pacini, in the process or means by which the electricity is excited —and lastly, in being supplied with nerves proceeding

from different sources, and this is perhaps the most
important of all the differences.   Hence in the several
fishes furnished with electric organs, these cannot be
considered as homologous, but only as analogous in
function.   Consequently there is no reason to suppose
that they have been inherited from a common progenitor; for had this been the case they would have
closely resembled each other in all respects.   Thus the
difficulty of an organ, apparently the same, arising in
several remotely allied species, disappears, leaving only
the lesser yet still great difficulty; namely, by what
graduated steps these organs have been developed in
each separate group of fishes.

The luminous organs which occur in a few insects,
belonging to widely different families, and which are
situated in different parts of the body, offer, under our
present state of ignorance, a difficulty almost exactly
parallel with that of the electric organs.   Other similar
cases could be given; for instance in plants, the very
curious contrivance of a mass of pollen-grains, borne on
a foot-stalk with an adhesive gland, is apparently the
same in Orchis and Asclepias,—genera almost as
remote as is possible amongst flowering plants; but
here again the parts are not homologous.   In all cases
of beings, far removed from each other in the scale of
organisation, which are furnished with similar and
peculiar organs, it will be found that although the
general appearance and function of the organs may be
the same, yet fundamental differences between them
can always be detected.   For instance, the eyes of
cephalopods or cuttle-fish and of vertebrate animals
appear wonderfully alike; and in such widely sundered groups no part of this resemblance can be due

to inheritance from a common progenitor. Mr. Mivart has advanced this case as one of special difficulty, but I am unable to see the force of his argument. An organ for vision must be formed of transparent tissue, and must include some sort of lens for throwing an image at the back of a darkened chamber. Beyond this superficial resemblance, there is hardly any real similarity between the eyes of cuttle-fish and vertebrates, as may be seen by consulting Hensen's admirable memoir on these organs in the Cephalopoda. It is impossible for me here to enter on details, but I may specify a few of the points of difference. The crystalline lens in the higher cuttle-fish consists of two parts, placed one behind the other like two lenses, both having a very different structure and disposition to what occurs in the vertebrata. The retina is wholly different, with an actual inversion of the elemental parts, and with a large nervous ganglion included within the membranes of the eye. The relations of the muscles are as different as it is possible to conceive, and so in other points. Hence it is not a little difficult to decide how far even the same terms ought to be employed in describing the eyes of the Cephalopoda and Vertebrata. It is, of course, open to any one to deny that the eye in either case could have been developed through the natural selection of successive slight variations; but if this be admitted in the one case, it is clearly possible in the other; and fundamental differences of structure in the visual organs of two groups might have been anticipated, in accordance with this view of their manner of formation. As two men have sometimes independently hit on the same invention, so in the several foregoing cases it appears that natural selection, working for the good of each being, and taking advantage of all favourable variations, has produced similar organs, as far as function is concerned, in distinct organic beings, which owe none of their structure in common to inheritance from a common progenitor.

Fritz Müller, in order to test the conclusions arrived at in this volume, has followed out with much care a nearly similar line of argument. Several families of crustaceans include a few species, possessing an air-breathing apparatus and fitted to live out of the water. In two of these families, which were more especially examined by Müller, and which are nearly related to each other, the species agree most closely in all important characters; namely in their sense organs, circulating system, in the position of the tufts of hair within their complex stomachs, and lastly in the whole structure of the water-breathing branchiæ, even to the microscopical hooks by which they are cleansed. Hence it might have been expected that in the few species belonging to both families which live on the land, the equally-important air-breathing apparatus would have been the same ; for why should this one apparatus, given for the same purpose, have been made to differ, whilst all the other important organs were closely similar or rather identical.

Fritz Müller argues that this close similarity in so many points of structure must, in accordance with the views advanced by me, be accounted for by inheritance from a common progenitor. But as the vast majority of the species in the above two families, as well as most other crustaceans, are aquatic in their habits, it is improbable in the highest degree, that their common progenitor should have been adapted for breathing air.

Müller was thus led carefully to examine the apparatus in the air-breathing species; and he found it to differ in each in several important points, as in the position of the orifices, in the manner in which they are opened and closed, and in some accessory details. Now such differences are intelligible, and might even have been expected, on the supposition that species belonging to distinct families had slowly become adapted to live more and more out of water, and to breathe the air. For these species, from belonging to distinct families, would have differed to a certain extent, and in accordance with the principle that the nature of each variation depends on two factors, viz., the nature of the organism and that of the surrounding conditions, their variability assuredly would not have been exactly the same. Consequently natural selection would have had different materials or variations to work on, in order to arrive at the same functional result; and the structures thus acquired would almost necessarily have differed. On the hypothesis of separate acts of creation the whole case remains unintelligible. This line of argument seems to have had great weight in leading Fritz Müller to accept the views maintained by me in this volume.

Another distinguished zoologist, the late Professor Claparède, has argued in the same manner, and has arrived at the same result. He shows that there are parasitic mites (Acaridæ), belonging to distinct sub-families and families, which are furnished with hair-claspers. These organs must have been independently developed, as they could not have been inherited from a common progenitor; and in the several groups they are formed by the modification of the fore-legs,—of the

hind-legs,—of the maxillæ or lips,—and of appendages
on the under side of the hind part of the body.

In the foregoing cases, we see the same end gained
and the same function performed, in beings not at all
or only remotely allied, by organs in appearance, though
not in development, closely similar. On the other hand,
it is a common rule throughout nature that the same end
should be gained, even sometimes in the case of closely-
related beings, by the most diversified means. How
differently constructed is the feathered wing of a bird
and the membrane-covered wing of a bat ; and still more
so the four wings of a butterfly, the two wings of a fly,
and the two wings with the elytra of a beetle. Bivalve
shells are made to open and shut, but on what a
number of patterns is the hinge constructed,—from
the long row of neatly interlocking teeth in a Nucula
to the simple ligament of a Mussel ! Seeds are disseminated by their minuteness, — by their capsule
being converted into a light balloon-like envelope,—
by being embedded in pulp or flesh, formed of the most
diverse parts, and rendered nutritious, as well as conspicuously coloured, so as to attract and be devoured by
birds,—by having hooks and grapnels of many kinds
and serrated awns, so as to adhere to the fur of
quadrupeds,—and by being furnished with wings and
plumes, as different in shape as they are elegant in
structure, so as to be wafted by every breeze. I will
give one other instance ; for this subject of the same
end being gained by the most diversified means well
deserves attention. Some authors maintain that
organic beings have been formed in many ways for
the sake of mere variety, almost like toys in a shop,

but such a view of nature is incredible. With plants having separated sexes, and with those in which, though hermaphrodites, the pollen does not spontaneously fall on the stigma, some aid is necessary for their fertilisation. With several kinds this is effected by the pollen-grains, which are light and incoherent, being blown by the wind through mere chance on to the stigma; and this is the simplest plan which can well be conceived. An almost equally simple, though very different, plan occurs in many plants in which a symmetrical flower secretes a few drops of nectar, and is consequently visited by insects; and these carry the pollen from the anthers to the stigma.

From this simple stage we may pass through an inexhaustible number of contrivances, all for the same purpose and effected in essentially the same manner, but entailing changes in every part of the flower. The nectar may be stored in variously shaped receptacles, with the stamens and pistils modified in many ways, sometimes forming trap-like contrivances, and sometimes capable of neatly adapted movements through irritability or elasticity. From such structures we may advance till we come to such a case of extraordinary adaptation as that lately described by Dr. Crüger in the Coryanthes. This orchid has part of its labellum or lower lip hollowed out into a great bucket, into which drops of almost pure water continually fall from two secreting horns which stand above it; and when the bucket is half full, the water overflows by a spout on one side. The basal part of the labellum stands over the bucket, and is itself hollowed out into a sort of chamber with two lateral entrances; within this chamber there are curious fleshy ridges. The most

R

ingenious man, if he had not witnessed what takes
place, could never have imagined what purpose all
these parts serve.   But Dr. Crüger saw crowds of large
humble-bees visiting the gigantic flowers of this orchid,
not in order to suck nectar, but to gnaw off the ridges
within the chamber above the bucket; in doing this
they frequently pushed each other into the bucket, and
their wings being thus wetted they could not fly away,
but were compelled to crawl out through the passage
formed by the spout or overflow.   Dr. Crüger saw a
"continual procession" of bees thus crawling out of
their involuntary bath.   The passage is narrow, and is
roofed over by the column, so that a bee, in forcing its
way out, first rubs its back against the viscid stigma
and then against the viscid glands of the pollen-masses.
The pollen-masses are thus glued to the back of the bee
which first happens to crawl out through the passage of
a lately expanded flower, and are thus carried away.
Dr. Crüger sent me a flower in spirits of wine, with a
bee which he had killed before it had quite crawled out
with a pollen-mass still fastened to its back.   When the
bee, thus provided, flies to another flower, or to the
same flower a second time, and is pushed by its
comrades into the bucket and then crawls out by the
passage, the pollen-mass necessarily comes first into
contact with the viscid stigma, and adheres to it, and
the flower is fertilised.   Now at last we see the full
use of every part of the flower, of the water-secreting
horns, of the bucket half full of water, which prevents
the bees from flying away, and forces them to crawl
out through the spout, and rub against the properly
placed viscid pollen-masses and the viscid stigma.

The construction of the flower in another closely

allied orchid, namely the Catasetum, is widely different, though serving the same end; and is equally curious. Bees visit these flowers, like those of the Coryanthes, in order to gnaw the labellum; in doing this they inevitably touch a long, tapering, sensitive projection, or, as I have called it, the antenna. This antenna, when touched, transmits a sensation or vibration to a certain membrane which is instantly ruptured; this sets free a spring by which the pollen-mass is shot forth, like an arrow, in the right direction, and adheres by its viscid extremity to the back of the bee. The pollen-mass of the male plant (for the sexes are separate in this orchid) is thus carried to the flower of the female plant, where it is brought into contact with the stigma, which is viscid enough to break certain elastic threads, and retaining the pollen, fertilisation is effected.

How, it may be asked, in the foregoing and in innumerable other instances, can we understand the graduated scale of complexity and the multifarious means for gaining the same end. The answer no doubt is, as already remarked, that when two forms vary, which already differ from each other in some slight degree, the variability will not be of the same exact nature, and consequently the results obtained through natural selection for the same general purpose will not be the same. We should also bear in mind that every highly developed organism has passed through many changes; and that each modified structure tends to be inherited, so that each modification will not readily be quite lost, but may be again and again further altered. Hence the structure of each part of each species, for whatever purpose it may serve, is the sum of many inherited changes, through which the species has passed during its successive adaptations to changed habits and conditions of life.

Finally then, although in many cases it is most difficult even to conjecture by what transitions organs have arrived at their present state; yet, considering how small the proportion of living and known forms is to the extinct and unknown, I have been astonished how rarely an organ can be named, towards which no transitional grade is known to lead. It certainly is true, that new organs appearing as if created for some special purpose, rarely or never appear in any being;— as indeed is shown by that old, but somewhat exaggerated, canon in natural history of "Natura non facit saltum." We meet with this admission in the writings of almost every experienced naturalist; or as Milne Edwards has well expressed it, Nature is prodigal in variety, but niggard in innovation. Why, on the theory of Creation, should there be so much variety and so little real novelty? Why should all the parts and organs of many independent beings, each supposed to have been separately created for its proper place in nature, be so commonly linked together by graduated steps? Why should not Nature take a sudden leap from structure to structure? On the theory of natural selection, we can clearly understand why she should not; for natural selection acts only by taking advantage of slight successive variations; she can never take a great and sudden leap, but must advance by short and sure, though slow steps.

Organs of little apparent Importance, as affected by Natural Selection.

As natural selection acts by life and death,—by the survival of the fittest, and by the destruction of the less well-fitted individuals,—I have sometimes felt great difficulty in understanding the origin or formation of parts of little importance; almost as great, though of a very different kind, as in the case of the most perfect and complex organs.

In the first place, we are much too ignorant in regard to the whole economy of any one organic being, to say what slight modifications would be of importance or not. In a former chapter I have given instances of very trifling characters, such as the down on fruit and the colour of its flesh, the colour of the skin and hair of quadrupeds, which, from being correlated with constitutional differences or from determining the attacks of insects, might assuredly be acted on by natural selection. The tail of the giraffe looks like an artificially constructed fly-flapper; and it seems at first incredible that this could have been adapted for its present purpose by successive slight modifications, each better and better fitted, for so trifling an object as to drive away flies; yet we should pause before being too positive even in this case, for we know that the distribution and existence of cattle and other animals in South America absolutely depend on their power of resisting the attacks of insects: so that individuals which could by any means defend themselves from these small enemies, would be able to range into new pastures and thus gain a great advantage. It is not that the larger quadrupeds are actually destroyed (except in some rare cases) by flies, but they are incessantly harassed and their strength reduced, so that they are more subject to disease, or not so well enabled in a coming dearth to search for food, or to escape from beasts of prey.

Organs now of trifling importance have probably in some cases been of high importance to an early progenitor, and, after having been slowly perfected at a former period, have been transmitted to existing species in nearly the same state, although now of very slight use; but any actually injurious deviations in their structure would of course have been checked by natural selection. Seeing how important an organ of locomotion the tail is in most aquatic animals, its general presence and use for many purposes in so many land animals, which in their lungs or modified swim-bladders betray their aquatic origin, may perhaps be thus accounted for. A well-developed tail having been formed in an aquatic animal, it might subsequently come to be worked in for all sorts of purposes,—as a fly-flapper, an organ of prehension, or as an aid in turning, as in the case of the dog, though the aid in this latter respect must be slight, for the hare, with hardly any tail, can double still more quickly.

In the second place, we may easily err in attributing importance to characters, and in believing that they have been developed through natural selection. We must by no means overlook the effects of the definite action of changed conditions of life,—of so-called spontaneous variations, which seem to depend in a quite subordinate degree on the nature of the conditions,—of the tendency to reversion to long-lost characters,—of the complex laws of growth, such as of correlation, compensation, of the pressure of one part on another, &c.,—and finally of sexual selection, by which characters of use to one sex are often gained and then transmitted more or less perfectly to the other sex, though of no use to this sex. But structures thus indirectly gained although at first of no advantage to a species, may subsequently have been taken advantage of by its modified descendants, under new conditions of life and newly acquired habits.

If green woodpeckers alone had existed, and we did not know that there were many black and pied kinds, I dare say that we should have thought that the green colour was a beautiful adaptation to conceal this tree-frequenting bird from its enemies; and consequently that it was a character of importance, and had been acquired through natural selection; as it is, the colour is probably in chief part due to sexual selection. A trailing palm in the Malay Archipelago climbs the loftiest trees by the aid of exquisitely constructed hooks clustered around the ends of the branches, and this contrivance, no doubt, is of the highest service to the plant; but as we see nearly similar hooks on many trees which are not climbers, and which, as there is reason to believe from the distribution of the thorn-bearing species in Africa and South America, serve as a defence against browsing quadrupeds, so the spikes on the palm may at first have been developed for this object, and subsequently have been improved and taken advantage of by the plant, as it underwent further modification and became a climber. The naked skin on the head of a vulture is generally considered as a direct adaptation for wallowing in putridity; and so it may be, or it may possibly be due to the direct action

of putrid matter; but we should be very cautious in
drawing any such inference, when we see that the skin
on the head of the clean-feeding male Turkey is likewise
naked.   The sutures in the skulls of young mammals
have been advanced as a beautiful adaptation for aiding
parturition, and no doubt they facilitate, or may be
indispensable for this act; but as sutures occur in the
skulls of young birds and reptiles, which have only to
escape from a broken egg, we may infer that this
structure has arisen from the laws of growth, and has
been taken advantage of in the parturition of the higher
animals.

We are profoundly ignorant of the cause of each
slight variation or individual difference; and we are
immediately made conscious of this by reflecting on the
differences between the breeds of our domesticated
animals in different countries,—more especially in the
less civilised countries where there has been but little
methodical selection.   Animals kept by savages in
different countries often have to struggle for their own
subsistence, and are exposed to a certain extent to
natural selection, and individuals with slightly different
constitutions would succeed best under different
climates.   With cattle susceptibility to the attacks
of flies is correlated with colour, as is the liability to be
poisoned by certain plants; so that even colour would
be thus subjected to the action of natural selection.
Some observers are convinced that a damp climate
affects the growth of the hair, and that with the hair
the horns are correlated.   Mountain breeds always
differ from lowland breeds; and a mountainous country
would probably affect the hind limbs from exercising
them more, and possibly even the form of the pelvis;

and then by the law of homologous variation, the front limbs and the head would probably be affected. The shape, also, of the pelvis might affect by pressure the shape of certain parts of the young in the womb. The laborious breathing necessary in high regions tends, as we have good reason to believe, to increase the size of the chest; and again correlation would come into play. The effects of lessened exercise together with abundant food on the whole organisation is probably still more important; and this, as H. von Nathusius has lately shown in his excellent Treatise, is apparently one chief cause of the great modification which the breeds of swine have undergone. But we are far too ignorant to speculate on the relative importance of the several known and unknown causes of variation; and I have made these remarks only to show that, if we are unable to account for the characteristic differences of our several domestic breeds, which nevertheless are generally admitted to have arisen through ordinary generation from one or a few parent-stocks, we ought not to lay too much stress on our ignorance of the precise cause of the slight analogous differences between true species.

Utilitarian Doctrine, how far true: Beauty, how acquired.

The foregoing remarks lead me to say a few words on the protest lately made by some naturalists, against the utilitarian doctrine that every detail of structure has been produced for the good of its possessor. They believe that many structures have been created for the sake of beauty, to delight man or the Creator (but this

latter point is beyond the scope of scientific discussion),
or for the sake of mere variety, a view already discussed.
Such doctrines, if true, would be absolutely fatal to my
theory.   I fully admit that many structures are now of
no direct use to their possessors, and may never have
been of any use to their progenitors ; but this does not
prove that they were formed solely for beauty or variety.
No doubt the definite action of changed conditions, and
the various causes of modifications, lately specified,
have all produced an effect, probably a great effect,
independently of any advantage thus gained.   But a
still more important consideration is that the chief
part of the organisation of every living creature is due
to inheritance ; and consequently, though each being
assuredly is well fitted for its place in nature, many
structures have now no very close and direct relation to
present habits of life.   Thus, we can hardly believe
that the webbed feet of the upland goose or of the
frigate-bird are of special use to these birds ; we cannot
believe that the similar bones in the arm of the
monkey, in the fore-leg of the horse, in the wing of
the bat, and in the flipper of the seal, are of special
use to these animals.   We may safely attribute these
structures to inheritance.   But webbed feet no doubt
were as useful to the progenitor of the upland goose
and of the frigate-bird, as they now are to the most
aquatic of living birds.   So we may believe that the
progenitor of the seal did not possess a flipper, but a
foot with five toes fitted for walking or grasping ; and
we may further venture to believe that the several
bones in the limbs of the monkey, horse, and bat,
were originally developed, on the principle of utility,
probably through the reduction of more numerous

bones in the fin of some ancient fish-like progenitor of the whole class. It is scarcely possible to decide how much allowance ought to be made for such causes of change, as the definite action of external conditions, so-called spontaneous variations, and the complex laws of growth; but with these important exceptions, we may conclude that the structure of every living creature either now is, or was formerly, of some direct or indirect use to its possessor.

With respect to the belief that organic beings have been created beautiful for the delight of man,—a belief which it has been pronounced is subversive of my whole theory,—I may first remark that the sense of beauty obviously depends on the nature of the mind, irrespective of any real quality in the admired object; and that the idea of what is beautiful, is not innate or unalterable. We see this, for instance, in the men of different races admiring an entirely different standard of beauty in their women. If beautiful objects had been created solely for man's gratification, it ought to be shown that before man appeared, there was less beauty on the face of the earth than since he came on the stage. Were the beautiful volute and cone shells of the Eocene epoch, and the gracefully sculptured ammonites of the Secondary period, created that man might ages afterwards admire them in his cabinet? Few objects are more beautiful than the minute siliceous cases of the diatomaceæ: were these created that they might be examined and admired under the higher powers of the microscope? The beauty in this latter case, and in many others, is apparently wholly due to symmetry of growth. Flowers rank amongst the most beautiful productions of nature; but they

have been rendered conspicuous in contrast with the
green leaves, and in consequence at the same time
beautiful, so that they may be easily observed by
insects. I have come to this conclusion from finding it
an invariable rule that when a flower is fertilised by
the wind it never has a gaily-coloured corolla. Several
plants habitually produce two kinds of flowers; one
kind open and coloured so as to attract insects; the
other closed, not coloured, destitute of nectar, and never
visited by insects. Hence we may conclude that, if
insects had not been developed on the face of the earth,
our plants would not have been decked with beautiful
flowers, but would have produced only such poor flowers
as we see on our fir, oak, nut and ash trees, on grasses,
spinach, docks, and nettles, which are all fertilised
through the agency of the wind. A similar line of
argument holds good with fruits; that a ripe strawberry
or cherry is as pleasing to the eye as to the palate,—
that the gaily-coloured fruit of the spindle-wood tree
and the scarlet berries of the holly are beautiful objects,
—will be admitted by every one. But this beauty
serves merely as a guide to birds and beasts, in order
that the fruit may be devoured and the manured seeds
disseminated: I infer that this is the case from having
as yet found no exception to the rule that seeds are
always thus disseminated when embedded within a
fruit of any kind (that is within a fleshy or pulpy
envelope), if it be coloured of any brilliant tint, or
rendered conspicuous by being white or black.

On the other hand, I willingly admit that a great
number of male animals, as all our most gorgeous birds,
some fishes, reptiles, and mammals, and a host of
magnificently coloured butterflies, have been rendered

beautiful for beauty's sake; but this has been effected through sexual selection, that is, by the more beautiful males having been continually preferred by the females, and not for the delight of man. So it is with the music of birds. We may infer from all this that a nearly similar taste for beautiful colours and for musical sounds runs through a large part of the animal kingdom. When the female is as beautifully coloured as the male, which is not rarely the case with birds and butterflies, the cause apparently lies in the colours acquired through sexual selection having been transmitted to both sexes, instead of to the males alone. How the sense of beauty in its simplest form—that is, the reception of a peculiar kind of pleasure from certain colours, forms, and sounds—was first developed in the mind of man and of the lower animals, is a very obscure subject. The same sort of difficulty is presented, if we enquire how it is that certain flavours and odours give pleasure, and others displeasure. Habit in all these cases appears to have come to a certain extent into play; but there must be some fundamental cause in the constitution of the nervous system in each species.

Natural selection cannot possibly produce any modification in a species exclusively for the good of another species; though throughout nature one species incessantly takes advantage of, and profits by, the structures of others. But natural selection can and does often produce structures for the direct injury of other animals, as we see in the fang of the adder, and in the ovipositor of the ichneumon, by which its eggs are deposited in the living bodies of other insects. If it could be proved that any part of the structure of any one species had been formed for the exclusive good of another species, it would annihilate my theory, for such could not have been produced through natural selection. Although many statements may be found in works on natural history to this effect, I cannot find even one which seems to me of any weight.   It is admitted that the rattlesnake has a poison-fang for its own defence, and for the destruction of its prey; but some authors suppose that at the same time it is furnished with a rattle for its own injury, namely, to warn its prey.   I would almost as soon believe that the cat curls the end of its tail when preparing to spring, in order to warn the doomed mouse.   It is a much more probable view that the rattlesnake uses its rattle, the cobra expands its frill, and the puff-adder swells whilst hissing so loudly and harshly, in order to alarm the many birds and beasts which are known to attack even the most venomous species.   Snakes act on the same principle which makes the hen ruffle her feathers and expand her wings when a dog approaches her chickens; but I have not space here to enlarge on the many ways by which animals endeavour to frighten away their enemies.

Natural selection will never produce in a being any structure more injurious than beneficial to that being, for natural selection acts solely by and for the good of each.   No organ will be formed, as Paley has remarked, for the purpose of causing pain or for doing an injury to its possessor.   If a fair balance be struck between the good and evil caused by each part, each will be found on the whole advantageous.   After the lapse of time, under changing conditions of life, if any part comes to be injurious, it will be modified; or if it be not so, the being will become extinct as myriads have become extinct.

Natural selection tends only to make each organic being as perfect as, or slightly more perfect than, the other inhabitants of the same country with which it comes into competition. And we see that this is the standard of perfection attained under nature. The endemic productions of New Zealand, for instance, are perfect one compared with another; but they are now rapidly yielding before the advancing legions of plants and animals introduced from Europe. Natural selection will not produce absolute perfection, nor do we always meet, as far as we can judge, with this high standard under nature. The correction for the aberration of light is said by Müller not to be perfect even in that most perfect organ, the human eye. Helmholtz, whose judgment no one will dispute, after describing in the strongest terms the wonderful powers of the human eye, adds these remarkable words : " That which we have discovered in the way of inexactness and imperfection in the optical machine and in the image on the retina, is as nothing in comparison with the incongruities which we have just come across in the domain of the sensations. One might say that nature has taken delight in accumulating contradictions in order to remove all foundation from the theory of a pre-existing harmony between the external and internal worlds." If our reason leads us to admire with enthusiasm a multitude of inimitable contrivances in nature, this same reason tells us, though we may easily err on both sides, that some other contrivances are less perfect. Can we consider the sting of the bee as perfect, which, when used against many kinds of enemies, cannot be withdrawn, owing to the backward serratures, and thus inevitably causes the death of the insect by tearing out its viscera?

If we look at the sting of the bee, as having existed in a remote progenitor, as a boring and serrated instrument, like that in so many members of the same great order, and that it has since been modified but not perfected for its present purpose, with the poison originally adapted for some other object, such as to produce galls, since intensified, we can perhaps understand how it is that the use of the sting should so often cause the insect's own death: for if on the whole the power of stinging be useful to the social community, it will fulfil all the requirements of natural selection, though it may cause the death of some few members. If we admire the truly wonderful power of scent by which the males of many insects find their females, can we admire the production for this single purpose of thousands of drones, which are utterly useless to the community for any other purpose, and which are ultimately slaughtered by their industrious and sterile sisters?   It may be difficult, but we ought to admire the savage instinctive hatred of the queen-bee, which urges her to destroy the young queens, her daughters, as soon as they are born, or to perish herself in the combat; for undoubtedly this is for the good of the community; and maternal love or maternal hatred, though the latter fortunately is most rare, is all the same to the inexorable principle of natural selection.   If we admire the several ingenious contrivances, by which orchids and many other plants are fertilised through insect agency, can we consider as equally perfect the elaboration of dense clouds of pollen by our fir-trees, so that a few granules may be wafted by chance on to the ovules ?

Summary: the Law of Unity of Type and of the Conditions of Existence embraced by the Theory of Natural Selection.

We have in this chapter discussed some of the difficulties and objections which may be urged against the theory. Many of them are serious ; but I think that in the discussion light has been thrown on several facts, which on the belief of independent acts of creation are utterly obscure. We have seen that species at any one period are not indefinitely variable, and are not linked together by a multitude of intermediate gradations, partly because the process of natural selection is always very slow, and at any one time acts only on a few forms ; and partly because the very process of natural selection implies the continual supplanting and extinction of preceding and intermediate gradations. Closely allied species, now living on a continuous area, must often have been formed when the area was not continuous, and when the conditions of life did not insensibly graduate away from one part to another. When two varieties are formed in two districts of a continuous area, an intermediate variety will often be formed, fitted for an intermediate zone ; but from reasons assigned, the intermediate variety will usually exist in lesser numbers than the two forms which it connects ; consequently the two latter, during the course of further modification, from existing in greater numbers, will have a great advantage over the less numerous

s

intermediate variety, and will thus generally succeed in supplanting and exterminating it.

We have seen in this chapter how cautious we should be in concluding that the most different habits of life could not graduate into each other; that a bat, for instance, could not have been formed by natural selection from an animal which at first only glided through the air.

We have seen that a species under new conditions of life may change its habits; or it may have diversified habits, with some very unlike those of its nearest congeners. Hence we can understand, bearing in mind that each organic being is trying to live wherever it can live, how it has arisen that there are upland geese with webbed feet, ground woodpeckers, diving thrushes, and petrels with the habits of auks.

Although the belief that an organ so perfect as the eye could have been formed by natural selection, is enough to stagger any one; yet in the case of any organ, if we know of a long series of gradations in complexity, each good for its possessor, then, under changing conditions of life, there is no logical impossibility in the acquirement of any conceivable degree of perfection through natural selection. In the cases in which we know of no intermediate or transitional states, we should be extremely cautious in concluding that none can have existed, for the metamorphoses of many organs show what wonderful changes in function are at least possible. For instance, a swimbladder has apparently been converted into an air-breathing lung. The same organ having performed simultaneously very different functions. and then having been in part or in whole specialised for one function; and two distinct organs having performed at the same time the same function, the one having been perfected whilst aided by the other, must often have largely facilitated transitions.

We have seen that in two beings widely remote from each other in the natural scale, organs serving for the same purpose and in external appearance closely similar may have been separately and independently formed; but when such organs are closely examined, essential differences in their structure can almost always be detected; and this naturally follows from the principle of natural selection. On the other hand, the common rule throughout nature is infinite diversity of structure for gaining the same end; and this again naturally follows from the same great principle.

In many cases we are far too ignorant to be enabled to assert that a part or organ is so unimportant for the welfare of a species, that modifications in its structure could not have been slowly accumulated by means of natural selection. In many other cases, modifications are probably the direct result of the laws of variation or of growth, independently of any good having been thus gained. But even such structures have often, as we may feel assured, been subsequently taken advantage of, and still further modified, for the good of species under new conditions of life. We may, also, believe that a part formerly of high importance has frequently been retained (as the tail of an aquatic animal by its terrestrial descendants), though it has become of such small importance that it could not, in its present state, have been acquired by means of natural selection.

Natural selection can produce nothing in one species for the exclusive good or injury of another; though it may well produce parts, organs, and excretions highly useful or even indispensable, or again highly injurious to another species, but in all cases at the same time useful to the possessor. In each well-stocked country natural selection acts through the competition of the inhabitants, and consequently leads to success in the battle for life, only in accordance with the standard of that particular country. Hence the inhabitants of one country, generally the smaller one, often yield to the inhabitants of another and generally the larger country. For in the larger country there will have existed more individuals and more diversified forms, and the competition will have been severer, and thus the standard of perfection will have been rendered higher. Natural selection will not necessarily lead to absolute perfection; nor, as far as we can judge by our limited faculties, can absolute perfection be everywhere predicated.

On the theory of natural selection we can clearly understand the full meaning of that old canon in natural history, "Natura non facit saltum." This canon, if we look to the present inhabitants alone of the world, is not strictly correct; but if we include all those of past times, whether known or unknown, it must on this theory be strictly true.

It is generally acknowledged that all organic beings have been formed on two great laws—Unity of Type, and the Conditions of Existence. By unity of type is meant that fundamental agreement in structure which we see in organic beings of the same class, and which is quite independent of their habits of life. On my theory, unity of type is explained by unity of descent. The expression of conditions of existence, so often insisted on by the illustrious Cuvier, is fully embraced by the principle of natural selection. For natural selection acts by either now adapting the varying parts of each being to its organic and inorganic conditions of life; or by having adapted them during past periods of time: the adaptations being aided in many cases by the increased use or disuse of parts, being affected by the direct action of the external conditions of life, and subjected in all cases to the several laws of growth and variation. Hence, in fact, the law of the Conditions of Existence is the higher law; as it includes, through the inheritance of former variations and adaptations, that of **Unity of Type**.

CHAPTER VII.

MISCELLANEOUS OBJECTIONS TO THE THEORY OF NATURAL SELECTION.

Longevity—Modifications not necessarily simultaneous—Modifications apparently of no direct service—Progressive development—Characters of small functional importance, the most constant—Supposed incompetence of natural selection to account for the incipient stages of useful structures—Causes which interfere with the acquisition through natural selection of useful structures—Gradations of structure with changed functions—Widely different organs in members of the same class, developed from one and the same source—Reasons for disbelieving in great and abrupt modifications.

I WILL devote this chapter to the consideration of various miscellaneous objections which have been advanced against my views, as some of the previous discussions may thus be made clearer; but it would be useless to discuss all of them, as many have been made by writers who have not taken the trouble to understand the subject. Thus a distinguished German naturalist has asserted that the weakest part of my theory is, that I consider all organic beings as imperfect: what I have really said is, that all are not as perfect as they might have been in relation to their conditions; and this is shown to be the case by so many native forms in many quarters of the world having yielded their places to intruding foreigners. Nor can organic beings, even if they were at any one time perfectly adapted to their conditions of life, have remained so, when their conditions changed, unless they themselves likewise changed and no one will dispute that the physical conditions of each country, as well as the numbers and kinds of its inhabitants, have undergone many mutations.

A critic has lately insisted, with some parade of mathematical accuracy, that longevity is a great advantage to all species, so that he who believes in natural selection "must arrange his genealogical tree" in such a manner that all the descendants have longer lives than their progenitors! Cannot our critic conceive that a biennial plant or one of the lower animals might range into a cold climate and perish there every winter; and yet, owing to advantages gained through natural selection, survive from year to year by means of its seeds or ova? Mr. E. Ray Lankester has recently discussed this subject, and he concludes, as far as its extreme complexity allows him to form a judgment, that longevity is generally related to the standard of each species in the scale of organisation, as well as to the amount of expenditure in reproduction and in general activity. And these conditions have, it is probable, been largely determined through natural selection.

It has been argued that, as none of the animals and plants of Egypt, of which we know anything, have changed during the last three or four thousand years, so probably have none in any part of the world. But, as Mr. G. H. Lewes has remarked, this line of argument proves too much, for the ancient domestic races figured on the Egyptian monuments, or embalmed, are closely similar or even identical with those now living; yet all naturalists admit that such races have been produced

through the modification of their original types.   The
many animals which have remained unchanged since
the commencement of the glacial period, would have
been an incomparably stronger case, for these have been
exposed to great changes of climate and have migrated
over great distances ; whereas, in Egypt, during the last
several thousand years, the conditions of life, as far as we
know, have remained absolutely uniform. The fact of little
or no modification having been effected since the glacial
period would have been of some avail against those who
believe in an innate and necessary law of development,
but is powerless against the doctrine of natural selection or the survival of the fittest, which implies that
when variations or individual differences of a beneficial
nature happen to arise, these will be preserved; but
this will be effected only under certain favourable
circumstances.

The celebrated palæontologist, Bronn, at the close of
his German translation of this work, asks, how, on the
principle of natural selection, can a variety live side by
side with the parent species ?   If both have become
fitted for slightly different habits of life or conditions,
they might live together; and if we lay on one side
polymorphic species, in which the variability seems to
be of a peculiar nature, and all mere temporary variations, such as size, albinism, &c., the more permanent
varieties are generally found, as far as I can discover,
inhabiting distinct stations,—such as high land or low
land, dry or moist districts.   Moreover, in the case of
animals which wander much about and cross freely,
their varieties seem to be generally confined to distinct
regions.

Bronn also insists that distinct species never differ

from each other in single characters, but in many parts;
and he asks, how it always comes that many parts
of the organisation should have been modified at the
same time through variation and natural selection? But
there is no necessity for supposing that all the parts of
any being have been simultaneously modified. The
most striking modifications, excellently adapted for
some purpose, might, as was formerly remarked, be
acquired by successive variations, if slight, first in one
part and then in another; and as they would be transmitted all together, they would appear to us as if they
had been simultaneously developed. The best answer,
however, to the above objection is afforded by those
domestic races which have been modified, chiefly through
man's power of selection, for some special purpose. Look
at the race and dray horse, or at the greyhound and
mastiff. Their whole frames and even their mental
characteristics have been modified; but if we could
trace each step in the history of their transformation,
—and the latter steps can be traced,—we should not
see great and simultaneous changes, but first one part
and then another slightly modified and improved. Even
when selection has been applied by man to some one
character alone,—of which our cultivated plants offer
the best instances,—it will invariably be found that
although this one part, whether it be the flower, fruit,
or leaves, has been greatly changed, almost all the other
parts have been slightly modified. This may be attributed partly to the principle of correlated growth, and
partly to so-called spontaneous variation.

A much more serious objection has been urged by
Bronn, and recently by Broca, namely, that many characters appear to be of no service whatever to their

possessors, and therefore cannot have been influenced through natural selection. Bronn adduces the length of the ears and tails in the different species of hares and mice,—the complex folds of enamel in the teeth of many animals, and a multitude of analogous cases. With respect to plants, this subject has been discussed by Nägeli in an admirable essay. He admits that natural selection has effected much, but he insists that the families of plants differ chiefly from each other in morphological characters, which appear to be quite unimportant for the welfare of the species. He consequently believes in an innate tendency towards progressive and more perfect development. He specifies the arrangement of the cells in the tissues, and of the leaves on the axis, as cases in which natural selection could not have acted. To these may be added the numerical divisions in the parts of the flower, the position of the ovules, the shape of the seed, when not of any use for dissemination, &c.

There is much force in the above objection. Nevertheless, we ought, in the first place, to be extremely cautious in pretending to decide what structures now are, or have formerly been, of use to each species. In the second place, it should always be borne in mind that when one part is modified, so will be other parts, through certain dimly seen causes, such as an increased or diminished flow of nutriment to a part, mutual pressure, an early developed part affecting one subsequently developed, and so forth,—as well as through other causes which lead to the many mysterious cases of correlation, which we do not in the least understand. These agencies may be all grouped together, for the sake of brevity, under the expression of the laws of growth. In the third place, we have to allow for the direct and definite action of changed conditions of life, and for so-called spontaneous variations, in which the nature of the conditions apparently plays a quite subordinate part. Bud-variations, such as the appearance of a moss-rose on a common rose, or of a nectarine on a peach-tree, offer good instances of spontaneous variations; but even in these ·cases, if we bear in mind the power of a minute drop of poison in producing complex galls, we ought not to feel too sure that the above variations are not the effect of some local change in the nature of the sap, due to some change in the conditions. There must be some efficient cause for each slight individual difference, as well as for more strongly marked variations which occasionally arise; and if the unknown cause were to act persistently, it is almost certain that all the individuals of the species would be similarly modified.

In the earlier editions of this work I under-rated, as it now seems probable, the frequency and importance of modifications due to spontaneous variability. But it is impossible to attribute to this cause the innumerable structures which are so well adapted to the habits of life of each species. I can no more believe in this, than that the well-adapted form of a race-horse or greyhound, which before the principle of selection by man was well understood, excited so much surprise in the minds of the older naturalists, can thus be explained.

It may be worth while to illustrate some of the foregoing remarks. With respect to the assumed inutility of various parts and organs, it is hardly necessary to observe that even in the higher and best-known animals many structures exist, which are so highly developed that no one doubts that they are of importance, yet their use has not been, or has only recently been, ascertained. As Bronn gives the length of the ears and tail in the several species of mice as instances, though trifling ones, of differences in structure which can be of no special use, I may mention that, according to Dr. Schöbl, the external ears of the common mouse are supplied in an extraordinary manner with nerves, so that they no doubt serve as tactile organs; hence the length of the ears can hardly be quite unimportant. We shall, also, presently see that the tail is a highly useful prehensile organ to some of the species; and its use would be much influenced by its length.

With respect to plants, to which on account of Nägeli's essay I shall confine myself in the following remarks, it will be admitted that the flowers of orchids present a multitude of curious structures, which a few years ago would have been considered as mere morphological differences without any special function; but they are now known to be of the highest importance for the fertilisation of the species through the aid of insects, and have probably been gained through natural selection. No one until lately would have imagined that in dimorphic and trimorphic plants the different lengths of the stamens and pistils, and their arrangement, could have been of any service, but now we know this to be the case.

In certain whole groups of plants the ovules stand erect, and in others they are suspended; and within the same ovarium of some few plants, one ovule holds the former and a second ovule the latter position. These positions seem at first purely morphological, or of no physiological signification; but Dr. Hooker informs me that within the same ovarium, the upper ovules alone in some cases, and in other cases the lower ones alone are fertilised; and he suggests that this probably depends on the direction in which the pollen-tubes enter the ovarium. If so, the position of the ovules, even when one is erect and the other suspended within the same ovarium, would follow from the selection of any slight deviations in position which favoured their fertilisation, and the production of seed.

Several plants belonging to distinct orders habitually produce flowers of two kinds,—the one open of the ordinary structure, the other closed and imperfect. These two kinds of flowers sometimes differ wonderfully in structure, yet may be seen to graduate into each other on the same plant. The ordinary and open flowers can be intercrossed; and the benefits which certainly are derived from this process are thus secured. The closed and imperfect flowers are, however, manifestly of high importance, as they yield with the utmost safety a large stock of seed, with the expenditure of wonderfully little pollen. The two kinds of flowers often differ much, as just stated, in structure. The petals in the imperfect flowers almost always consist of mere rudiments, and the pollen-grains are reduced in diameter. In Ononis columnæ five of the alternate stamens are rudimentary; and in some species of Viola three stamens are in this state, two retaining their proper function, but being of very small size. In six out of thirty of the closed flowers in an Indian violet (name unknown, for the plants have never produced with me perfect flowers), the sepals are reduced from the normal number of five to three. In one section of the Malpighiaceæ the closed flowers, according to A. de Jussieu, are still further modified, for the five stamens which stand opposite to the sepals are all aborted, a

sixth stamen standing opposite to a petal being alone
developed; and this stamen is not present in the ordinary flowers of these species; the style is aborted; and
the ovaria are reduced from three to two. Now although
natural selection may well have had the power to prevent some of the flowers from expanding, and to reduce
the amount of pollen, when rendered by the closure of
the flowers superflous, yet hardly any of the above special
modifications can have been thus determined, but must
have followed from the laws of growth, including the
functional inactivity of parts, during the progress of the
reduction of the pollen and the closure of the flowers.

It is so necessary to appreciate the important effects
of the laws of growth, that I will give some additional
cases of another kind, namely of differences in the same
part or organ, due to differences in relative position on
the same plant. In the Spanish chestnut, and in certain
fir-trees, the angles of divergence of the leaves differ,
according to Schacht, in the nearly horizontal and in
the upright branches. In the common rue and some
other plants, one flower, usually the central or terminal
one, opens first, and has five sepals and petals, and five
divisions to the ovarium; whilst all the other flowers
on the plant are tetramerous. In the British Adoxa the
uppermost flower generally has two calyx-lobes with the
other organs tetramerous, whilst the surrounding flowers
generally have three calyx-lobes with the other organs
pentamerous. In many Compositæ and Umbelliferæ
(and in some other plants) the circumferential flowers
have their corollas much more developed than those of
the centre; and this seems often connected with the
abortion of the reproductive organs. It is a more
curious fact, previously referred to, that the achenes

or seeds of the circumference and centre sometimes differ greatly in form, colour, and other characters. In Carthamus and some other Compositæ the central achenes alone are furnished with a pappus; and in Hyoseris the same head yields achenes of three different forms. In certain Umbelliferæ the exterior seeds, according to Tausch, are orthospermous, and the central one cœlospermous, and this is a character which was considered by De Candolle to be in other species of the highest systematic importance. Prof. Braun mentions a Fumariaceous genus, in which the flowers in the lower part of the spike bear oval, ribbed, one-seeded nutlets; and in the upper part of the spike, lanceolate, two-valved, and two-seeded siliques. In these several cases, with the exception of that of the well developed ray-florets, which are of service in making the flowers conspicuous to insects, natural selection cannot, as far as we can judge, have come into play, or only in a quite subordinate manner. All these modifications follow from the relative position and inter-action of the parts; and it can hardly be doubted that if all the flowers and leaves on the same plant had been subjected to the same external and internal condition, as are the flowers and leaves in certain positions, all would have been modified in the same manner.

In numerous other cases we find modifications of structure, which are considered by botanists to be generally of a highly important nature, affecting only some of the flowers on the same plant, or occurring on distinct plants, which grow close together under the same conditions. As these variations seem of no special use to the plants, they cannot have been influenced by natural selection. Of their cause we are quite ignorant;

we cannot even attribute them, as in the last class of
cases, to any proximate agency, such as relative position.
I will give only a few instances. It is so common to
observe on the same plant, flowers indifferently tetramerous, pentamerous, &c., that I need not give examples;
but as numerical variations are comparatively rare when
the parts are few, I may mention that, according to De
Candolle, the flowers of Papaver bracteatum offer either
two sepals with four petals (which is the common type
with poppies), or three sepals with six petals. The
manner in which the petals are folded in the bud is in
most groups a very constant morphological character;
but Professor Asa Gray states that with some species
of Mimulus, the æstivation is almost as frequently that
of the Rhinanthideæ as of the Antirrhinideæ, to which
latter tribe the genus belongs. Aug. St. Hilaire gives
the following cases: the genus Zanthoxylon belongs to
a division of the Rutaceæ with a single ovary, but in
some species flowers may be found on the same plant,
and even in the same panicle, with either one or two
ovaries. In Helianthemum the capsule has been
described as unilocular or 3-locular; and in H. mutabile,
"Une lame, *plus ou moins large*, s'étend entre le pericarpe
et le placenta." In the flowers of Saponaria officinalis,
Dr. Masters has observed instances of both marginal
and free central placentation. Lastly, St. Hilaire found
towards the southern extreme of the range of Gomphia
oleæformis two forms which he did not at first doubt
were distinct species, but he subsequently saw them
growing on the same bush; and he then adds, "Voilà
donc dans un même individu des loges et un style qui
se rattachent tantôt à un axe verticale et tantôt à un
gynobase."

We thus see that with plants many morphological changes may be attributed to the laws of growth and the inter-action of parts, independently of natural selection. But with respect to Nägeli's doctrine of an innate tendency towards perfection or progressive development, can it be said in the case of these strongly pronounced variations, that the plants have been caught in the act of progressing towards a higher state of development? On the contrary, I should infer from the mere fact of the parts in question differing or varying greatly on the same plant, that such modifications were of extremely small importance to the plants themselves, of whatever importance they may generally be to us for our classifications. The acquisition of a useless part can hardly be said to raise an organism in the natural scale; and in the case of the imperfect, closed flowers above described, if any new principle has to be invoked, it must be one of retrogression rather than of progression; and so it must be with many parasitic and degraded animals. We are ignorant of the exciting cause of the above specified modifications; but if the unknown cause were to act almost uniformly for a length of time, we may infer that the result would be almost uniform; and in this case all the individuals of the species would be modified in the same manner.

From the fact of the above characters being unimportant for the welfare of the species, any slight variations which occurred in them would not have been accumulated and augmented through natural selection. A structure which has been developed through long-continued selection, when it ceases to be of service to a species, generally becomes variable, as we see with rudimentary organs; for it will no longer be regulated by

T

this same power of selection.   But when, from the nature of the organism and of the conditions, modifications have been induced which are unimportant for the welfare of the species, they may be, and apparently often have been, transmitted in nearly the same state to numerous, otherwise modified, descendants.   It cannot have been of much importance to the greater number of mammals, birds, or reptiles, whether they were clothed with hair, feathers, or scales; yet hair has been transmitted to almost all mammals, feathers to all birds, and scales to all true reptiles.   A structure, whatever it may be, which is common to many allied forms, is ranked by us as of high systematic importance, and consequently is often assumed to be of high vital importance to the species.   Thus, as I am inclined to believe, morphological differences, which we consider as important— such as the arrangement of the leaves, the divisions of the flower or of the ovarium, the position of the ovules, &c. —first appeared in many cases as fluctuating variations, which sooner or later became constant through the nature of the organism and of the surrounding conditions, as well as through the intercrossing of distinct individuals, but not through natural selection; for as these morphological characters do not affect the welfare of the species, any slight deviations in them could not have been governed or accumulated through this latter agency.   It is a strange result which we thus arrive at, namely that characters of slight vital importance to the species, are the most important to the systematist; but, as we shall hereafter see when we treat of the genetic principle of classification, this is by no means so paradoxical as it may at first appear.

Although we have no good evidence of the existence

in organic beings of an innate tendency towards progressive development, yet this necessarily follows, as I
have attempted to show in the fourth chapter, through
the continued action of natural selection.   For the best
definition which has ever been given of a high standard
of organisation, is the degree to which the parts have
been specialised or differentiated; and natural selection
tends towards this end, inasmuch as the parts are thus
enabled to perform their functions more efficiently.

A distinguished zoologist, Mr. St. George Mivart, has
recently collected all the objections which have ever
been advanced by myself and others against the theory
of natural selection, as propounded by Mr. Wallace and
myself, and has illustrated them with admirable art and
force.   When thus marshalled, they make a formidable
array; and as it forms no part of Mr. Mivart's plan to
give the various facts and considerations opposed to his
conclusions, no slight effort of reason and memory is
left to the reader, who may wish to weigh the evidence
on both sides.   When discussing special cases, Mr.
Mivart passes over the effects of the increased use and
disuse of parts, which I have always maintained to be
highly important, and have treated in my 'Variation
under Domestication' at greater length than, as I believe,
any other writer.   He likewise often assumes that I
attribute nothing to variation, independently of natural
selection, whereas in the work just referred to I have
collected a greater number of well-established cases than
can be found in any other work known to me.   My
judgment may not be trustworthy, but after reading with
care Mr. Mivart's book, and comparing each section with
what I have said on the same head, I never before felt

so strongly convinced of the general truth of the conclusions here arrived at, subject, of course, in so intricate a subject, to much partial error.

All Mr. Mivart's objections will be, or have been, considered in the present volume. The one new point which appears to have struck many readers is, "that natural selection is incompetent to account for the incipient stages of useful structures." This subject is intimately connected with that of the gradation of characters, often accompanied by a change of function,— for instance, the conversion of a swim-bladder into lungs,—points which were discussed in the last chapter under two headings. Nevertheless, I will here consider in some detail several of the cases advanced by Mr. Mivart, selecting those which are the most illustrative, as want of space prevents me from considering all.

The giraffe, by its lofty stature, much elongated neck, fore-legs, head and tongue, has its whole frame beautifully adapted for browsing on the higher branches of trees. It can thus obtain food beyond the reach of the other Ungulata or hoofed animals inhabiting the same country; and this must be a great advantage to it during dearths. The Niata cattle in S. America show us how small a difference in structure may make, during such periods, a great difference in preserving an animal's life. These cattle can browse as well as others on grass, but from the projection of the lower jaw they cannot, during the often recurrent droughts, browse on the twigs of trees, reeds, &c., to which food the common cattle and horses are then driven; so that at these times the Niatas perish, if not fed by their owners. Before coming to Mr. Mivart's objections, it may be well to explain once again how natural selection will act in all ordinary cases. Man

has modified some of his animals, without necessarily
having attended to special points of structure, by simply
preserving and breeding from the fleetest individuals, as
with the race-horse and greyhound, or as with the game-
cock, by breeding from the victorious birds. So under
nature with the nascent giraffe, the individuals which
were the highest browsers and were able during dearths
to reach even an inch or two above the others, will often
have been preserved; for they will have roamed over
the whole country in search of food. That the individuals of the same species often differ slightly in the
relative lengths of all their parts may be seen in many
works of natural history, in which careful measurements
are given. These slight proportional differences, due to
the laws of growth and variation, are not of the slightest
use or importance to most species. But it will have
been otherwise with the nascent giraffe, considering its
probable habits of life; for those individuals which had
some one part or several parts of their bodies rather more
elongated than usual, would generally have survived.
These will have intercrossed and left offspring, either
inheriting the same bodily peculiarities, or with a
tendency to vary again in the same manner; whilst the
individuals, less favoured in the same respects, will have
been the most liable to perish.

We here see that there is no need to separate single
pairs, as man does, when he methodically improves a
breed: natural selection will preserve and thus separate
all the superior individuals, allowing them freely to intercross, and will destroy all the inferior individuals. By
this process long-continued, which exactly corresponds
with what I have called unconscious selection by man,
combined no doubt in a most important manner with

the inherited effects of the increased use of parts, it seems to me almost certain that an ordinary hoofed quadruped might be converted into a giraffe.

To this conclusion Mr. Mivart brings forward two objections. One is that the increased size of the body would obviously require an increased supply of food, and he considers it as "very problematical whether the disadvantages thence arising would not, in times of scarcity, more than counterbalance the advantages." But as the giraffe does actually exist in large numbers in S. Africa, and as some of the largest antelopes in the world, taller than an ox, abound there, why should we doubt that, as far as size is concerned, intermediate gradations could formerly have existed there, subjected as now to severe dearths. Assuredly the being able to reach, at each stage of increased size, to a supply of food, left untouched by the other hoofed quadrupeds of the country, would have been of some advantage to the nascent giraffe. Nor must we overlook the fact, that increased bulk would act as a protection against almost all beasts of prey excepting the lion; and against this animal, its tall neck,—and the taller the better,—would, as Mr. Chauncey Wright has remarked, serve as a watchtower. It is from this cause, as Sir S. Baker remarks, that no animal is more difficult to stalk than the giraffe. This animal also uses its long neck as a means of offence or defence, by violently swinging its head armed with stump-like horns. The preservation of each species can rarely be determined by any one advantage, but by the union of all, great and small.

Mr. Mivart then asks (and this is his second objection), if natural selection be so potent, and if high browsing be so great an advantage, why has not any other hoofed

quadruped acquired a long neck and lofty stature, besides
the giraffe, and, in a lesser degree, the camel, guanaco,
and macrauchenia ? Or, again, why has not any member
of the group acquired a long proboscis ? With respect
to S. Africa, which was formerly inhabited by numerous
herds of the giraffe, the answer is not difficult, and can
best be given by an illustration. In every meadow in
England in which trees grow, we see the lower branches
trimmed or planed to an exact level by the browsing of
the horses or cattle; and what advantage would it be,
for instance, to sheep, if kept there, to acquire slightly
longer necks ? In every district some one kind of animal
will almost certainly be able to browse higher than the
others; and it is almost equally certain that this one
kind alone could have its neck elongated for this purpose,
through natural selection and the effects of increased use.
In S. Africa the competition for browsing on the higher
branches of the acacias and other trees must be between
giraffe and giraffe, and not with the other ungulate animals.

Why, in other quarters of the world, various animals
belonging to this same order have not acquired either
an elongated neck or a proboscis, cannot be distinctly
answered; but it is as unreasonable to expect a distinct
answer to such a question, as why some event in the
history of mankind did not occur in one country, whilst
it did in another. We are ignorant with respect to the
conditions which determine the numbers and range of
each species; and we cannot even conjecture what
changes of structure would be favourable to its increase
in some new country. We can, however, see in a general
manner that various causes might have interfered with
the development of a long neck or proboscis. To reach
the foliage at a considerable height (without climbing, for

which hoofed animals are singularly ill-constructed) implies greatly increased bulk of body; and we know that some areas support singularly few large quadrupeds, for instance S. America, though it is so luxuriant; whilst S. Africa abounds with them to an unparalleled degree. Why this should be so, we do not know; nor why the later tertiary periods should have been much more favourable for their existence than the present time. Whatever the causes may have been, we can see that certain districts and times would have been much more favourable than others for the development of so large a quadruped as the giraffe.

In order that an animal should acquire some structure specially and largely developed, it is almost indispensable that several other parts should be modified and co-adapted. Although every part of the body varies slightly, it does not follow that the necessary parts should always vary in the right direction and to the right degree. With the different species of our domesticated animals we know that the parts vary in a different manner and degree; and that some species are much more variable than others. Even if the fitting variations did arise, it does not follow that natural selection would be able to act on them, and produce a structure which apparently would be beneficial to the species. For instance, if the number of individuals existing in a country is determined chiefly through destruction by beasts of prey,—by external or internal parasites, &c.,—as seems often to be the case, then natural selection will be able to do little, or will be greatly retarded, in modifying any particular structure for obtaining food. Lastly, natural selection is a slow process, and the same favourable conditions must long endure in order that any marked effect should thus be produced.

Except by assigning such general and vague reasons, we cannot explain why, in many quarters of the world, hoofed quadrupeds have not acquired much elongated necks or other means for browsing on the higher branches of trees.

Objections of the same nature as the foregoing have been advanced by many writers. In each case various causes, besides the general ones just indicated, have probably interfered with the acquisition through natural selection of structures, which it is thought would be beneficial to certain species. One writer asks, why has not the ostrich acquired the power of flight? But a moment's reflection will show what an enormous supply of food would be necessary to give to this bird of the desert force to move its huge body through the air. Oceanic islands are inhabited by bats and seals, but by no terrestrial mammals; yet as some of these bats are peculiar species, they must have long inhabited their present homes. Therefore Sir C. Lyell asks, and assigns certain reasons in answer, why have not seals and bats given birth on such islands to forms fitted to live on the land? But seals would necessarily be first converted into terrestrial carnivorous animals of considerable size, and bats into terrestrial insectivorous animals; for the former there would be no prey; for the bats ground-insects would serve as food, but these would already be largely preyed on by the reptiles or birds, which first colonise and abound on most oceanic islands. Gradations of structure, with each stage beneficial to a changing species, will be favoured only under certain peculiar conditions. A strictly terrestrial animal, by occasionally hunting for food in shallow water, then in streams or lakes, might at last be converted into an animal so throughly aquatic as to brave the open ocean. But seals would not find on oceanic islands the conditions favourable to their gradual reconversion into a terrestrial form. Bats, as formerly shown, probably acquired their wings by at first gliding through the air from tree to tree, like the so-called flying squirrels, for the sake of escaping from their enemies, or for avoiding falls; but when the power of true flight had once been acquired, it would never be reconverted back, at least for the above purposes, into the less efficient power of gliding through the air. Bats might, indeed, like many birds, have had their wings greatly reduced in size, or completely lost, through disuse; but in this case it would be necessary that they should first have acquired the power of running quickly on the ground, by the aid of their hind legs alone, so as to compete with birds or other ground animals; and for such a change a bat seems singularly ill-fitted. These conjectural remarks have been made merely to show that a transition of structure, with each step beneficial, is a highly complex affair; and that there is nothing strange in a transition not having occurred in any particular case.

Lastly, more than one writer has asked, why have some animals had their mental powers more highly developed than others, as such development would be advantageous to all? Why have not apes acquired the intellectual powers of man? Various causes could be assigned; but as they are conjectural, and their relative probability cannot be weighed, it would be useless to give them. A definite answer to the latter question ought not to be expected, seeing that no one can solve the simpler problem why, of two races of savages, one has risen higher in the scale of civilisation than the other; and this apparently implies increased brain-power.

We will return to Mr. Mivart's other objections. Insects often resemble for the sake of protection various
objects, such as green or decayed leaves, dead twigs,
bits of lichen, flowers, spines, excrement of birds, and
living insects; but to this latter point I shall hereafter
recur. The resemblance is often wonderfully close, and
is not confined to colour, but extends to form, and even
to the manner in which the insects hold themselves.
The caterpillars which project motionless like dead twigs
from the bushes on which they feed, offer an excellent
instance of a resemblance of this kind. The cases of
the imitation of such objects as the excrement of birds,
are rare and exceptional. On this head, Mr. Mivart
remarks, " As, according to Mr. Darwin's theory, there
is a constant tendency to indefinite variation, and as
the minute incipient variations will be in *all directions*,
they must tend to neutralize each other, and at first to
form such unstable modifications that it is difficult, if
not impossible, to see how such indefinite oscillations
of infinitesimal beginnings can ever build up a sufficiently appreciable resemblance to a leaf, bamboo, or
other object, for Natural Selection to seize upon and
perpetuate."

But in all the foregoing cases the insects in their
original state no doubt presented some rude and
accidental resemblance to an object commonly found
in the stations frequented by them. Nor is this at all
improbable, considering the almost infinite number of
surrounding objects and the diversity in form and colour
of the hosts of insects which exist. As some rude
resemblance is necessary for the first start, we can
understand how it is that the larger and higher animals
do not (with the exception, as far as I know, of one fish)

resemble for the sake of protection special objects, but only the surface which commonly surrounds them, and this chiefly in colour. Assuming that an insect originally happened to resemble in some degree a dead twig or a decayed leaf, and that it varied slightly in many ways, then all the variations which rendered the insect at all more like any such object, and thus favoured its escape, would be preserved, whilst other variations would be neglected and ultimately lost; or, if they rendered the insect at all less like the imitated object, they would be eliminated. There would indeed be force in Mr. Mivart's objection, if we were to attempt to account for the above resemblances, independently of natural selection, through mere fluctuating variability; but as the case stands there is none.

Nor can I see any force in Mr. Mivart's difficulty with respect to "the last touches of perfection in the mimicry;" as in the case given by Mr. Wallace, of a walking-stick insect (Ceroxylus laceratus), which resembles "a stick grown over by a creeping moss or jungermannia." So close was this resemblance, that a native Dyak maintained that the foliaceous excrescences were really moss. Insects are preyed on by birds and other enemies, whose sight is probably sharper than ours, and every grade in resemblance which aided an insect to escape notice or detection, would tend towards its preservation; and the more perfect the resemblance so much the better for the insect. Considering the nature of the differences between the species in the group which includes the above Ceroxylus, there is nothing improbable in this insect having varied in the irregularities on its surface, and in these having become more or less green-coloured; for in every group the characters which differ in the several species are the most apt to vary, whilst the generic characters, or those common to all the species, are the most constant.

The Greenland whale is one of the most wonderful animals in the world, and the baleen, or whale-bone, one of its greatest peculiarities. The baleen consists of a row, on each side, of the upper jaw, of about 300 plates or laminæ, which stand close together transversely to the longer axis of the mouth. Within the main row there are some subsidiary rows. The extremities and inner margins of all the plates are frayed into stiff bristles, which clothe the whole gigantic palate, and serve to strain or sift the water, and thus to secure the minute prey on which these great animals subsist. The middle and longest lamina in the Greenland whale is ten, twelve, or even fifteen feet in length; but in the different species of Cetaceans there are gradations in length; the middle lamina being in one species, according to Scoresby, four feet, in another three, in another eighteen inches, and in the Balænoptera rostrata only about nine inches in length. The quality of the whale-bone also differs in the different species.

With respect to the baleen, Mr. Mivart remarks that if it "had once attained such a size and development as to be at all useful, then its preservation and augmentation within serviceable limits would be promoted by natural selection alone. But how to obtain the beginning of such useful development?" In answer, it may be asked, why should not the early progenitors of the whales with baleen have possessed a mouth constructed something like the lamellated beak of a duck? Ducks, like whales, subsist by sifting the mud and water; and the

family has sometimes been called *Criblatores*, or sifters.
I hope that I may not be misconstrued into saying that
the progenitors of whales did actually possess mouths
lamellated like the beak of a duck.   I wish only to show
that this is not incredible, and that the immense plates
of baleen in the Greenland whale might have been
developed from such lamellæ by finely graduated steps,
each of service to its possessor.

The beak of a shoveller-duck (Spatula clypeata) is
a more beautiful and complex structure than the mouth
of a whale.   The upper mandible is furnished on each
side (in the specimen examined by me) with a row or
comb formed of 188 thin, elastic lamellæ, obliquely bevelled so as to be pointed, and placed transversely to the
longer axis of the mouth.   They arise from the palate, and
are attached by flexible membrane to the sides of the
mandible.   Those standing towards the middle are the
longest, being about one-third of an inch in length, and
they project ·14 of an inch beneath the edge.   At their
bases there is a short subsidiary row of obliquely transverse lamellæ.   In these several respects they resemble
the plates of baleen in the mouth of a whale.   But
towards the extremity of the beak they differ much, as
they project inwards, instead of straight downwards.
The entire head of the shoveller, though incomparably
less bulky, is about one-eighteenth of the length of the
head of a moderately large Balænoptera rostrata, in
which species the baleen is only nine inches long; so
that if we were to make the head of the shoveller as
long as that of the Balænoptera, the lamellæ would be
six inches in length,—that is, two-thirds of the length
of the baleen in this species of whale.   The lower
mandible of the shoveller-duck is furnished with lamellæ

of equal length with those above, but finer; and in being thus furnished it differs conspicuously from the lower jaw of a whale, which is destitute of baleen. On the other hand, the extremities of these lower lamellæ are frayed into fine bristly points, so that they thus curiously resemble the plates of baleen. In the genus Prion, a member of the distinct family of the Petrels, the upper mandible alone is furnished with lamellæ, which are well developed and project beneath the margin; so that the beak of this bird resembles in this respect the mouth of a whale.

From the highly developed structure of the shoveller's beak we may proceed (as I have learnt from information and specimens sent to me by Mr. Salvin), without any great break, as far as fitness for sifting is concerned, through the beak of the Merganetta armata, and in some respects through that of the Aix sponsa, to the beak of the common duck. In this latter species, the lamellæ are much coarser than in the shoveller, and are firmly attached to the sides of the mandible; they are only about 50 in number on each side, and do not project at all beneath the margin. They are square-topped, and are edged with translucent hardish tissue, as if for crushing food. The edges of the lower mandible are crossed by numerous fine ridges, which project very little. Although the beak is thus very inferior as a sifter to that of the shoveller, yet this bird, as every one knows, constantly uses it for this purpose. There are other species, as I hear from Mr. Salvin, in which the lamellæ are considerably less developed than in the common duck; but I do not know whether they use their beaks for sifting the water.

Turning to another group of the same family. In the

Egyptian goose (Chenalopex) the beak closely resembles that of the common duck; but the lamellæ are not so numerous, nor so distinct from each other, nor do they project so much inwards; yet this goose, as I am informed by Mr. E. Bartlett, "uses its bill like a duck by throwing the water out at the corners." Its chief food, however, is grass, which it crops like the common goose. In this latter bird, the lamellæ of the upper mandible are much coarser than in the common duck, almost confluent, about 27 in number on each side, and terminating upwards in teeth-like knobs. The palate is also covered with hard rounded knobs. The edges of the lower mandible are serrated with teeth much more prominent, coarser, and sharper than in the duck. The common goose does not sift the water, but uses its beak exclusively for tearing or cutting herbage, for which purpose it is so well fitted, that it can crop grass closer than almost any other animal. There are other species of geese, as I hear from Mr. Bartlett, in which the lamellæ are less developed than in the common goose.

We thus see that a member of the duck family, with a beak constructed like that of the common goose and adapted solely for grazing, or even a member with a beak having less well-developed lamellæ, might be converted by small changes into a species like the Egyptian goose, —this into one like the common duck,—and, lastly, into one like the shoveller, provided with a beak almost exclusively adapted for sifting the water; for this bird could hardly use any part of its beak, except the hooked tip, for seizing or tearing solid food. The beak of a goose, as I may add, might also be converted by small changes into one provided with prominent, recurved teeth, like those of the Merganser (a member of the same family), serving for the widely different purpose of securing live fish.

Returning to the whales. The Hyperoodon bidens is destitute of true teeth in an efficient condition, but its palate is roughened, according to Lacepède, with small, unequal, hard points of horn. There is, therefore, nothing improbable in supposing that some early Cetacean form was provided with similar points of horn on the palate, but rather more regularly placed, and which. like the knobs on the beak of the goose, aided it in seizing or tearing its food. If so, it will hardly be denied that the points might have been converted through variation and natural selection into lamellæ as well-developed as those of the Egyptian goose, in which case they would have been used both for seizing objects and for sifting the water; then into lamellæ like those of the domestic duck; and so onwards, until they became as well constructed as those of the shoveller, in which case they would have served exclusively as a sifting apparatus. From this stage, in which the lamellæ would be two-thirds of the length of the plates of baleen in the Balænoptera rostrata, gradations, which may be observed in still-existing Cetaceans, lead us onwards to the enormous plates of baleen in the Greenland whale. Nor is there the least reason to doubt that each step in this scale might have been as serviceable to certain ancient Cetaceans, with the functions of the parts slowly changing during the progress of development, as are the gradations in the beaks of the different existing members of the duck-family. We should bear in mind that each species of duck is subjected to a severe struggle for existence, and that the structure of every part of its frame must be well adapted to its conditions of life.

The Pleuronectidæ, or Flat-fish, are remarkable for their asymmetrical bodies. They rest on one side,—in the greater number of species on the left, but in some on the right side; and occasionally reversed adult specimens occur. The lower, or resting-surface, resembles at first sight the ventral surface of an ordinary fish: it is of a white colour, less developed in many ways than the upper side, with the lateral fins often of smaller size. But the eyes offer the most remarkable peculiarity; for they are both placed on the upper side of the head. During early youth, however, they stand opposite to each other, and the whole body is then symmetrical, with both sides equally coloured. Soon the eye proper to the lower side begins to glide slowly round the head to the upper side; but does not pass right through the skull, as was formerly thought to be the case. It is obvious that unless the lower eye did thus travel round, it could not be used by the fish whilst lying in its habitual position on one side. The lower eye would, also, have been liable to be abraded by the sandy bottom. That the Pleuronectidæ are admirably adapted by their flattened and asymmetrical structure for their habits of life, is manifest from several species, such as soles, flounders, &c., being extremely common. The chief advantages thus gained seem to be protection from their enemies, and facility for feeding on the ground. The different members, however, of the family present, as Schiödte remarks, "a long series of forms exhibiting a gradual transition from Hippoglossus pinguis, which does not in any considerable degree alter the shape in which it leaves the ovum, to the soles, which are entirely thrown to one side."

Mr. Mivart has taken up this case, and remarks that

a sudden spontaneous transformation in the position of the eyes is hardly conceivable, in which I quite agree with him. He then adds: "if the transit was gradual, then how such transit of one eye a minute fraction of the journey towards the other side of the head could benefit the individual is, indeed, far from clear. It seems, even, that such an incipient transformation must rather have been injurious." But he might have found an answer to this objection in the excellent observations published in 1867 by Malm. The Pleuronectidæ, whilst very young and still symmetrical, with their eyes standing on opposite sides of the head, cannot long retain a vertical position, owing to the excessive depth of their bodies, the small size of their lateral fins, and to their being destitute of a swimbladder. Hence soon growing tired, they fall to the bottom on one side. Whilst thus at rest they often twist, as Malm observed, the lower eye upwards, to see above them; and they do this so vigorously that the eye is pressed hard against the upper part of the orbit. The forehead between the eyes consequently becomes, as could be plainly seen, temporarily contracted in breadth. On one occasion Malm saw a young fish raise and depress the lower eye through an angular distance of about seventy degrees.

We should remember that the skull at this early age is cartilaginous and flexible, so that it readily yields to muscular action. It is also known with the higher animals, even after early youth, that the skull yields and is altered in shape, if the skin or muscles be permanently contracted through disease or some accident. With long-eared rabbits, if one ear lops forwards and downwards, its weight drags forward all the bones of the skull on the same side, of which I have given a.

figure. Malm states that the newly-hatched young of perches, salmon, and several other symmetrical fishes, have the habit of occasionally resting on one side at the bottom; and he has observed that they often then strain their lower eyes so as to look upwards; and their skulls are thus rendered rather crooked. These fishes, however, are soon able to hold themselves in a vertical position, and no permanent effect is thus produced. With the Pleuronectidæ, on the other hand, the older they grow the more habitually they rest on one side, owing to the increasing flatness of their bodies, and a permanent effect is thus produced on the form of the head, and on the position of the eyes. Judging from analogy, the tendency to distortion would no doubt be increased through the principle of inheritance. Schiödte believes, in opposition to some other naturalists, that the Pleuronectidæ are not quite symmetrical even in the embryo; and if this be so, we could understand how it is that certain species, whilst young, habitually fall over and rest on the left side, and other species on the right side. Malm adds, in confirmation of the above view, that the adult Trachypterus arcticus, which is not a member of the Pleuronectidæ, rests on its left side at the bottom, and swims diagonally through the water; and in this fish, the two sides of the head are said to be somewhat dissimilar. Our great authority on Fishes, Dr. Günther, concludes his abstract of Malm's paper, by remarking that "the author gives a very simple explanation of the abnormal condition of the Pleuronectoids."

We thus see that the first stages of the transit of the eye from one side of the head to the other, which Mr. Mivart considers would be injurious, may be attributed to the habit, no doubt beneficial to the individual and to the species, of endeavouring to look upwards with both eyes, whilst resting on one side at the bottom. We may also attribute to the inherited effects of use the fact of the mouth in several kinds of flat-fish being bent towards the lower surface, with the jaw bones stronger and more effective on this, the eyeless side of the head, than on the other, for the sake, as Dr. Traquair supposes, of feeding with ease on the ground. Disuse, on the other hand, will account for the less developed condition of the whole inferior half of the body, including the lateral fins; though Yarrell thinks that the reduced size of these fins is advantageous to the fish, as "there is so much less room for their action, than with the larger fins above." Perhaps the lesser number of teeth in the proportion of four to seven in the upper halves of the two jaws of the plaice, to twenty-five to thirty in the lower halves, may likewise be accounted for by disuse. From the colour-less state of the ventral surface of most fishes and of many other animals, we may reasonably suppose that the absence of colour in flat-fish on the side, whether it be the right or left, which is undermost, is due to the exclu-sion of light. But it cannot be supposed that the peculiar speckled appearance of the upper side of the sole, so like the sandy bed of the sea, or the power in some species, as recently shown by Pouchet, of changing their colour in accordance with the surrounding surface, or the presence of bony tubercles on the upper side of the turbot, are due to the action of the light. Here natural selection has probably come into play, as well as in adapting the general shape of the body of these fishes, and many other peculiarities, to their habits of life. We should keep in mind, as I have before insisted, that the inherited effects of the increased use of parts, and perhaps of their disuse, will be strengthened by natural selection. For all spontaneous variations in the right direction will thus be preserved; as will those individuals which inherit in the highest degree the effects of the increased and beneficial use of any part. How much to attribute in each particular case to the effects of use, and how much to natural selection, it seems impossible to decide.

I may give another instance of a structure which apparently owes its origin exclusively to use or habit. The extremity of the tail in some American monkeys has been converted into a wonderfully perfect prehensile organ, and serves as a fifth hand. A reviewer who agrees with Mr. Mivart in every detail, remarks on this structure: "It is impossible to believe that in any number of ages the first slight incipient tendency to grasp could preserve the lives of the individuals possessing it, or favour their chance of having and of rearing offspring." But there is no necessity for any such belief. Habit, and this almost implies that some benefit great or small is thus derived, would in all probability suffice for the work. Brehm saw the young of an African monkey (Cercopithecus) clinging to the under surface of their mother by their hands, and at the same time they hooked their little tails round that of their mother. Professor Henslow kept in confinement some harvest mice (Mus messorius) which do not possess a structurally prehensile tail; but he frequently observed that they curled their tails round the branches of a bush placed in the cage, and thus aided themselves in climbing. I have received an analogous account from Dr. Günther, who has seen a mouse thus suspend itself. If the harvest mouse had been more strictly arboreal, it would perhaps have had its tail rendered structurally prehensile, as is the case with some members of the same order.   Why Cercopithecus, considering its habits whilst young, has not become thus provided, it would be difficult to say. It is, however, possible that the long tail of this monkey may be of more service to it as a balancing organ in making its prodigious leaps, than as a prehensile organ.

The mammary glands are common to the whole class of mammals, and are indispensable for their existence ; they must, therefore, have been developed at an extremely remote period, and we can know nothing positively about their manner of development.   Mr. Mivart asks : " Is it conceivable that the young of any animal was ever saved from destruction by accidentally sucking a drop of scarcely nutritious fluid from an accidentally hypertrophied cutaneous gland of its mother ?   And even if one was so, what chance was there of the perpetuation of such a variation ? "   But the case is not here put fairly.   It is admitted by most evolutionists that mammals are descended from a marsupial form ; and if so, the mammary glands will have been at first developed within the marsupial sack.   In the case of the fish (Hippocampus) the eggs are hatched, and the young are reared for a time, within a sack of this nature ; and an American naturalist, Mr. Lockwood, believes from what he has seen of the development of the young, that they are nourished by a secretion from the cutaneous glands of the sack.   Now with the early progenitors of mammals, almost before they deserved to be thus designated, is it not at least possible that the young might have been similarly nourished ? And in this case, the individuals which secreted a fluid, in some degree or manner the most nutritious, so as to partake of the nature of milk. would in the long run have reared a larger number of well-nourished offspring, than would the individuals which secreted a poorer fluid; and thus the cutaneous glands, which are the homologues of the mammary glands, would have been improved or rendered more effective. It accords with the widely extended principle of specialisation, that the glands over a certain space of the sack should have become more highly developed than the remainder; and they would then have formed a breast, but at first without a nipple, as we see in the Ornithorhyncus, at the base of the mammalian series. Through what agency the glands over a certain space became more highly specialised than the others, I will not pretend to decide, whether in part through compensation of growth, the effects of use, or of natural selection.

The development of the mammary glands would have been of no service, and could not have been effected through natural selection, unless the young at the same time were able to partake of the secretion. There is no greater difficulty in understanding how young mammals have instinctively learnt to suck the breast, than in understanding how unhatched chickens have learnt to break the egg-shell by tapping against it with their specially adapted beaks; or how a few hours after leaving the shell they have learnt to pick up grains of food. In such cases the most probable solution seems to be, that the habit was at first acquired by practice at a more advanced age, and afterwards transmitted to the offspring at an earlier age. But the young kangaroo is said not to suck, only to cling to the nipple of its mother, who has the power of injecting milk into the mouth of her helpless, half-formed offspring. On this head Mr. Mivart remarks: "Did no special provision exist, the young one must infallibly be choked by the intrusion of the milk into the windpipe. But there *is* a special provision. The larynx is so elongated that it rises up into the posterior end of the nasal passage, and is thus enabled to give free entrance to the air for the lungs, while the milk passes harmlessly on each side of this elongated larynx, and so safely attains the gullet behind it." Mr. Mivart then asks how did natural selection remove in the adult kangaroo (and in most other mammals, on the assumption that they are descended from a marsupial form), "this at least perfectly innocent and harmless structure?" It may be suggested in answer that the voice, which is certainly of high importance to many animals, could hardly have been used with full force as long as the larynx entered the nasal passage; and Professor Flower has suggested to me that this structure would have greatly interfered with an animal swallowing solid food.

We will now turn for a short space to the lower divisions of the animal kingdom. The Echinodermata (starfishes, sea-urchins, &c.) are furnished with remarkable organs, called pedicellariæ, which consist, when well developed, of a tridactyle forceps—that is, of one formed of three serrated arms, neatly fitting together and placed on the summit of a flexible stem, moved by muscles. These forceps can seize firmly hold of any object; and Alexander Agassiz has seen an Echinus or sea-urchin rapidly passing particles of excrement from forceps to forceps down certain lines of its body, in order that its shell should not be fouled. But there is no doubt that besides removing dirt of all kinds, they subserve other functions; and one of these apparently is defence.

With respect to these organs, Mr. Mivart, as on so

many previous occasions, asks: "What would be the utility of the *first rudimentary beginnings* of such structures, and how could such incipient buddings have ever preserved the life of a single Echinus?" He adds, "not even the *sudden* development of the snapping action could have been beneficial without the freely moveable stalk, nor could the latter have been efficient without the snapping jaws, yet no minute merely indefinite variations could simultaneously evolve these complex co-ordinations of structure; to deny this seems to do no less than to affirm a startling paradox." Paradoxical as this may appear to Mr. Mivart, tridactyle forcepses, immovably fixed at the base, but capable of a snapping action, certainly exist on some star-fishes; and this is intelligible if they serve, at least in part, as a means of defence. Mr. Agassiz, to whose great kindness I am indebted for much information on the subject, informs me that there are other star-fishes, in which one of the three arms of the forceps is reduced to a support for the other two; and again, other genera in which the third arm is completely lost. In Echinoneus, the shell is described by M. Perrier as bearing two kinds of pedicellariæ, one resembling those of Echinus, and the other those of Spatangus; and such cases are always interesting as affording the means of apparently sudden transitions, through the abortion of one of the two states of an organ.

With respect to the steps by which these curious organs have been evolved, Mr. Agassiz infers from his own researches and those of Müller, that both in star-fishes and sea-urchins the pedicellariæ must undoubtedly be looked at as modified spines. This may be inferred from their manner of development in the individual,

as well as from a long and perfect series of gradations
in different species and genera, from simple granules to
ordinary spines, to perfect tridactyle pedicellariæ.   The
gradation extends even to the manner in which ordinary
spines and the pedicellariæ with their supporting calcareous rods are articulated to the shell.   In certain
genera of star-fishes, "the very combinations needed to
show that the pedicellariæ are only modified branching
spines" may be found.   Thus we have fixed spines,
with three equi-distant, serrated, moveable branches,
articulated to near their bases; and higher up, on the
same spine, three other moveable branches.   Now when
the latter arise from the summit of a spine they form in
fact a rude tridactyle pedicellaria, and such may be seen
on the same spine together with the three lower branches.
In this case the identity in nature between the arms of
the pedicellariæ and the moveable branches of a spine, is
unmistakable.   It is generally admitted that the ordinary
spines serve as a protection; and if so, there can be no
reason to doubt that those furnished with serrated and
moveable branches likewise serve for the same purpose;
and they would thus serve still more effectively as soon
as by meeting together they acted as a prehensile or
snapping apparatus.   Thus every gradation, from an
ordinary fixed spine to a fixed pedicellaria, would be of
service.

In certain genera of star-fishes these organs, instead
of being fixed or borne on an immovable support, are
placed on the summit of a flexible and muscular, though
short, stem; and in this case they probably subserve
some additional function besides defence.   In the sea-
urchins the steps can be followed by which a fixed spine
becomes articulated to the shell, and is thus rendered

moveable. I wish I had space here to give a fuller abstract of Mr. Agassiz's interesting observations on the development of the pedicellariæ. All possible gradations, as he adds, may likewise be found between the pedicellariæ of the star-fishes and the hooks of the Ophiurians, another group of the Echinodermata ; and again between the pedicellariæ of sea-urchins and the anchors of the Holothuriæ, also belonging to the same great class.

Certain compound animals, or zoophytes as they have been termed, namely the Polyzoa, are provided with curious organs called avicularia. These differ much in structure in the different species. In their most perfect condition, they curiously resemble the head and beak of a vulture in miniature, seated on a neck and capable of movement, as is likewise the lower jaw or mandible. In one species observed by me all the avicularia on the same branch often moved simultaneously backwards and forwards, with the lower jaw widely open, through an angle of about 90°, in the course of five seconds; and their movement caused the whole polyzoary to tremble. When the jaws are touched with a needle they seize it so firmly that the branch can thus be shaken.

Mr. Mivart adduces this case, chiefly on account of the supposed difficulty of organs, namely the avicularia of the Polyzoa and the pedicellariæ of the Echinodermata, which he considers as "essentially similar," having been developed through natural selection in widely distinct divisions of the animal kingdom. But, as far as structure is concerned, I can see no similarity between tridactyle pedicellariæ and avicularia. The latter resemble somewhat more closely the chelæ or pincers of Crustaceans ; and Mr. Mivart might have adduced with equal appropriateness this resemblance as a special difficulty ; or even their resemblance to the head and beak of a bird. The avicularia are believed by Mr. Busk, Dr. Smitt, and Dr. Nitsche—naturalists who have carefully studied this group—to be homologous with the zooids and their cells which compose the zoophyte; the moveable lip or lid of the cell corresponding with the lower and moveable mandible of the avicularium. Mr. Busk, however, does not know of any gradations now existing between a zooid and an avicularium. It is therefore impossible to con-jecture by what serviceable gradations the one could have been converted into the other : but it by no means follows from this that such gradations have not existed.

As the chelæ of Crustaceans resemble in some degree the avicularia of Polyzoa, both serving as pincers, it may be worth while to show that with the former a long series of serviceable gradations still exists. In the first and simplest stage, the terminal segment of a limb shuts down either on the square summit of the broad penultimate segment, or against one whole side; and is thus enabled to catch hold of an object ; but the limb still serves as an organ of locomotion. We next find one corner of the broad penultimate segment slightly prominent, sometimes furnished with irregular teeth; and against these the terminal segment shuts down. By an increase in the size of this projection, with its shape, as well as that of the terminal segment, slightly modified and improved, the pincers are rendered more and more perfect, until we have at last an instrument as efficient as the chelæ of a lobster; and all these gradations can be actually traced.

Besides the avicularia, the Polyzoa possess curious organs called vibracula. These generally consist of long bristles, capable of movement and easily excited. In one species examined by me the vibracula were slightly curved and serrated along the outer margin; and all of them on the same polyzoary often moved simultaneously; so that, acting like long oars, they swept a branch rapidly across the object-glass of my microscope. When a branch was placed on its face, the vibracula became entangled, and they made violent efforts to free themselves. They are supposed to serve as a defence, and may be seen, as Mr. Busk remarks, "to sweep slowly and carefully over the surface of the polyzoary, removing what might be noxious to the delicate inhabitants of the cells when their tentacula are protruded." The avicularia, like the vibracula, probably serve for defence, but they also catch and kill small living animals, which it is believed are afterwards swept by the currents within reach of the tentacula of the zooids. Some species are provided with avicularia and vibracula; some with avicularia alone, and a few with vibracula alone.

It is not easy to imagine two objects more widely different in appearance than a bristle or vibraculum, and an avicularium like the head of a bird; yet they are almost certainly homologous and have been developed from the same common source, namely a zooid with its cell. Hence we can understand how it is that these organs graduate in some cases, as I am informed by Mr. Busk, into each other. Thus with the avicularia of several species of Lepralia, the moveable mandible is so much produced and is so like a bristle, that the presence of the upper or fixed beak alone serves to determine its avicularian nature. The vibracula may have been directly developed from the lips of the cells, without having passed through the avicularian stage; but it seems more probable that they have passed through this stage, as during the early stages of the transformation, the other parts of the cell with the included zooid could hardly have disappeared at once. In many cases the vibracula have a grooved support at the base, which seems to represent the fixed beak; though this support in some species is quite absent. This view of the development of the vibracula, if trustworthy, is interesting; for supposing that all the species provided with avicularia had become extinct, no one with the most vivid imagination would ever have thought that the vibracula had originally existed as part of an organ, resembling a bird's head or an irregular box or hood. It is interesting to see two such widely different organs developed from a common origin; and as the moveable lip of the cell serves as a protection to the zooid, there is no difficulty in believing that all the gradations, by which the lip became converted first into the lower mandible of an avicularium and then into an elongated bristle, likewise served as a protection in different ways and under different circumstances.

In the vegetable kingdom Mr. Mivart only alludes to two cases, namely the structure of the flowers of orchids, and the movements of climbing plants. With respect to the former, he says, "the explanation of their *origin* is deemed throughly unsatisfactory—utterly insufficient to explain the incipient, infinitesimal beginnings of structures which are of utility only when they are considerably developed." As I have fully treated this subject in another work, I will here give only a few details on one alone of the most striking peculiarities of the flowers of orchids, namely their pollinia. A pollinium when highly developed consists of a mass of pollen-grains, affixed to an elastic foot-stalk or caudicle, and this to a little mass of extremely viscid matter. The pollinia are by this means transported by insects from one flower to the stigma of another. In some orchids there is no caudicle to the pollen-masses, and the grains are merely tied together by fine threads; but as these are not confined to orchids, they need not here be considered; yet I may mention that at the base of the orchidaceous series, in Cypripedium, we can see how the threads were probably first developed. In other orchids the threads cohere at one end of the pollen-masses; and this forms the first or nascent trace of a caudicle. That this is the origin of the caudicle, even when of considerable length and highly developed, we have good evidence in the aborted pollen-grains which can sometimes be detected embedded within the central and solid parts.

With respect to the second chief peculiarity, namely the little mass of viscid matter attached to the end of the caudicle, a long series of gradations can be specified, each of plain service to the plant. In most flowers belonging to other orders the stigma secretes a little viscid matter. Now in certain orchids similar viscid matter is secreted, but in much larger quantities by one alone of the three stigmas; and this stigma, perhaps in consequence of the copious secretion, is rendered sterile. When an insect visits a flower of this kind, it rubs off some of the viscid matter and thus at the same time drags away some of the pollen-grains. From this simple condition, which differs but little from that of a multitude of common flowers, there are endless gradations,— to species in which the pollen-mass terminates in a very short, free caudicle,—to others in which the caudicle becomes firmly attached to the viscid matter, with the sterile stigma itself much modified. In this latter case we have a pollinium in its most highly developed and perfect condition. He who will carefully examine the flowers of orchids for himself will not deny the existence of the above series of gradations—from a mass of pollen-grains merely tied together by threads, with the stigma differing but little from that of an ordinary flower, to a highly complex pollinium, admirably adapted for transportal by insects; nor will he deny that all the gradations in the several species are admirably adapted in relation to the general structure of each flower for its fertilisation by different insects. In this, and in almost every other case, the enquiry may be pushed further backwards; and it may be asked how did the stigma of an ordinary flower become viscid, but as we do not know the full history of any one group of beings, it is as useless to ask, as it is hopeless to attempt answering, such questions.

We will now turn to climbing plants. These can be arranged in a long series, from those which simply twine round a support, to those which I have called leaf-climbers, and to those provided with tendrils. In these two latter classes the stems have generally, but not always, lost the power of twining, though they retain the power of revolving, which the tendrils likewise possess. The gradations from leaf-climbers to tendril-bearers are wonderfully close, and certain plants may be indifferently placed in either class. But in ascending the series from simple twiners to leaf-climbers, an important quality is added, namely sensitiveness to a touch, by which means the foot-stalks of the leaves or flowers, or these modified and converted into tendrils, are excited to bend round and clasp the touching object. He who will read my memoir on these plants will, I think, admit that all the many gradations in function and structure between simple twiners and tendril-bearers are in each case beneficial in a high degree to the species. For instance, it is clearly a great advantage to a twining plant to become a leaf-climber; and it is probable that every twiner which possessed leaves with long foot-stalks would have been developed into a leaf-climber, if the foot-stalks had possessed in any slight degree the requisite sensitiveness to a touch.

As twining is the simplest means of ascending a support, and forms the basis of our series, it may naturally be asked how did plants acquire this power in an incipient degree, afterwards to be improved and increased through natural selection. The power of twining depends, firstly, on the stems whilst young being extremely flexible (but this is a character common to many plants which are not climbers); and, secondly, on their continually bending to all points of the compass, one after the other in succession, in the same order. By this movement the stems are inclined to all sides, and are made to move round and round. As soon as the lower part of a stem strikes against any object and is stopped, the upper part still goes on bending and revolving, and thus necessarily twines round and up the support. The revolving movement ceases after the early growth of each shoot. As in many widely separated families of plants, single species and single genera possess the power of revolving, and have thus become twiners, they must have independently acquired it, and cannot have inherited it from a common progenitor. Hence I was led to predict that some slight tendency to a movement of this kind would be found to be far from uncommon with plants which did not climb; and that this had afforded the basis for natural selection to work on and improve.   When I made this prediction, I knew of only one imperfect case, namely of the young flower-peduncles of a Maurandia which revolved slightly and irregularly, like the stems of twining plants, but without making any use of this habit.   Soon afterwards Fritz Müller discovered that the young stems of an Alisma and of a Linum,—plants which do not climb and are widely separated in the natural system,—revolved plainly, though irregularly; and he states that he has reason to suspect that this occurs with some other plants.   These slight movements appear to be of no service to the plants in question; anyhow, they are not of the least use in the way of climbing, which is the point that concerns us.   Nevertheless we can see that if the stems of these plants had been flexible, and if under the conditions to which they are exposed it had profited them to ascend to a height, then the habit of slightly and irregularly revolving might have been increased and utilised through natural selection, until they had become converted into well-developed twining species.

With respect to the sensitiveness of the foot-stalks of the leaves and flowers, and of tendrils, nearly the same remarks are applicable as in the case of the revolving movements of twining plants.   As a vast number of species, belonging to widely distinct groups, are endowed with this kind of sensitiveness, it ought to be found in a nascent condition in many plants which have not become climbers.   This is the case: I observed that the young flower-peduncles of the above Maurandia curved themselves a little towards the side which was touched

Morren found in several species of Oxalis that the leaves and their foot-stalks moved, especially after exposure to a hot sun, when they were gently and repeatedly touched, or when the plant was shaken. I repeated these observations on some other species of Oxalis with the same result; in some of them the movement was distinct, but was best seen in the young leaves; in others it was extremely slight. It is a more important fact that according to the high authority of Hofmeister, the young shoots and leaves of all plants move after being shaken; and with climbing plants it is, as we know, only during the early stages of growth that the foot-stalks and tendrils are sensitive.

It is scarcely possible that the above slight movements, due to a touch or shake, in the young and growing organs of plants, can be of any functional importance to them. But plants possess, in obedience to various stimuli, powers of movement, which are of manifest importance to them; for instance, towards and more rarely from the light,—in opposition to, and more rarely in the direction of, the attraction of gravity. When the nerves and muscles of an animal are excited by galvanism or by the absorption of strychnine, the consequent movements may be called an incidental result, for the nerves and muscles have not been rendered specially sensitive to these stimuli. So with plants it appears that, from having the power of movement in obedience to certain stimuli, they are excited in an incidental manner by a touch, or by being shaken. Hence there is no great difficulty in admitting that in the case of leaf-climbers and tendril-bearers, it is this tendency which has been taken advantage of and increased through natural selection. It is, however, probable, from reasons which I have assigned in my memoir, that this will have occurred only with plants which had already acquired the power of revolving, and had thus become twiners.

I have already endeavoured to explain how plants became twiners, namely, by the increase of a tendency to slight and irregular revolving movements, which were at first of no use to them; this movement, as well as that due to a touch or shake, being the incidental result of the power of moving, gained for other and beneficial purposes. Whether, during the gradual development of climbing plants, natural selection has been aided by the inherited effects of use, I will not pretend to decide; but we know that certain periodical movements, for instance the so-called sleep of plants, are governed by habit.

I have now considered enough, perhaps more than enough, of the cases, selected with care by a skilful naturalist, to prove that natural selection is incompetent to account for the incipient stages of useful structures; and I have shown, as I hope, that there is no great difficulty on this head. A good opportunity has thus been afforded for enlarging a little on gradations of structure, often associated with changed functions,—an important subject, which was not treated at sufficient length in the former editions of this work. I will now briefly recapitulate the foregoing cases.

With the giraffe, the continued preservation of the individuals of some extinct high-reaching ruminant, which had the longest necks, legs, &c., and could browse a little above the average height, and the continued destruction of those which could not browse so high, would have sufficed for the production of this remarkable quadruped; but the prolonged use of all the parts together with inheritance will have aided in an important manner in their co-ordination. With the many insects which imitate various objects, there is no improbability in the belief that an accidental resemblance to some common object was in each case the foundation for the work of natural selection, since perfected through the occasional preservation of slight variations which made the resemblance at all closer; and this will have been carried on as long as the insect continued to vary, and as long as a more and more perfect resemblance led to its escape from sharp-sighted enemies. In certain species of whales there is a tendency to the formation of irregular little points of horn on the palate; and it seems to be quite within the scope of natural selection to preserve all favourable variations, until the points were converted first into lamellated knobs or teeth, like those on the beak of a goose,—then into short lamellæ, like those of the domestic ducks,—and then into lamellæ, as perfect as those of the shoveller-duck,—and finally into the gigantic plates of baleen, as in the mouth of the Greenland whale. In the family of the ducks, the lamellæ are first used as teeth, then partly as teeth and partly as a sifting apparatus, and at last almost exclusively for this latter purpose.

With such structures as the above lamellæ of horn or whalebone, habit or use can have done little or nothing, as far as we can judge, towards their development. On the other hand, the transportal of the lower eye of a flat-fish to the upper side of the head, and the formation of a prehensile tail, may be attributed almost wholly to continued use, together with inheritance. With respect to the mammæ of the higher animals, the most probable conjecture is that primordially the cutaneous glands

over the whole suface of a marsupial sack secreted a
nutritious fluid; and that these glands were improved
in function through natural selection, and concentrated
into a confined area, in which case they would have
formed a mamma.   There is no more difficulty in understanding how the branched spines of some ancient Echi-
noderm, which served as a defence, became developed
through natural selection into tridactyle pedicellariæ,
than in understanding the development of the pincers
of crustaceans, through slight, serviceable modifications
in the ultimate and penultimate segments of a limb,
which was at first used solely for locomotion.   In the
avicularia and vibracula of the Polyzoa we have organs
widely different in appearance developed from the same
source; and with the vibracula we can understand how
the successive gradations might have been of service.
With the pollinia of orchids, the threads which originally served to tie together the pollen-grains, can be
traced cohering into caudicles; and the steps can likewise be followed by which viscid matter, such as that
secreted by the stigmas of ordinary flowers, and still
subserving nearly but not quite the same purpose,
became attached to the free ends of the caudicles;—all
these gradations being of manifest benefit to the plants
in question.   With respect to climbing plants, I need
not repeat what has been so lately said.

It has often been asked, if natural selection be so
potent, why has not this or that structure been gained
by certain species, to which it would apparently have
been advantageous?   But it is unreasonable to expect a
precise answer to such questions, considering our ignorance of the past history of each species, and of the conditions which at the present day determine its numbers and

range. In most cases only general reasons, but in some few cases special reasons, can be assigned. Thus to adapt a species to new habits of life, many co-ordinated modifications are almost indispensable, and it may often have happened that the requisite parts did not vary in the right manner or to the right degree. Many species must have been prevented from increasing in numbers through destructive agencies, which stood in no relation to certain structures, which we imagine would have been gained through natural selection from appearing to us advantageous to the species. In this case, as the struggle for life did not depend on such structures, they could not have been acquired through natural selection. In many cases complex and long-enduring conditions, often of a peculiar nature, are necessary for the development of a structure; and the requisite conditions may seldom have concurred. The belief that any given structure, which we think, often erroneously, would have been beneficial to a species, would have been gained under all circumstances through natural selection, is opposed to what we can understand of its manner of action. Mr. Mivart does not deny that natural selection has effected something; but he considers it as "demonstrably insufficient" to account for the phenomena which I explain by its agency. His chief arguments have now been considered, and the others will hereafter be considered. They seem to me to partake little of the character of demonstration, and to have little weight in comparison with those in favour of the power of natural selection, aided by the other agencies often specified. I am bound to add, that some of the facts and arguments here used by me, have been advanced for the same purpose in an able article lately published in the 'Medico-Chirurgical Review.'

At the present day almost all naturalists admit evolution under some form. Mr. Mivart believes that species change through "an internal force or tendency," about which it is not pretended that anything is known. That species have a capacity for change will be admitted by all evolutionists; but there is no need, as it seems to me, to invoke any internal force beyond the tendency to ordinary variability, which through the aid of selection by man has given rise to many well-adapted domestic races, and which through the aid of natural selection would equally well give rise by graduated steps to natural races or species. The final result will generally have been, as already explained, an advance, but in some few cases a retrogression, in organisation.

Mr. Mivart is further inclined to believe, and some naturalists agree with him, that new species manifest themselves "with suddenness and by modifications appearing at once." For instance, he supposes that the differences between the extinct three-toed Hipparion and the horse arose suddenly. He thinks it difficult to believe that the wing of a bird "was developed in any other way than by a comparatively sudden modification of a marked and important kind;" and apparently he would extend the same view to the wings of bats and pterodactyles. This conclusion, which implies great breaks or discontinuity in the series, appears to me improbable in the highest degree.

Every one who believes in slow and gradual evolution, will of course admit that specific changes may have been as abrupt and as great as any single variation which we meet with under nature, or even under domestication. But as species are more variable when domesticated or cultivated than under their natural conditions, it is not probable that such great and abrupt variations have often occurred under nature, as are known occasionally to arise under domestication. Of these latter variations several may be attributed to reversion; and the characters which thus reappear were, it is probable, in many cases at first gained in a gradual manner. A still greater number must be called monstrosities, such as six-fingered men, porcupine men, Ancon sheep, Niata cattle, &c.; and as they are widely different in character from natural species, they throw very little light on our subject. Excluding such cases of abrupt variations, the few which remain would at best constitute, if found in a state of nature, doubtful species, closely related to their parental types.

My reasons for doubting whether natural species have changed as abruptly as have occasionally domestic races, and for entirely disbelieving that they have changed in the wonderful manner indicated by Mr. Mivart, are as follows. According to our experience, abrupt and strongly marked variations occur in our domesticated productions, singly and at rather long intervals of time. If such occurred under nature, they would be liable, as formerly explained, to be lost by accidental causes of destruction and by subsequent inter-crossing; and so it is known to be under domestication, unless abrupt variations of this kind are specially preserved and separated by the care of man. Hence in order that a new species should suddenly appear in the manner supposed by Mr. Mivart, it is almost necessary to believe, in opposition to all analogy, that several wonderfully changed individuals appeared simultaneously within the same district. This difficulty, as in the case of unconscious selection by man, is avoided on the theory of gradual evolution, through the preservation of a large number of individuals, which varied more or less in any favourable direction, and of the destruction of a large number which varied in an opposite manner.

That many species have been evolved in an extremely gradual manner, there can hardly be a doubt. The species and even the genera of many large natural families are so closely allied together, that it is difficult to distinguish not a few of them. On every continent in proceeding from north to south, from lowland to upland, &c., we meet with a host of closely related or representative species; as we likewise do on certain distinct continents, which we have reason to believe were formerly connected. But in making these and the following remarks, I am compelled to allude to subjects hereafter to be discussed. Look at the many outlying islands round a continent, and see how many of their inhabitants can be raised only to the rank of doubtful species. So it is if we look to past times, and compare the species which have just passed away with those still living within the same areas; or if we compare the fossil species embedded in the sub-stages of the same geological formation. It is indeed manifest that multitudes of species are related in the closest manner to other species that still exist, or have lately existed; and it will hardly be maintained that such species have been developed in an abrupt or sudden manner. Nor should it be forgotten, when we look to the special parts of allied species, instead of to distinct species, that numerous and wonderfully fine gradations can be traced, connecting together widely different structures.

Many large groups of facts are intelligible only on the principle that species have been evolved by very small steps. For instance, the fact that the species included in the larger genera are more closely related to each other, and present a greater number of varieties than do the species in the smaller genera. The former are also grouped in little clusters, like varieties round species; and they present other analogies with varieties, as was shown in our second chapter. On this same principle we can understand how it is that specific characters are more variable than generic characters; and how the parts which are developed in an extraordinary degree or manner are more variable than other parts of the same species. Many analogous facts, all pointing in the same direction, could be added.

Although very many species have almost certainly been produced by steps not greater than those separating fine varieties; yet it may be maintained that some have been developed in a different and abrupt manner. Such an admission, however, ought not to be made without strong evidence being assigned. The vague and in some respects false analogies, as they have been shown to be by Mr. Chauncey Wright, which have been advanced in favour of this view, such as the sudden crystallisation of inorganic substances, or the falling of a facetted spheroid from one facet to another, hardly deserve consideration. One class of facts, however, namely, the sudden appearance of new and distinct forms of life in our geological formations supports at first sight the belief in abrupt development. But the value of this evidence depends entirely on the perfection of the geological record, in relation to periods remote in the history of the world. If the record is as fragmentary as many geologists strenuously assert, there is nothing strange in new forms appearing as if suddenly developed.

Unless we admit transformations as prodigious as

those advocated by Mr. Mivart, such as the sudden development of the wings of birds or bats, or the sudden conversion of a Hipparion into a horse, hardly any light is thrown by the belief in abrupt modifications on the deficiency of connecting links in our geological formations. But against the belief in such abrupt changes, embryology enters a strong protest. It is notorious that the wings of birds and bats, and the legs of horses or other quadrupeds, are undistinguishable at an early embryonic period, and that they become differentiated by insensibly fine steps. Embryological resemblances of all kinds can be accounted for, as we shall hereafter see, by the progenitors of our existing species having varied after early youth, and having transmitted their newly acquired characters to their offspring, at a corresponding age. The embryo is thus left almost unaffected, and serves as a record of the past condition of the species. Hence it is that existing species during the early stages of their development so often resemble ancient and extinct forms belonging to the same class. On this view of the meaning of embryological resemblances, and indeed on any view, it is incredible that an animal should have undergone such momentous and abrupt transformations, as those above indicated; and yet should not bear even a trace in its embryonic condition of any sudden modification; every detail in its structure being developed by insensibly fine steps.

He who believes that some ancient form was transformed suddenly through an internal force or tendency into, for instance, one furnished with wings, will be almost compelled to assume, in opposition to all analogy, that many individuals varied simultaneously. It cannot be denied that such abrupt and great changes of structure are

widely different from those which most species apparently have undergone.   He will further be compelled to
believe that many structures beautifully adapted to all
the other parts of the same creature and to the surrounding conditions, have ' been suddenly produced; and of
such complex and wonderful co-adaptations, he will not
be able to assign a shadow of an explanation.   He will be
forced to admit that these great and sudden transformations have left no trace of their action on the embryo.
To admit all this is, as it seems to me, to enter into the
realms of miracle, and to leave those of Science.

CHAPTER VIII.

INSTINCT.

Instincts comparable with habits, but different in their origin—Instincts graduated—Aphides and ants—Instincts variable—Domestic instincts, their origin—Natural instincts of the cuckoo, molothrus, ostrich, and parasitic bees—Slave-making ants—Hive-bee, its cell-making instinct—Changes of instinct and structure not necessarily simultaneous—Difficulties of the theory of the Natural Selection of instincts—Neuter or sterile insects—Summary.

MANY instincts are so wonderful that their development will probably appear to the reader a difficulty sufficient to overthrow my whole theory. I may here premise, that I have nothing to do with the origin of the mental powers, any more than I have with that of life itself. We are concerned only with the diversities of instinct and of the other mental faculties in animals of the same class.

I will not attempt any definition of instinct. It would be easy to show that several distinct mental actions are commonly embraced by this term; but every one understands what is meant, when it is said that instinct impels the cuckoo to migrate and to lay her eggs in other birds' nests. An action, which we ourselves require experience to enable us to perform, when performed by an animal, more especially by a very young one, without experience, and when performed by many individuals in the same way, without their knowing for what purpose it is performed, is usually said to be instinctive. But I could show that none of these characters are universal. A little dose of judgment or reason, as Pierre Huber expresses it, often comes into play, even with animals low in the scale of nature.

Frederick Cuvier and several of the older metaphysicians have compared instinct with habit. This comparison gives, I think, an accurate notion of the frame of mind under which an instinctive action is performed, but not necessarily of its origin. How unconsciously many habitual actions are performed, indeed not rarely in direct opposition to our conscious will! yet they may be modified by the will or reason. Habits easily become associated with other habits, with certain periods of time, and states of the body. When once acquired, they often remain constant throughout life. Several other points of resemblance between instincts and habits could be pointed out. As in repeating a well-known song, so in instincts, one action follows another by a sort of rhythm; if a person be interrupted in a song, or in repeating anything by rote, he is generally forced to go back to recover the habitual train of thought; so P. Huber found it was with a caterpillar, which makes a very complicated hammock; for if he took a caterpillar which had completed its hammock up to, say, the sixth stage of construction, and put it into a hammock completed up only to the third stage, the caterpillar simply re-performed the fourth, fifth, and sixth stages of construction. If, however, a caterpillar were taken out of a hammock made up, for instance, to the third stage, and were put into one finished up to the sixth stage, so that much of its work was already done for it, far from deriving any

benefit from this, it was much embarrassed, and in order
to complete its hammock, seemed forced to start from
the third stage, where it had left off, and thus tried to
complete the already finished work.

If we suppose any habitual action to become inherited
—and it can be shown that this does sometimes happen
—then the resemblance between what originally was a
habit and an instinct becomes so close as not to be distinguished. If Mozart, instead of playing the pianoforte
at three years old with wonderfully little practice, had
played a tune with no practice at all, he might truly be
said to have done so instinctively. But it would be a
serious error to suppose that the greater number of
instincts have been acquired by habit in one generation, and then transmitted by inheritance to succeeding
generations. It can be clearly shown that the most
wonderful instincts with which we are acquainted,
namely, those of the hive-bee and of many ants, could
not possibly have been acquired by habit.

It will be universally admitted that instincts are as
important as corporeal structures for the welfare of each
species, under its present conditions of life. Under
changed conditions of life, it is at least possible that
slight modifications of instinct might be profitable to a
species; and if it can be shown that instincts do vary
ever so little, then I can see no difficulty in natural
selection preserving and continually accumulating variations of instinct to any extent that was profitable. It
is thus, as I believe, that all the most complex and
wonderful instincts have originated. As modifications
of corporeal structure arise from, and are increased by,
use or habit, and are diminished or lost by disuse, so I
do not doubt it has been with instincts. But I believe

Y

that the effects of habit are in many cases of subordinate importance to the effects of the natural selection of what may be called spontaneous variations of instincts;—that is of variations produced by the same unknown causes which produce slight deviations of bodily structure.

No complex instinct can possibly be produced through natural selection, except by the slow and gradual accumulation of numerous slight, yet profitable, variations. Hence, as in the case of corporeal structures, we ought to find in nature, not the actual transitional gradations by which each complex instinct has been acquired—for these could be found only in the lineal ancestors of each species—but we ought to find in the collateral lines of descent some evidence of such gradations; or we ought at least to be able to show that gradations of some kind are possible; and this we certainly can do. I have been surprised to find, making allowance for the instincts of animals having been but little observed except in Europe and North America, and for no instinct being known amongst extinct species, how very generally gradations, leading to the most complex instincts, can be discovered. Changes of instinct may sometimes be facilitated by the same species having different instincts at different periods of life, or at different seasons of the year, or when placed under different circumstances, &c.; in which case either the one or the other instinct might be preserved by natural selection. And such instances of diversity of instinct in the same species can be shown to occur in nature.

Again, as in the case of corporeal structure, and conformably to my theory, the instinct of each species is good for itself, but has never, as far as we can judge, been produced for the exclusive good of others. One of the strongest instances of an animal apparently perform-

ing an action for the sole good of another, with which I
am acquainted, is that of aphides voluntarily yielding,
as was first observed by Huber, their sweet excretion to
ants: that they do so voluntarily, the following facts
show. I removed all the ants from a group of about
a dozen aphides on a dock-plant, and prevented their
attendance during several hours. After this interval, I
felt sure that the aphides would want to excrete. I
watched them for some time through a lens, but not one
excreted; I then tickled and stroked them with a hair
in the same manner, as well as I could, as the ants do
with their antennæ; but not one excreted. Afterwards
I allowed an ant to visit them, and it immediately
seemed, by its eager way of running about, to be well
aware what a rich flock it had discovered; it then began
to play with its antennæ on the abdomen first of one
aphis and then of another; and each, as soon as it felt
the antennæ, immediately lifted up its abdomen and
excreted a limpid drop of sweet juice, which was eagerly
devoured by the ant. Even the quite young aphides
behaved in this manner, showing that the action was
instinctive, and not the result of experience. It is
certain, from the observations of Huber, that the aphides
show no dislike to the ants: if the latter be not present
they are at last compelled to eject their excretion. But
as the excretion is extremely viscid, it is no doubt a
convenience to the aphides to have it removed; therefore probably they do not excrete solely for the good of
the ants. Although there is no evidence that any animal
performs an action for the exclusive good of another
species, yet each tries to take advantage of the instincts
of others, as each takes advantage of the weaker bodily
structure of other species. So again certain instincts

cannot be considered as absolutely perfect; but as details on this and other such points are not indispensable, they may be here passed over.

As some degree of variation in instincts under a state of nature, and the inheritance of such variations, are indispensable for the action of natural selection, as many instances as possible ought to be given; but want of space prevents me. I can only assert that instincts certainly do vary—for instance, the migratory instinct, both in extent and direction, and in its total loss. So it is with the nests of birds, which vary partly in dependence on the situations chosen, and on the nature and temperature of the country inhabited, but often from causes wholly unknown to us: Audubon has given several remarkable cases of differences in the nests of the same species in the northern and southern United States. Why, it has been asked, if instinct be variable, has it not granted to the bee "the ability to use some other material when wax was deficient"? But what other natural material could bees use? They will work, as I have seen, with wax hardened with vermilion or softened with lard. Andrew Knight observed that his bees, instead of laboriously collecting propolis, used a cement of wax and turpentine, with which he had covered decorticated trees. It has lately been shown that bees, instead of searching for pollen, will gladly use a very different substance, namely oatmeal. Fear of any particular enemy is certainly an instinctive quality, as may be seen in nestling birds, though it is strengthened by experience, and by the sight of fear of the same enemy in other animals. The fear of man is slowly acquired, as I have elsewhere shown, by the various animals which inhabit desert islands; and we see an instance

of this even in England, in the greater wildness of all
our large birds in comparison with our small birds;
for the large birds have been most persecuted by man.
We may safely attribute the greater wildness of our large
birds to this cause; for in uninhabited islands large birds
are not more fearful than small; and the magpie, so wary
in England, is tame in Norway, as is the hooded crow
in Egypt.

That the mental qualities of animals of the same kind,
born in a state of nature, vary much, could be shown
by many facts.   Several cases could also be adduced of
occasional and strange habits in wild animals, which, if
advantageous to the species, might have given rise,
through natural selection, to new instincts.   But I am
well aware that these general statements, without the
facts in detail, will produce but a feeble effect on the
reader's mind.   I can only repeat my assurance, that I do
not speak without good evidence.

Inherited Changes of Habit or Instinct in Domesticated Animals.

The possibility, or even probability, of inherited variations of instinct in a state of nature will be strengthened
by briefly considering a few cases under domestication.
We shall thus be enabled to see the part which habit
and the selection of so-called spontaneous variations have
played in modifying the mental qualities of our domestic
animals.   It is notorious how much domestic animals
vary in their mental qualities.   With cats, for instance,
one naturally takes to catching rats, and another mice,
and these tendencies are known to be inherited.   One
cat, according to Mr. St. John, always brought home
game-birds, another hares or rabbits, and another hunted

on marshy ground and almost nightly caught woodcocks or snipes. A number of curious and authentic instances could be given of various shades of disposition and of taste, and likewise of the oddest tricks, associated with certain frames of mind or periods of time, being inherited. But let us look to the familiar case of the breeds of the dogs : it cannot be doubted that young pointers (I have myself seen a striking instance) will sometimes point and even back other dogs the very first time that they are taken out; retrieving is certainly in some degree inherited by retrievers; and a tendency to run round, instead of at, a flock of sheep, by shepherd dogs. I cannot see that these actions, performed without experience by the young, and in nearly the same manner by each individual, performed with eager delight by each breed, and without the end being known—for the young pointer can no more know that he points to aid his master, than the white butterfly knows why she lays her eggs on the leaf of the cabbage—I cannot see that these actions differ essentially from true instincts. If we were to behold one kind of wolf, when young and without any training, as soon as it scented its prey, stand motionless like a statue, and then slowly crawl forward with a peculiar gait; and another kind of wolf rushing round, instead of at, a herd of deer, and driving them to a distant point, we should assuredly call these actions instinctive. Domestic instincts, as they may be called, are certainly far less fixed than natural instincts; but they have been acted on by far less rigorous selection, and have been transmitted for an incomparably shorter period, under less fixed conditions of life.

How strongly these domestic instincts, habits, and dispositions are inherited, and how curiously they

become mingled, is well shown when different breeds of
dogs are crossed. Thus it is known that a cross with a
bull-dog has affected for many generations the courage
and obstinacy of greyhounds; and a cross with a greyhound has given to a whole family of shepherd-dogs a
tendency to hunt hares. These domestic instincts, when
thus tested by crossing, resemble natural instincts, which
in a like manner become curiously blended together,
and for a long period exhibit traces of the instincts of
either parent: for example, Le Roy describes a dog,
whose great-grandfather was a wolf, and this dog showed
a trace of its wild parentage only in one way, by not
coming in a straight line to his master, when called.

Domestic instincts are sometimes spoken of as actions
which have become inherited solely from long-continued
and compulsory habit; but this is not true. No one
would ever have thought of teaching, or probably could
have taught, the tumbler-pigeon to tumble,—an action
which, as I have witnessed, is performed by young birds,
that have never seen a pigeon tumble. We may believe
that some one pigeon showed a slight tendency to this
strange habit, and that the long-continued selection of
the best individuals in successive generations made
tumblers what they now are; and near Glasgow there
are house-tumblers, as I hear from Mr. Brent, which
cannot fly eighteen inches high without going head over
heels. It may be doubted whether any one would have
thought of training a dog to point, had not some one dog
naturally shown a tendency in this line; and this is
known occasionally to happen, as I once saw, in a pure
terrier: the act of pointing is probably, as many have
thought, only the exaggerated pause of an animal preparing to spring on its prey. When the first tendency

to point was once displayed, methodical selection and the inherited effects of compulsory training in each successive generation would soon complete the work; and unconscious selection is still in progress, as each man tries to procure, without intending to improve the breed, dogs which stand and hunt best. On the other hand, habit alone in some cases has sufficed; hardly any animal is more difficult to tame than the young of the wild rabbit; scarcely any animal is tamer than the young of the tame rabbit; but I can hardly suppose that domestic rabbits have often been selected for tameness alone; so that we must attribute at least the greater part of the inherited change from extreme wildness to extreme tameness, to habit and long-continued close confinement.

Natural instincts are lost under domestication: a remarkable instance of this is seen in those breeds of fowls which very rarely or never become "broody," that is, never wish to sit on their eggs. Familiarity alone prevents our seeing how largely and how permanently the minds of our domestic animals have been modified. It is scarcely possible to doubt that the love of man has become instinctive in the dog. All wolves, foxes, jackals, and species of the cat genus, when kept tame, are most eager to attack poultry, sheep, and pigs; and this tendency has been found incurable in dogs which have been brought home as puppies from countries such as Tierra del Fuego and Australia, where the savages do not keep these domestic animals. How rarely, on the other hand, do our civilised dogs, even when quite young, require to be taught not to attack poultry, sheep, and pigs! No doubt they occasionally do make an attack, and are then beaten; and if not cured, they are de stroyed; so that habit and some degree of selection have probably concurred in civilising by inheritance our dogs. On the other hand, young chickens have lost, wholly by habit, that fear of the dog and cat which no doubt was originally instinctive in them; for I am informed by Captain Hutton that the young chickens of the parent-stock, the Gallus bankiva, when reared in India under a hen, are at first excessively wild. So it is with young pheasants reared in England under a hen. It is not that chickens have lost all fear, but fear only of dogs and cats, for if the hen gives the danger-chuckle, they will run (more especially young turkeys) from under her, and conceal themselves in the surrounding grass or thickets; and this is evidently done for the instinctive purpose of allowing, as we see in wild ground-birds, their mother to fly away. But this instinct retained by our chickens has become useless under domestication, for the mother-hen has almost lost by disuse the power of flight.

Hence, we may conclude, that under domestication instincts have been acquired, and natural instincts have been lost, partly by habit, and partly by man selecting and accumulating, during successive generations, peculiar mental habits and actions, which at first appeared from what we must in our ignorance call an accident. In some cases compulsory habit alone has sufficed to produce inherited mental changes; in other cases, compulsory habit has done nothing, and all has been the result of selection, pursued both methodically and unconsciously: but in most cases habit and selection have probably concurred.

Special Instincts.

We shall, perhaps, best understand how instincts in a state of nature have become modified by selection, by considering a few cases. I will select only three,—namely, the instinct which leads the cuckoo to lay her eggs in other birds' nests; the slave-making instinct of certain ants; and the cell-making power of the hive-bee. These two latter instincts have generally and justly been ranked by naturalists as the most wonderful of all known instincts.

Instincts of the Cuckoo. — It is supposed by some naturalists that the more immediate cause of the instinct of the cuckoo is, that she lays her eggs, not daily, but at intervals of two or three days; so that, if she were to make her own nest and sit on her own eggs, those first laid would have to be left for some time unincubated, or there would be eggs and young birds of different ages in the same nest. If this were the case, the process of laying and hatching might be inconveniently long, more especially as she migrates at a very early period; and the first hatched young would probably have to be fed by the male alone. But the American cuckoo is in this predicament; for she makes her own nest, and has eggs and young successively hatched, all at the same time. It has been both asserted and denied that the American cuckoo occasionally lays her eggs in other birds' nests; but I have lately heard from Dr. Merrell, of Iowa, that he once found in Illinois a young cuckoo together with a young jay in the nest of a Blue jay (Garrulus cristatus); and as both were nearly full feathered, there could be no mistake in their identification. I could also give several instances of various birds which have been known occasionally to lay their eggs in other birds' nests.   Now let us suppose that the ancient progenitor of our European cuckoo had the habits of the American cuckoo, and that she occasionally laid an egg in another bird's nest.   If the old bird profited by this occasional habit through being enabled to migrate earlier or through any other cause; or if the young were made more vigorous by advantage being taken of the mistaken instinct of another species than when reared by their own mother, encumbered as she could hardly fail to be by having eggs and young of different ages at the same time; then the old birds or the fostered young would gain an advantage.   And analogy would lead us to believe, that the young thus reared would be apt to follow by inheritance the occasional and aberrant habit of their mother, and in their turn would be apt to lay their eggs in other birds' nests, and thus be more successful in rearing their young.   By a continued process of this nature, I believe that the strange instinct of our cuckoo has been generated.   It has, also, recently been ascertained on sufficient evidence, by Adolf Müller, that the cuckoo occasionally lays her eggs on the bare ground, sits on them, and feeds her young.   This rare event is probably a case of reversion to the long-lost, aboriginal instinct of nidification.

It has been objected that I have not noticed other related instincts and adaptations of structure in the cuckoo, which are spoken of as necessarily co-ordinated. But in all cases, speculation on an instinct known to us only in a single species, is useless, for we have hitherto had no facts to guide us.   Until recently the instincts of the European and of the non-parasitic American cuckoo alone were known; now, owing to Mr. Ramsay's observations, we have learnt something about three Australian species, which lay their eggs in other birds' nests. The chief points to be referred to are three: first, that the common cuckoo, with rare exceptions, lays only one egg in a nest, so that the large and voracious young bird receives ample food. Secondly, that the eggs are remarkably small, not exceeding those of the skylark, —a bird about one-fourth as large as the cuckoo. That the small size of the egg is a real case of adaptation we may infer from the fact of the non-parasitic American cuckoo laying full-sized eggs. Thirdly, that the young cuckoo, soon after birth, has the instinct, the strength, and a properly shaped back for ejecting its foster-brothers, which then perish from cold and hunger. This has been boldly called a beneficent arrangement, in order that the young cuckoo may get sufficient food, and that its foster-brothers may perish before they had acquired much feeling!

Turning now to the Australian species; though these birds generally lay only one egg in a nest, it is not rare to find two and even three eggs in the same nest. In the Bronze cuckoo the eggs vary greatly in size, from eight to ten lines in length. Now if it had been of an advantage to this species to have laid eggs even smaller than those now laid, so as to have deceived certain foster-parents, or, as is more probable, to have been hatched within a shorter period (for it is asserted that there is a relation between the size of eggs and the period of their incubation), then there is no difficulty in believing that a race or species might have been formed which would have laid smaller and smaller eggs; for these would have been more safely hatched and reared. Mr. Ramsay remarks that two of the Australian cuckoos, when they

lay their eggs in an open nest, manifest a decided preference for nests containing eggs similar in colour to
their own.   The European species apparently manifests
some tendency towards a similar instinct, but not rarely
departs from it, as is shown by her laying her dull and
pale-coloured eggs in the nest of the Hedge-warbler
with bright greenish-blue eggs.   Had our cuckoo invariably displayed the above instinct, it would assuredly
have been added to those which it is assumed must all
have been acquired together.   The eggs of the Australian
Bronze cuckoo vary, according to Mr. Ramsay, to an
extraordinary degree in colour; so that in this respect,
as well as in size, natural selection might have secured
and fixed any advantageous variation.

In the case of the European cuckoo, the offspring of
the foster-parents are commonly ejected from the nest
within three days after the cuckoo is hatched; and as
the latter at this age is in a most helpless condition, Mr.
Gould was formerly inclined to believe that the act of
ejection was performed by the foster-parents themselves.
But he has now received a trustworthy account of a
young cuckoo which was actually seen, whilst still blind
and not able even to hold up its own head, in the act of
ejecting its foster-brothers.   One of these was replaced
in the nest by the observer, and was again thrown out.
With respect to the means by which this strange and
odious instinct was acquired, if it were of great importance for the young cuckoo, as is probably the case, to
receive as much food as possible soon after birth, I can
see no special difficulty in its having gradually acquired,
during successive generations, the blind desire, the
strength, and structure necessary for the work of ejection; for those young cuckoos which had such habits

and structure best developed would be the most securely reared. The first step towards the acquisition of the proper instinct might have been mere unintentional restlessness on the part of the young bird, when somewhat advanced in age and strength; the habit having been afterwards improved, and transmitted to an earlier age. I can see no more difficulty in this, than in the unhatched young of other birds acquiring the instinct to break through their own shells;—or than in young snakes acquiring in their upper jaws, as Owen has remarked, a transitory sharp tooth for cutting through the tough egg-shell. For if each part is liable to individual variations at all ages, and the variations tend to be inherited at a corresponding or earlier age,—propositions which cannot be disputed,—then the instincts and structure of the young could be slowly modified as surely as those of the adult; and both cases must stand or fall together with the whole theory of natural selection.

Some species of Molothrus, a widely distinct genus of American birds, allied to our starlings, have parasitic habits like those of the cuckoo; and the species present an interesting gradation in the perfection of their instincts. The sexes of Molothrus badius are stated by an excellent observer, Mr. Hudson, sometimes to live promiscuously together in flocks, and sometimes to pair. They either build a nest of their own, or seize on one belonging to some other bird, occasionally throwing out the nestlings of the stranger. They either lay their eggs in the nest thus appropriated, or oddly enough build one for themselves on the top of it. They usually sit on their own eggs and rear their own young; but Mr. Hudson says it is probable that they are occasionally parasitic, for he has seen the young of this species

following old birds of a distinct kind and clamouring to
be fed by them. The parasitic habits of another species
of Molothrus, the M. bonariensis, are much more highly
developed than those of the last, but are still far from
perfect. This bird, as far as it is known, invariably lays
its eggs in the nests of strangers; but it is remarkable
that several together sometimes commence to build an
irregular untidy nest of their own, placed in singularly
ill-adapted situations, as on the leaves of a large thistle.
They never, however, as far as Mr. Hudson has ascertained, complete a nest for themselves. They often lay
so many eggs—from fifteen to twenty—in the same
foster-nest, that few or none can possibly be hatched.
They have, moreover, the extraordinary habit of pecking
holes in the eggs, whether of their own species or of their
foster-parents, which they find in the appropriated nests.
They drop also many eggs on the bare ground, which
are thus wasted. A third species, the M. pecoris of
North America, has acquired instincts as perfect as
those of the cuckoo, for it never lays more than one egg
in a foster-nest, so that the young bird is securely reared.
Mr. Hudson is a strong disbeliever in evolution, but he
appears to have been so much struck by the imperfect
instincts of the Molothrus bonariensis that he quotes
my words, and asks, "Must we consider these habits,
not as especially endowed or created instincts, but
as small consequences of one general law, namely,
transition?"

Various birds, as has already been remarked, occasionally lay their eggs in the nests of other birds. This
habit is not very uncommon with the Gallinaceæ, and
throws some light on the singular instinct of the ostrich.
In this family several hen-birds unite and lay first a

few eggs in one nest and then in another; and these
are hatched by the males. This instinct may probably
be accounted for by the fact of the hens laying a large
number of eggs, but, as with the cuckoo, at intervals
of two or three days. The instinct, however, of the
American ostrich, as in the case of the Molothrus bonariensis, has not as yet been perfected; for a surprising
number of eggs lie strewed over the plains, so that in
one day's hunting I picked up no less than twenty lost
and wasted eggs.

Many bees are parasitic, and regularly lay their eggs
in the nests of other kinds of bees. This case is more
remarkable than that of the cuckoo; for these bees have
not only had their instincts but their structure modified
in accordance with their parasitic habits; for they do
not possess the pollen-collecting apparatus which would
have been indispensable if they had stored up food for
their own young. Some species of Sphegidæ (wasp-like
insects) are likewise parasitic; and M. Fabre has lately
shown good reason for believing that, although the
Tachytes nigra generally makes its own burrow and
stores it with paralysed prey for its own larvæ, yet that,
when this insect finds a burrow already made and stored
by another sphex, it takes advantage of the prize, and
becomes for the occasion parasitic. In this case, as with
that of the Molothrus or cuckoo, I can see no difficulty
in natural selection making an occasional habit permanent, if of advantage to the species, and if the insect
whose nest and stored food are feloniously appropriated,
be not thus exterminated.

Slave-making instinct.—This remarkable instinct was
first discovered in the Formica (Polyerges) rufescens by
Pierre Huber, a better observer even than his celebrated

father. This ant is absolutely dependent on its slaves ;
without their aid, the species would certainly become
extinct in a single year. The males and fertile females
do no work of any kind, and the workers or sterile
females, though most energetic and courageous in
capturing slaves, do no other work. They are incapable of making their own nests, or of feeding their
own larvæ. When the old nest is found inconvenient,
and they have to migrate, it is the slaves which determine the migration, and actually carry their masters in
their jaws. So utterly helpless are the masters, that
when Huber shut up thirty of them without a
slave, but with plenty of the food which they like
best, and with their own larvæ and pupæ to stimu-
late them to work, they did nothing; they could not
even feed themselves, and many perished of hunger.
Huber then introduced a single slave (F. fusca), and she
instantly set to work, fed and saved the survivors ;
made some cells and tended the larvæ, and put all to
rights. What can be more extraordinary than these
well-ascertained facts ? If we had not known of any
other slave-making ant, it would have been hopeless
to speculate how so wonderful an instinct could have
been perfected.

Another species, Formica sanguinea, was likewise first
discovered by P. Huber to be a slave-making ant. This
species is found in the southern parts of England, and
its habits have been attended to by Mr. F. Smith, of
the British Museum, to whom I am much indebted for
information on this and other subjects. Although fully
trusting to the statements of Huber and Mr. Smith, I
tried to approach the subject in a sceptical frame of
mind, as any one may well be excused for doubting the

z

existence of so extraordinary an instinct as that of making
slaves.　Hence, I will give the observations which I
made in some little detail.　I opened fourteen nests of
F. sanguinea, and found a few slaves in all.　Males and
fertile females of the slave species (F. fusca) are found
only in their own proper communities, and have never
been observed in the nests of F. sanguinea.　The slaves
are black and not above half the size of their red masters,
so that the contrast in their appearance is great.　When
the nest is slightly disturbed, the slaves occasionally
come out, and like their masters are much agitated and
defend the nest: when the nest is much disturbed, and
the larvæ and pupæ are exposed, the slaves work ener-
getically together with their masters in carrying them
away to a place of safety.　Hence, it is clear, that the
slaves feel quite at home.　During the months of June
and July, on three successive years, I watched for many
hours several nests in Surrey and Sussex, and never saw
a slave either leave or enter a nest.　As, during these
months, the slaves are very few in number, I thought
that they might behave differently when more numerous;
but Mr. Smith informs me that he has watched the nests
at various hours during May, June, and August, both
in Surrey and Hampshire, and has never seen the slaves,
though present in large numbers in August, either leave
or enter the nest.　Hence he considers them as strictly
household slaves.　The masters, on the other hand, may
be constantly seen bringing in materials for the nest,
and food of all kinds.　During the year 1860, however,
in the month of July, I came across a community with an
unusually large stock of slaves, and I observed a few
slaves mingled with their masters leaving the nest, and
marching along the same road to a tall Scotch-fir-tree,

twenty-five yards distant, which they ascended together, probably in search of aphides or cocci. According to Huber, who had ample opportunities for observation, the slaves in Switzerland habitually work with their masters in making the nest, and they alone open and close the doors in the morning and evening; and, as Huber expressly states, their principal office is to search for aphides. This difference in the usual habits of the masters and slaves in the two countries, probably depends merely on the slaves being captured in greater numbers in Switzerland than in England.

One day I fortunately witnessed a migration of F. sanguinea from one nest to another, and it was a most interesting spectacle to behold the masters carefully carrying their slaves in their jaws instead of being carried by them, as in the case of F. rufescens. Another day my attention was struck by about a score of the slave-makers haunting the same spot, and evidently not in search of food; they approached and were vigorously repulsed by an independent community of the slave-species (F. fusca); sometimes as many as three of these ants clinging to the legs of the slave-making F. sanguinea. The latter ruthlessly killed their small opponents, and carried their dead bodies as food to their nest, twenty-nine yards distant; but they were prevented from getting any pupæ to rear as slaves. I then dug up a small parcel of the pupæ of F. fusca from another nest, and put them down on a bare spot near the place of combat; they were eagerly seized and carried off by the tyrants, who perhaps fancied that, after all, they had been victorious in their late combat.

At the same time I laid on the same place a small parcel of the pupæ of another species, F. flava, with a

few of these little yellow ants still clinging to the
fragments of their nest. This species is sometimes,
though rarely, made into slaves, as has been described
by Mr. Smith. Although so small a species, it is very
courageous, and I have seen it ferociously attack other
ants. In one instance I found to my surprise an independent community of F. flava under a stone beneath
a nest of the slave-making F. sanguinea; and when I
had accidentally disturbed both nests, the little ants
attacked their big neighbours with surprising courage.
Now I was curious to ascertain whether F. sanguinea
could distinguish the pupæ of F. fusca, which they
habitually make into slaves, from those of the little
and furious F. flava, which they rarely capture, and it
was evident that they did at once distinguish them;
for we have seen that they eagerly and instantly seized
the pupæ of F. fusca, whereas they were much terrified
when they came across the pupæ, or even the earth from
the nest, of F. flava, and quickly ran away; but in
about a quarter of an hour, shortly after all the little
yellow ants had crawled away, they took heart and
carried off the pupæ.

One evening I visited another community of F.
sanguinea, and found a number of these ants returning
home and entering their nests, carrying the dead bodies
of F. fusca (showing that it was not a migration) and
numerous pupæ. I traced a long file of ants burthened
with booty, for about forty yards back, to a very thick
clump of heath, whence I saw the last individual of F.
sanguinea emerge, carrying a pupa; but I was not able
to find the desolated nest in the thick heath. The nest,
however, must have been close at hand, for two or three
individuals of F. fusca were rushing about in the greatest

agitation, and one was perched motionless with its own
pupa in its mouth on the top of a spray of heath, an
image of despair over its ravaged home.

Such are the facts, though they did not need confirmation by me, in regard to the wonderful instinct of
making slaves.   Let it be observed what a contrast the
instinctive habits of F. sanguinea present with those of
the continental F. rufescens.   The latter does not build
its own nest, does not determine its own migrations, does
not collect food for itself or its young, and cannot even
feed itself: it is absolutely dependent on its numerous
slaves.   Formica sanguinea, on the other hand, possesses much fewer slaves, and in the early part of the
summer extremely few: the masters determine when
and where a new nest shall be formed, and when they
migrate, the masters carry the slaves.   Both in Switzerland and England the slaves seem to have the exclusive
care of the larvæ, and the masters alone go on slave-
making expeditions.   In Switzerland the slaves and
masters work together, making and bringing materials
for the nest; both, but chiefly the slaves, tend, and milk,
as it may be called, their aphides; and thus both collect
food for the community.   In England the masters alone
usually leave the nest to collect building materials and
food for themselves, their slaves and larvæ.   So that the
masters in this country receive much less service from
their slaves than they do in Switzerland.

By what steps the instinct of F. sanguinea originated
I will not pretend to conjecture.   But as ants which
are not slave-makers will, as I have seen, carry off the
pupæ of other species, if scattered near their nests, it is
possible that such pupæ originally stored as food might
become developed; and the foreign ants thus uninten-

tionally reared would then follow their proper instincts, and do what work they could. If their presence proved useful to the species which had seized them—if it were more advantageous to this species to capture workers than to procreate them—the habit of collecting pupæ, originally for food, might by natural selection be strengthened and rendered permanent for the very different purpose of raising slaves. When the instinct was once acquired, if carried out to a much less extent even than in our British F. sanguinea, which, as we have seen, is less aided by its slaves than the same species in Switzerland, natural selection might increase and modify the instinct—always supposing each modification to be of use to the species—until an ant was formed as abjectly dependent on its slaves as is the Formica rufescens.

Cell-making instinct of the Hive-Bee.—I will not here enter on minute details on this subject, but will merely give an outline of the conclusions at which I have arrived. He must be a dull man who can examine the exquisite structure of a comb, so beautifully adapted to its end, without enthusiastic admiration. We hear from mathematicians that bees have practically solved a recondite problem, and have made their cells of the proper shape to hold the greatest possible amount of honey, with the least possible consumption of precious wax in their construction. It has been remarked that a skilful workman with fitting tools and measures, would find it very difficult to make cells of wax of the true form, though this is effected by a crowd of bees working in a dark hive. Granting whatever instincts you please, it seems at first quite inconceivable how they can make all the necessary angles and planes, or even perceive when they are correctly made. But the difficulty is not nearly so great as it at first appears : all this beautiful work can be shown, I think, to follow from a few simple instincts.

I was led to investigate this subject by Mr. Waterhouse, who has shown that the form of the cell stands in close relation to the presence of adjoining cells ; and the following view may, perhaps, be considered only as a modification of his theory. Let us look to the great principle of gradation, and see whether Nature does not reveal to us her method of work. At one end of a short series we have humble-bees, which use their old cocoons to hold honey, sometimes adding to them short tubes of wax, and likewise making separate and very irregular rounded cells of wax. At the other end of the series we have the cells of the hive-bee, placed in a double layer : each cell, as is well known, is an hexagonal prism, with the basal edges of its six sides bevelled so as to join an inverted pyramid, of three rhombs. These rhombs have certain angles, and the three which form the pyramidal base of a single cell on one side of the comb enter into the composition of the bases of three adjoining cells on the opposite side. In the series between the extreme perfection of the cells of the hive-bee and the simplicity of those of the humble-bee we have the cells of the Mexican Melipona domestica, carefully described and figured by Pierre Huber. The Melipona itself is intermediate in structure between the hive and humble bee, but more nearly related to the latter ; it forms a nearly regular waxen comb of cylindrical cells, in which the young are hatched, and, in addition, some large cells of wax for holding honey. These latter cells are nearly spherical and of nearly equal sizes, and are aggregated into an irregular mass. But the important point to notice is, that these cells are always made at that degree of nearness to each other that they would have intersected or broken into each other if the spheres had been completed; but this is never permitted, the bees building perfectly flat walls of wax between the spheres which thus tend to intersect. Hence, each cell consists of an outer spherical portion, and of two, three, or more flat surfaces, according as the cell adjoins two, three, or more other cells. When one cell rests on three other cells, which, from the spheres being nearly of the same size, is very frequently and necessarily the case, the three flat surfaces are united into a pyramid; and this pyramid as Huber has remarked, is manifestly a gross imitation of the three-sided pyramidal base of the cell of the hive-bee. As in the cells of the hive-bee, so here, the three plane surfaces in any one cell necessarily enter into the construction of three adjoining cells. It is obvious that the Melipona saves wax, and what is more important, labour, by this manner of building; for the flat walls between the adjoining cells are not double, but are of the same thickness as the outer spherical portions, and yet each flat portion forms a part of two cells.

Reflecting on this case, it occurred to me that if the Melipona had made its spheres at some given distance from each other, and had made them of equal sizes and had arranged them symmetrically in a double layer, the resulting structure would have been as perfect as the comb of the hive-bee. Accordingly I wrote to Professor Miller of Cambridge, and this geometer has kindly read over the following statement, drawn up from his information, and tells me that it is strictly correct:—

If a number of equal spheres be described with their centres placed in two parallel layers; with the centre of each sphere at the distance of radius $\times \sqrt{2}$, or radius $\times$ 1·41421 (or at some lesser distance), from the centres of the six surrounding spheres in the same layer; and at the same distance from the centres of the adjoining spheres in the other and parallel layer; then, if planes of intersection between the several spheres in both layers be formed, there will result a double layer of hexagonal prisms united together by pyramidal bases formed of three rhombs; and the rhombs and the sides of the hexagonal prisms will have every angle identically the same with the best measurements which have been made of the cells of the hive-bee. But I hear from Prof. Wyman, who has made numerous careful measurements, that the accuracy of the workmanship of the bee has been greatly exaggerated; so much so, that whatever the typical form of the cell may be, it is rarely, if ever, realised.

Hence we may safely conclude that, if we could slightly modify the instincts already possessed by the Melipona, and in themselves not very wonderful, this bee would make a structure as wonderfully perfect as that of the hive-bee. We must suppose the Melipona to have the power of forming her cells truly spherical, and of equal sizes; and this would not be very surprising, seeing that she already does so to a certain extent, and seeing what perfectly cylindrical burrows many insects make in wood, apparently by turning round on a fixed point. We must suppose the Melipona to arrange her cells in level layers, as she already does her cylindrical cells; and we must further suppose, and this is the greatest difficulty, that she can somehow judge accurately at what distance to stand from her fellow-labourers when several are making their spheres; but she is already so far enabled to judge of distance, that she always describes her spheres so as to intersect to a certain extent; and then she unites the points of intersection by perfectly flat surfaces. By such modifications of instincts which in themselves are not very wonderful,—hardly more wonderful than those which guide a bird to make its nest,—I believe that the hive-bee has acquired, through natural selection, her inimitable architectural powers.

But this theory can be tested by experiment. Following the example of Mr. Tegetmeier, I separated two combs, and put between them a long, thick, rectangular strip of wax: the bees instantly began to excavate minute circular pits in it; and as they deepened these little pits, they made them wider and wider until they were converted into shallow basins, appearing to the eye perfectly true or parts of a sphere, and of about the diameter of a cell. It was most interesting to observe that, wherever several bees had begun to excavate these basins near together, they had begun their work at such a distance from each other, that by the time the basins had acquired the above-stated width (*i.e.* about the width of an ordinary cell), and were in depth about one sixth of the diameter of the sphere of which they formed a part, the rims of the basins intersected or broke into each other. As soon as this occurred, the bees ceased to excavate, and began to build up flat walls of wax on the lines of intersection between the basins, so that each hexagonal prism was built upon the scalloped edge of a smooth basin, instead of on the straight edges of a three-sided pyramid as in the case of ordinary cells.

I then put into the hive, instead of a thick, rectangular piece of wax, a thin and narrow, knife-edged ridge, coloured with vermilion. The bees instantly began on both sides to excavate little basins near to each other, in the same way as before; but the ridge of wax was so thin, that the bottoms of the basins, if they had been excavated to the same depth as in the former experiment, would have broken into each other from the opposite sides. The bees, however, did not suffer this to happen, and they stopped their excavations in due time; so that the basins, as soon as they had been a little deepened, came to have flat bases; and these flat bases, formed by thin little plates of the vermilion wax left ungnawed, were situated, as far as the eye could judge, exactly along the planes of imaginary intersection between the basins on the opposite sides of the ridge of wax. In some parts, only small portions, in other parts, large portions of a rhombic plate were thus left between the opposed basins, but the work, from the unnatural state of things, had not been neatly performed. The bees must have worked at very nearly the same rate in circularly gnawing away and deepening the basins on both sides of the ridge of vermilion wax, in order to have thus succeeded in leaving flat plates between the basins, by stopping work at the planes of intersection.

Considering how flexible thin wax is, I do not see that there is any difficulty in the bees, whilst at work on the two sides of a strip of wax, perceiving when they have gnawed the wax away to the proper thinness, and then stopping their work. In ordinary combs it has appeared to me that the bees do not always succeed in working at exactly the same rate from the opposite sides; for I have noticed half-completed rhombs at the base of a just commenced cell, which were slightly concave on one side, where I suppose that the bees had excavated too quickly, and convex on the opposed side where the bees had worked less quickly. In one well marked instance, I put the comb back into the hive, and allowed the bees to go on working for a short time, and again examined the cell, and I found that the rhombic plate had been completed, and had become *perfectly flat*: it was absolutely impossible, from the extreme thinness of the little plate, that they could have effected this by gnawing away the convex side; and I suspect that the bees in such cases stand on opposite sides and push and bend the ductile and warm wax (which as I have tried is easily done) into its proper intermediate plane, and thus flatten it.

From the experiment of the ridge of vermilion wax we can see that, if the bees were to build for themselves a thin wall of wax, they could make their cells of the proper shape, by standing at the proper distance from each other, by excavating at the same rate, and by endeavouring to make equal spherical hollows, but never allowing the spheres to break into each other. Now bees, as may be clearly seen by examining the edge of a growing comb, do make a rough, circumferential wall or rim all round the comb; and they gnaw this away from the opposite sides, always working circularly as they deepen each cell. They do not make the whole three-sided pyramidal base of any one cell at the same time, but only that one rhombic plate which stands on the extreme growing margin, or the two plates, as the case may be; and they never complete the upper edges of the rhombic plates, until the hexagonal walls are commenced. Some of these statements differ from

those made by the justly celebrated elder Huber, but
I am convinced of their accuracy; and if I had
space, I could show that they are conformable with
my theory.

Huber's statement, that the very first cell is excavated
out of a little parallel-sided wall of wax, is not, so far as I
have seen, strictly correct; the first commencement having
always been a little hood of wax; but I will not here
enter on details. We see how important a part excavation plays in the construction of the cells; but it would
be a great error to suppose that the bees cannot build
up a rough wall of wax in the proper position—that is,
along the plane of intersection between two adjoining
spheres. I have several specimens showing clearly that
they can do this. Even in the rude circumferential rim
or wall of wax round a growing comb, flexures may
sometimes be observed, corresponding in position to the
planes of the rhombic basal plates of future cells. But
the rough wall of wax has in every case to be finished
off, by being largely gnawed away on both sides. The
manner in which the bees build is curious; they always
make the first rough wall from ten to twenty times thicker
than the excessively thin finished wall of the cell, which
will ultimately be left. We shall understand how they
work, by supposing masons first to pile up a broad ridge
of cement, and then to begin cutting it away equally
on both sides near the ground, till a smooth, very thin
wall is left in the middle; the masons always piling up
the cut-away cement, and adding fresh cement on the
summit of the ridge. We shall thus have a thin wall
steadily growing upward but always crowned by a
gigantic coping. From all the cells, both those just
commenced and those completed, being thus crowned

by a strong coping of wax, the bees can cluster and crawl over the comb without injuring the delicate hexagonal walls. These walls, as Professor Miller has kindly ascertained for me, vary greatly in thickness; being, on an average of twelve measurements made near the border of the comb, $\frac{1}{352}$ of an inch in thickness; whereas the basal rhomboidal plates are thicker, nearly in the proportion of three to two, having a mean thickness, from twenty-one measurements, of $\frac{1}{229}$ of an inch. By the above singular manner of building, strength is continually given to the comb, with the utmost ultimate economy of wax.

It seems at first to add to the difficulty of understanding how the cells are made, that a multitude of bees all work together; one bee after working a short time at one cell going to another, so that, as Huber has stated, a score of individuals work even at the commencement of the first cell. I was able practically to show this fact, by covering the edges of the hexagonal walls of a single cell, or the extreme margin of the circumferential rim of a growing comb, with an extremely thin layer of melted vermilion wax; and I invariably found that the colour was most delicately diffused by the bees—as delicately as a painter could have done it with his brush —by atoms of the coloured wax having been taken from the spot on which it had been placed, and worked into the growing edges of the cells all round. The work of construction seems to be a sort of balance struck between many bees, all instinctively standing at the same relative distance from each other, all trying to sweep equal spheres, and then building up, or leaving ungnawed, the planes of intersection between these spheres. It was really curious to note in cases of difficulty, as when two pieces of comb met at an angle, how often the bees would pull down and rebuild in different ways the same cell, sometimes recurring to a shape which they had at first rejected.

When bees have a place on which they can stand in their proper positions for working,—for instance, on a slip of wood, placed directly under the middle of a comb growing downwards, so that the comb has to be built over one face of the slip—in this case the bees can lay the foundations of one wall of a new hexagon, in its strictly proper place, projecting beyond the other completed cells. It suffices that the bees should be enabled to stand at their proper relative distances from each other and from the walls of the last completed cells, and then, by striking imaginary spheres, they can build up a wall intermediate between two adjoining spheres; but, as far as I have seen, they never gnaw away and finish off the angles of a cell till a large part both of that cell and of the adjoining cells has been built. This capacity in bees of laying down under certain circumstances a rough wall in its proper place between two just-commenced cells, is important, as it bears on a fact, which seems at first subversive of the foregoing theory; namely, that the cells on the extreme margin of wasp-combs are sometimes strictly hexagonal; but I have not space here to enter on this subject. Nor does there seem to me any great difficulty in a single insect (as in the case of a queen-wasp) making hexagonal cells, if she were to work alternately on the inside and outside of two or three cells commenced at the same time, always standing at the proper relative distance from the parts of the cells just begun, sweeping spheres or cylinders, and building up intermediate planes.

As natural selection acts only by the accumulation of slight modifications of structure or instinct, each profitable to the individual under its conditions of life, it may reasonably be asked, how a long and graduated succession of modified architectural instincts, all tending towards the present perfect plan of construction, could have profited the progenitors of the hive-bee? I think the answer is not difficult: cells constructed like those of the bee or the wasp gain in strength, and save much in labour and space, and in the materials of which they are constructed. With respect to the formation of wax, it is known that bees are often hard pressed to get sufficient nectar, and I am informed by Mr. Tegetmeier that it has been experimentally proved that from twelve to fifteen pounds of dry sugar are consumed by a hive of bees for the secretion of a pound of wax; so that a prodigious quantity of fluid nectar must be collected and consumed by the bees in a hive for the secretion of the wax necessary for the construction of their combs. Moreover, many bees have to remain idle for many days during the process of secretion. A large store of honey is indispensable to support a large stock of bees during the winter; and the security of the hive is known mainly to depend on a large number of bees being supported. Hence the saving of wax by largely saving honey and the time consumed in collecting the honey must be an important element of success to any family of bees. Of course the success of the species may be dependent on the number of its enemies, or parasites, or on quite distinct causes, and so be altogether independent of the quantity of honey which the bees can collect. But let us suppose that this latter circumstance determined, as it probably often has determined, whether a bee allied

to our humble-bees could exist in large numbers in any
country; and let us further suppose that the community
lived through the winter, and consequently required a
store of honey: there can in this case be no doubt that
it would be an advantage to our imaginary humble-bee
if a slight modification in her instincts led her to make
her waxen cells near together, so as to intersect a little;
for a wall in common even to two adjoining cells would
save some little labour and wax. Hence it would continually be more and more advantageous to our humble-
bees, if they were to make their cells more and more
regular, nearer together, and aggregated into a mass, like
the cells of the Melipona; for in this case a large part
of the bounding surface of each cell would serve to bound
the adjoining cells, and much labour and wax would be
saved. Again, from the same cause, it would be advantageous to the Melipona, if she were to make her cells
closer together, and more regular in every way than at
present; for then, as we have seen, the spherical surfaces
would wholly disappear and be replaced by plane surfaces; and the Melipona would make a comb as perfect
as that of the hive-bee. Beyond this stage of perfection
in architecture, natural selection could not lead; for the
comb of the hive-bee, as far as we can see, is absolutely
perfect in economising labour and wax.

Thus, as I believe, the most wonderful of all known
instincts, that of the hive-bee, can be explained by
natural selection having taken advantage of numerous,
successive, slight modifications of simpler instincts;
natural selection having, by slow degrees, more and
more perfectly led the bees to sweep equal spheres at a
given distance from each other in a double layer, and to
build up and excavate the wax along the planes of inter-

section; the bees, of course, no more knowing that they swept their spheres at one particular distance from each other, than they know what are the several angles of the hexagonal prisms and of the basal rhombic plates; the motive power of the process of natural selection having been the construction of cells of due strength and of the proper size and shape for the larvæ, this being effected with the greatest possible economy of labour and wax; that individual swarm which thus made the best cells with least labour, and least waste of honey in the secretion of wax, having succeeded best, and having transmitted their newly-acquired economical instincts to new swarms, which in their turn will have had the best chance of succeeding in the struggle for existence.

Objections to the Theory of Natural Selection as applied to Instincts: Neuter and Sterile Insects.

It has been objected to the foregoing view of the origin of instincts that " the variations of structure and of instinct must have been simultaneous and accurately adjusted to each other, as a modification in the one without an immediate corresponding change in the other would have been fatal." The force of this objection rests entirely on the assumption that the changes in the instincts and structure are abrupt. To take as an illustration the case of the larger titmouse (Parus major) alluded to in a previous chapter; this bird often holds the seeds of the yew between its feet on a branch, and hammers with its beak till it gets at the kernel. Now what special difficulty would there be in natural selection preserving all the slight individual variations in the shape of the beak, which were better and better adapted to break open the seeds, until a beak was formed, as well constructed for this purpose as that of the nut-hatch, at the same time that habit, or compulsion, or spontaneous variations of taste, led the bird to become more and more of a seed-eater ? In this case the beak is supposed to be slowly modified by natural selection, subsequently to, but in accordance with, slowly changing habits or taste ; but let the feet of the titmouse vary and grow larger from correlation with the beak, or from any other unknown cause, and it is not improbable that such larger feet would lead the bird to climb more and more until it acquired the remarkable climbing instinct and power of the nuthatch. In this case a gradual change of structure is supposed to lead to changed instinctive habits. To take one more case : few instincts are more remarkable than that which leads the swift of the Eastern Islands to make its nest wholly of inspissated saliva. Some birds build their nests of mud, believed to be moistened with saliva; and one of the swifts of North America makes its nest (as I have seen) of sticks agglutinated with saliva, and even with flakes of this substance. Is it then very improbable that the natural selection of individual swifts, which secreted more and more saliva, should at last produce a species with instincts leading it to neglect other materials, and to make its nest exclusively of inspissated saliva ? And so in other cases. It must, however, be admitted that in many instances we cannot conjecture whether it was instinct or structure which first varied.

No doubt many instincts of very difficult explanation could be opposed to the theory of natural selection— cases, in which we cannot see how an instinct could have originated ; cases, in which no intermediate gradations are known to exist; cases of instincts of such trifling importance, that they could hardly have been acted on by natural selection; cases of instincts almost identically the same in animals so remote in the scale of nature, that we cannot account for their similarity by inheritance from a common progenitor, and consequently must believe that they were independently acquired through natural selection. I will not here enter on these several cases, but will confine myself to one special difficulty, which at first appeared to me insuperable, and actually fatal to the whole theory. I allude to the neuters or sterile females in insect-communities; for these neuters often differ widely in instinct and in structure from both the males and fertile females, and yet, from being sterile, they cannot propagate their kind.

The subject well deserves to be discussed at great length, but I will here take only a single case, that of working or sterile ants. How the workers have been rendered sterile is a difficulty; but not much greater than that of any other striking modification of structure; for it can be shown that some insects and other articulate animals in a state of nature occasionally become sterile; and if such insects had been social, and it had been profitable to the community that a number should have been annually born capable of work, but incapable of procreation, I can see no especial difficulty in this having been effected through natural selection. But I must pass over this preliminary difficulty. The great difficulty lies in the working ants differing widely from both the males and the fertile females in structure, as in the shape of the thorax, and in being destitute of wings and sometimes of eyes, and in instinct. As far

as instinct alone is concerned, the wonderful difference
in this respect between the workers and the perfect
females, would have been better exemplified by the hive-
bee.   If a working ant or other neuter insect had been
an ordinary animal, I should have unhesitatingly assumed that all its characters had been slowly acquired
through natural selection; namely, by individuals having
been born with slight profitable modifications, which
were inherited by the offspring; and that these again
varied and again were selected, and so onwards.   But
with the working ant we have an insect differing greatly
from its parents, yet absolutely sterile; so that it could
never have transmitted successively acquired modifications of structure or instinct to its progeny.   It may
well be asked how is it possible to reconcile this case
with the theory of natural selection?

First, let it be remembered that we have innumerable
instances, both in our domestic productions and in those
in a state of nature, of all sorts of differences of inherited
structure which are correlated with certain ages, and
with either sex.   We have differences correlated not
only with one sex, but with that short period when the
reproductive system is active, as in the nuptial plumage
of many birds, and in the hooked jaws of the male salmon.
We have even slight differences in the horns of different
breeds of cattle in relation to an artificially imperfect
state of the male sex; for oxen of certain breeds have
longer horns than the oxen of other breeds, relatively to
the length of the horns in both the bulls and cows of
these same breeds.   Hence I can see no great difficulty
in any character becoming correlated with the sterile
condition of certain members of insect-communities:
the difficulty lies in understanding how such correlated

modifications of structure could have been slowly accumulated by natural selection.

This difficulty, though appearing insuperable, is lessened, or, as I believe, disappears, when it is remembered that selection may be applied to the family, as well as to the individual, and may thus gain the desired end. Breeders of cattle wish the flesh and fat to be well marbled together: an animal thus characterised has been slaughtered, but the breeder has gone with confidence to the same stock and has succeeded. Such faith may be placed in the power of selection, that a breed of cattle, always yielding oxen with extraordinarily long horns, could, it is probable, be formed by carefully watching which individual bulls and cows, when matched, produced oxen with the longest horns; and yet no one ox would ever have propagated its kind. Here is a better and real illustration: according to M. Verlot, some varieties of the double annual Stock from having been long and carefully selected to the right degree, always produce a large proportion of seedlings bearing double and quite sterile flowers; but they likewise yield some single and fertile plants. These latter, by which alone the variety can be propagated, may be compared with the fertile male and female ants, and the double sterile plants with the neuters of the same community. As with the varieties of the stock, so with social insects, selection has been applied to the family, and not to the individual, for the sake of gaining a serviceable end. Hence we may conclude that slight modifications of structure or of instinct, correlated with the sterile condition of certain members of the community, have proved advantageous: consequently the fertile males and females have flourished, and transmitted to their fertile offspring a tendency to produce sterile members with the same modifications. This process must have been repeated many times, until that prodigious amount of difference between the fertile and sterile females of the same species has been produced, which we see in many social insects.

But we have not as yet touched on the acme of the difficulty; namely, the fact that the neuters of several ants differ, not only from the fertile females and males, but from each other, sometimes to an almost incredible degree, and are thus divided into two or even three castes. The castes, moreover, do not commonly graduate into each other, but are perfectly well defined; being as distinct from each other as are any two species of the same genus, or rather as any two genera of the same family. Thus in Eciton, there are working and soldier neuters, with jaws and instincts extraordinarily different: in Cryptocerus, the workers of one caste alone carry a wonderful sort of shield on their heads, the use of which is quite unknown: in the Mexican Myrmecocystus, the workers of one caste never leave the nest; they are fed by the workers of another caste, and they have an enormously developed abdomen which secretes a sort of honey, supplying the place of that excreted by the aphides, or the domestic cattle as they may be called, which our European ants guard and imprison.

It will indeed be thought that I have an overweening confidence in the principle of natural selection, when I do not admit that such wonderful and well-established facts at once annihilate the theory. In the simpler case of neuter insects all of one caste, which, as I believe, have been rendered different from the fertile males and females through natural selection, we may conclude

from the analogy of ordinary variations, that the successive, slight, profitable modifications did not first arise
in all the neuters in the same nest, but in some few
alone; and that by the survival of the communities
with females which produced most neuters having the
advantageous modification, all the neuters ultimately
came to be thus characterised.   According to this view
we ought occasionally to find in the same nest neuter
insects, presenting gradations of structure; and this we
do find, even not rarely, considering how few neuter
insects out of Europe have been carefully examined.
Mr. F. Smith has shown that the neuters of several British
ants differ surprisingly from each other in size and sometimes in colour; and that the extreme forms can be
linked together by individuals taken out of the same
nest: I have myself compared perfect gradations of this
kind.   It sometimes happens that the larger or the smaller
sized workers are the most numerous; or that both large
and small are numerous, whilst those of an intermediate
size are scanty in numbers.   Formica flava has larger
and smaller workers, with some few of intermediate size;
and, in this species, as Mr. F. Smith has observed, the
larger workers have simple eyes (ocelli), which though
small can be plainly distinguished, whereas the smaller
workers have their ocelli rudimentary.   Having carefully dissected several specimens of these workers, I
can affirm that the eyes are far more rudimentary in
the smaller workers than can be accounted for merely
by their proportionally lesser size; and I fully believe,
though I dare not assert so positively, that the workers
of intermediate size have their ocelli in an exactly intermediate condition.   So that here we have two bodies of
sterile workers in the same nest, differing not only in

size, but in their organs of vision, yet connected by some
few members in an intermediate condition.   I may
digress by adding, that if the smaller workers had been
the most useful to the community, and those males and
females had been continually selected, which produced
more and more of the smaller workers, until all the
workers were in this condition; we should then have
had a species of ant with neuters in nearly the same
condition as those of Myrmica.   For the workers of
Myrmica have not even rudiments of ocelli, though
the male and female ants of this genus have well-
developed ocelli.

I may give one other case: so confidently did I
expect occasionally to find gradations of important structures between the different castes of neuters in the same
species, that I gladly availed myself of Mr. F. Smith's
offer of numerous specimens from the same nest of the
driver ant (Anomma) of West Africa.   The reader will
perhaps best appreciate the amount of difference in these
workers, by my giving not the actual measurements,
but a strictly accurate illustration: the difference was
the same as if we were to see a set of workmen building
a house, of whom many were five feet four inches high,
and many sixteen feet high; but we must in addition
suppose that the larger workmen had heads four instead
of three times as big as those of the smaller men, and
jaws nearly five times as big.   The jaws, moreover, of
the working ants of the several sizes differed wonderfully
in shape, and in the form and number of the teeth.   But
the important fact for us is, that, though the workers can
be grouped into castes of different sizes, yet they graduate
insensibly into each other, as does the widely-different
structure of their jaws.   I speak confidently on this latter

point, as Sir J. Lubbock made drawings for me, with
the camera lucida, of the jaws which I dissected from
the workers of the several sizes. Mr. Bates, in his
interesting 'Naturalist on the Amazons,' has described
analogous cases.

With these facts before me, I believe that natural
selection, by acting on the fertile ants or parents, could
form a species which should regularly produce neuters,
all of large size with one form of jaw, or all of small
size with widely different jaws; or lastly, and this is
the greatest difficulty, one set of workers of one size
and structure, and simultaneously another set of workers
of a different size and structure;—a graduated series
having first been formed, as in the case of the driver
ant, and then the extreme forms having been produced
in greater and greater numbers, through the survival of
the parents which generated them, until none with an
intermediate structure were produced.

An analogous explanation has been given by Mr.
Wallace, of the equally complex case, of certain Malayan
Butterflies regularly appearing under two or even three
distinct female forms; and by Fritz Müller, of certain
Brazilian crustaceans likewise appearing under two
widely distinct male forms. But this subject need not
here be discussed.

I have now explained how, as I believe, the wonderful
fact of two distinctly defined castes of sterile workers
existing in the same nest, both widely different from
each other and from their parents, has originated. We
can see how useful their production may have been to a
social community of ants, on the same principle that
the division of labour is useful to civilised man. Ants,
however, work by inherited instincts and by inherited

organs or tools, whilst man works by acquired know-ledge and manufactured instruments. But I must confess, that, with all my faith in natural selection, I should never have anticipated that this principle could have been efficient in so high a degree, had not the case of these neuter insects led me to this conclusion. I have, therefore, discussed this case, at some little but wholly insufficient length, in order to show the power of natural selection, and likewise because this is by far the most serious special difficulty which my theory has encountered. The case, also, is very interesting, as it proves that with animals, as with plants, any amount of modification may be effected by the accumulation of numerous, slight, spontaneous variations, which are in any way profitable, without exercise or habit having been brought into play. For peculiar habits confined to the workers or sterile females, however long they might be followed, could not possibly affect the males and fertile females, which alone leave descendants. I am surprised that no one has hitherto advanced this demonstrative case of neuter insects, against the well-known doctrine of inherited habit, as advanced by Lamarck.

Summary.

I have endeavoured in this chapter briefly to show that the mental qualities of our domestic animals vary, and that the variations are inherited. Still more briefly I have attempted to show that instincts vary slightly in a state of nature. No one will dispute that instincts are of the highest importance to each animal. Therefore there is no real difficulty, under changing conditions of life, in natural selection accumulating to any extent slight modifica-tions of instinct which are in any way useful. In many

cases habit or use and disuse have probably come into play.
I do not pretend that the facts given in this chapter
strengthen in any great degree my theory : but none of the
cases of difficulty, to the best of my judgment, annihilate
it.   On the other hand, the fact that instincts are not
always absolutely perfect and are liable to mistakes :—that
no instinct can be shown to have been produced for the
good of other animals, though animals take advantage of
the instincts of others ; that the canon in natural history,
of "Natura non facit saltum," is applicable to instincts as
well as to corporeal structure, and is plainly explicable on
the foregoing views, but is otherwise inexplicable,—all tend
to corroborate the theory of natural selection.

This theory is also strengthened by some few other facts in
regard to instincts ; as by that common case of closely
allied, but distinct, species, when inhabiting distant parts
of the world and living under considerably different conditions of life, yet often retaining nearly the same instincts.
For instance, we can understand, on the principle of inheritance, how it is that the thrush of tropical South
America lines its nest with mud, in the same peculiar
manner as does our British thrush ; how it is that the
Hornbills of Africa and India have the same extraordinary
instinct of plastering up and imprisoning the females in a
hole in a tree, with only a small hole left in the plaster
through which the males feed them and their young when
hatched : how it is that the male wrens (Troglodytes) of
North America build "cock-nests," to roost in, like the
males of our Kitty-wrens,—a habit wholly unlike that of
any other known bird.   Finally, it may not be a logical
deduction, but to my imagination it is far more satisfactory
to look at such instincts as the young cuckoo ejecting its
foster-brothers,—ants making slaves,—the larvæ of ichneumonidæ feeding within the live bodies of caterpillars,—not
as specially endowed or created instincts, but as small consequences of one general law leading to the advancement of
all organic beings,—namely, multiply, vary, let the strongest
live and the weakest die.

CHAPTER IX.

HYBRIDISM.

THE view commonly entertained by naturalists is that
species, when intercrossed, have been specially endowed
with sterility, in order to prevent their confusion. This
view certainly seems at first highly probable, for species
living together could hardly have been kept distinct had
they been capable of freely crossing. The subject is in
many ways important for us, more especially as the
sterility of species when first crossed, and that of their
hybrid offspring, cannot have been acquired, as I shall
show, by the preservation of successive profitable

degrees of sterility. It is an incidental result of differences in the reproductive systems of the parent-species.

In treating this subject, two classes of facts, to a large extent fundamentally different, have generally been confounded; namely, the sterility of species when first crossed, and the sterility of the hybrids produced from them.

Pure species have of course their organs of reproduction in a perfect condition, yet when intercrossed they produce either few or no offspring. Hybrids, on the other hand, have their reproductive organs functionally impotent, as may be clearly seen in the state of the male element in both plants and animals; though the formative organs themselves are perfect in structure, as far as the microscope reveals. In the first case the two sexual elements which go to form the embryo are perfect; in the second case they are either not at all developed, or are imperfectly developed. This distinction is important, when the cause of the sterility, which is common to the two cases, has to be considered. The distinction probably has been slurred over, owing to the sterility in both cases being looked on as a special endowment, beyond the province of our reasoning powers.

The fertility of varieties, that is of the forms known or believed to be descended from common parents, when crossed, and likewise the fertility of their mongrel offspring, is, with reference to my theory, of equal importance with the sterility of species; for it seems to make a broad and clear distinction between varieties and species.

Degrees of Sterility.—First, for the sterility of species

when crossed and of their hybrid offspring. It is impossible to study the several memoirs and works of those two conscientious and admirable observers, Kölreuter and Gärtner, who almost devoted their lives to this subject, without being deeply impressed with the high generality of some degree of sterility. Kölreuter makes the rule universal; but then he cuts the knot, for in ten cases in which he found two forms, considered by most authors as distinct species, quite fertile together, he unhesitatingly ranks them as varieties. Gärtner, also, makes the rule equally universal; and he disputes the entire fertility of Kölreuter's ten cases. But in these and in many other cases, Gärtner is obliged carefully to count the seeds, in order to show that there is any degree of sterility. He always compares the maximum number of seeds produced by two species when first crossed, and the maximum produced by their hybrid offspring, with the average number produced by both pure parent-species in a state of nature. But causes of serious error here intervene: a plant, to be hybridised, must be castrated, and, what is often more important, must be secluded in order to prevent pollen being brought to it by insects from other plants. Nearly all the plants experimented on by Gärtner were potted, and were kept in a chamber in his house. That these processes are often injurious to the fertility of a plant cannot be doubted; for Gärtner gives in his table about a score of cases of plants which he castrated, and artificially fertilised with their own pollen, and (excluding all cases such as the Leguminosæ, in which there is an acknowledged difficulty in the manipulation) half of these twenty plants had their fertility in some degree impaired. Moreover, as

Gärtner repeatedly crossed some forms, such as the common red and blue pimpernels (Anagallis arvensis and coerulea), which the best botanists rank as varieties, and found them absolutely sterile, we may doubt whether many species are really so sterile, when intercrossed, as he believed.

It is certain, on the one hand, that the sterility of various species when crossed is so different in degree and graduates away so insensibly, and, on the other hand, that the fertility of pure species is so easily affected by various circumstances, that for all practical purposes it is most difficult to say where perfect fertility ends and sterility begins. I think no better evidence of this can be required than that the two most experienced observers who have ever lived, namely Kölreuter and Gärtner, arrived at diametrically opposite conclusions in regard to some of the very same forms. It is also most instructive to compare—but I have not space here to enter on details—the evidence advanced by our best botanists on the question whether certain doubtful forms should be ranked as species or varieties, with the evidence from fertility adduced by different hybridisers, or by the same observer from experiments made during different years. It can thus be shown that neither sterility nor fertility affords any certain distinction between species and varieties. The evidence from this source graduates away, and is doubtful in the same degree as is the evidence derived from other constitutional and structural differences.

In regard to the sterility of hybrids in successive generations; though Gärtner was enabled to rear some hybrids, carefully guarding them from a cross with either pure parent, for six or seven, and in one case for ten generations, yet he asserts positively that their fertility never increases, but generally decreases greatly and suddenly. With respect to this decrease, it may first be noticed that when any deviation in structure or constitution is common to both parents, this is often transmitted in an augmented degree to the offspring; and both sexual elements in hybrid plants are already affected in some degree. But I believe that their fertility has been diminished in nearly all these cases by an independent cause, namely, by too close interbreeding. I have made so many experiments and collected so many facts, showing on the one hand that an occasional cross with a distinct individual or variety increases the vigour and fertility of the offspring, and on the other hand that very close interbreeding lessens their vigour and fertility, that I cannot doubt the correctness of this conclusion. Hybrids are seldom raised by experimentalists in great numbers; and as the parent-species, or other allied hybrids, generally grow in the same garden, the visits of insects must be carefully prevented during the flowering season: hence hybrids, if left to themselves, will generally be fertilised during each generation by pollen from the same flower; and this would probably be injurious to their fertility, already lessened by their hybrid origin. I am strengthened in this conviction by a remarkable statement repeatedly made by Gärtner, namely, that if even the less fertile hybrids be artificially fertilised with hybrid pollen of the same kind, their fertility, notwithstanding the frequent ill effects from manipulation, sometimes decidedly increases, and goes on increasing. Now, in the process of artificial fertilisation, pollen is as often taken by chance (as I know from my own experience)

2 B

from the anthers of another flower, as from the anthers
of the flower itself which is to be fertilised; so that a
cross between two flowers, though probably often on the
same plant, would be thus effected.   Moreover, whenever complicated experiments are in progress, so careful
an observer as Gärtner would have castrated his
hybrids, and this would have ensured in each generation
a cross with pollen from a distinct flower, either from
the same plant or from another plant of the same hybrid
nature.   And thus, the strange fact of an increase of fertility in the successive generations of *artificially fertilised*
hybrids, in contrast with those spontaneously self-
fertilised, may, as I believe, be accounted for by too close
interbreeding having been avoided.

Now let us turn to the results arrived at by a third
most experienced hybridiser, namely, the Hon. and Rev.
W. Herbert. He is as emphatic in his conclusion that
some hybrids are perfectly fertile—as fertile as the pure
parent-species—as are Kölreuter and Gärtner that some
degree of sterility between distinct species is a universal
law of nature.   He experimented on some of the very
same species as did Gärtner.   The difference in their
results may, I think, be in part accounted for by
Herbert's great horticultural skill, and by his having
hot-houses at his command.   Of his many important
statements I will here give only a single one as an
example, namely, that "every ovule in a pod of Crinum
capense fertilised by C. revolutum produced a plant,
which I never saw to occur in a case of its natural
fecundation."   So that here we have perfect or even
more than commonly perfect fertility, in a first cross
between two distinct species.

This case of the Crinum leads me to refer to a

singular fact, namely, that individual plants of certain
species of Lobelia, Verbascum and Passiflora, can easily
be fertilised by pollen from a distinct species, but not
by pollen from the same plant, though this pollen can
be proved to be perfectly sound by fertilising other
plants or species. In the genus Hippeastrum, in
Corydalis as shown by Professor Hildebrand, in various
orchids as shown by Mr. Scott and Fritz Müller, all the
individuals are in this peculiar condition. So that with
some species, certain abnormal individuals, and in other
species all the individuals, can actually be hybridised
much more readily than they can be fertilised by pollen
from the same individual plant! To give one instance,
a bulb of Hippeastrum aulicum produced four flowers;
three were fertilised by Herbert with their own pollen,
and the fourth was subsequently fertilised by the pollen
of a compound hybrid descended from three distinct
species: the result was that " the ovaries of the three
first flowers soon ceased to grow, and after a few days
perished entirely, whereas the pod impregnated by the
pollen of the hybrid made vigorous growth and rapid
progress to maturity, and bore good seed, which
vegetated freely." Mr. Herbert tried similar experiments during many years, and always with the same
result. These cases serve to show on what slight and
mysterious causes the lesser or greater fertility of a
species sometimes depends.

The practical experiments of horticulturists, though
not made with scientific precision, deserve some notice.
It is notorious in how complicated a manner the
species of Pelargonium, Fuchsia, Calceolaria, Petunia,
Rhododendron, &c., have been crossed, yet many of
these hybrids seed freely. For instance, Herbert

asserts that a hybrid from Calceolaria integrifolia and plantaginea, species most widely dissimilar in general habit, "reproduces itself as perfectly as if it had been a natural species from the mountains of Chili." I have taken some pains to ascertain the degree of fertility of some of the complex crosses of Rhododendrons, and I am assured that many of them are perfectly fertile. Mr. C. Noble, for instance, informs me that he raises stocks for grafting from a hybrid between Rhod. ponticum and catawbiense, and that this hybrid "seeds as freely as it is possible to imagine." Had hybrids when fairly treated, always gone on decreasing in fertility in each successive generation, as Gärtner believed to be the case, the fact would have been notorious to nurserymen. Horticulturists raise large beds of the same hybrid, and such alone are fairly treated, for by insect agency the several individuals are allowed to cross freely with each other, and the injurious influence of close interbreeding is thus prevented. Any one may readily convince himself of the efficiency of insect-agency by examining the flowers of the more sterile kinds of hybrid Rhododendrons, which produce no pollen, for he will find on their stigmas plenty of pollen brought from other flowers.

In regard to animals, much fewer experiments have been carefully tried than with plants. If our systematic arrangements can be trusted, that is, if the genera of animals are as distinct from each other as are the genera of plants, then we may infer that animals more widely distinct in the scale of nature can be crossed more easily than in the case of plants; but the hybrids themselves are, I think, more sterile. It should, however, be borne in mind that, owing to few animals

breeding freely under confinement, few experiments
have been fairly tried: for instance, the canary-bird has
been crossed with nine distinct species of finches, but,
as not one of these breeds freely in confinement, we
have no right to expect that the first crosses between
them and the canary, or that their hybrids, should be
perfectly fertile. Again, with respect to the fertility
in successive generations of the more fertile hybrid
animals, I hardly know of an instance in which two
families of the same hybrid have been raised at the
same time from different parents, so as to avoid the ill
effects of close interbreeding. On the contrary, brothers
and sisters have usually been crossed in each successive
generation, in opposition to the constantly repeated
admonition of every breeder. And in this case, it is
not at all surprising that the inherent sterility in the
hybrids should have gone on increasing.

Although I know of hardly any thoroughly well-
authenticated cases of perfectly fertile hybrid animals, I
have reason to believe that the hybrids from Cervulus
vaginalis and Reevesii, and from Phasianus colchicus
with P. torquatus, are perfectly fertile. M. Quatrefages
states that the hybrids from two moths (Bombyx
cynthia and arrindia) were proved in Paris to be fertile
inter se for eight generations. It has lately been
asserted that two such distinct species as the hare and
rabbit, when they can be got to breed together, produce
offspring, which are highly fertile when crossed with
one of the parent-species. The hybrids from the
common and Chinese geese (A. cygnoides), species
which are so different that they are generally ranked in
distinct genera, have often bred in this country with
either pure parent, and in one single instance they have

bred *inter se*. This was effected by Mr. Eyton, who raised two hybrids from the same parents, but from different hatches; and from these two birds he raised no less than eight hybrids (grandchildren of the pure geese) from one nest. In India, however, these cross-bred geese must be far more fertile; for I am assured by two eminently capable judges, namely Mr. Blyth and Capt. Hutton, that whole flocks of these crossed geese are kept in various parts of the country; and as they are kept for profit, where neither pure parent-species exists, they must certainly be highly or perfectly fertile.

With our domesticated animals, the various races when crossed together are quite fertile; yet in many cases they are descended from two or more wild species. From this fact we must conclude either that the aboriginal parent-species at first produced perfectly fertile hybrids, or that the hybrids subsequently reared under domestication became quite fertile. This latter alternative, which was first propounded by Pallas, seems by far the most probable, and can, indeed, hardly be doubted. It is, for instance, almost certain that our dogs are descended from several wild stocks; yet. with perhaps the exception of certain indigenous domestic dogs of South America, all are quite fertile together; but analogy makes me greatly doubt, whether the several aboriginal species would at first have freely bred together and have produced quite fertile hybrids. So again I have lately acquired decisive evidence that the crossed offspring from the Indian humped and common cattle are *inter se* perfectly fertile; and from the observations by Rütimeyer on their important osteo-logical differences, as well as from those by Mr. Blyth on their differences in habits, voice, constitution, &c.,

these two forms must be regarded as good and distinct
species. The same remarks may be extended to the two
chief races of the pig. We must, therefore, either give
up the belief of the universal sterility of species when
crossed; or we must look at this sterility in animals,
not as an indelible characteristic, but as one capable of
being removed by domestication.

Finally, considering all the ascertained facts on the
intercrossing of plants and animals, it may be concluded
that some degree of sterility, both in first crosses and in
hybrids, is an extremely general result; but that it
cannot, under our present state of knowledge, be considered as absolutely universal.

Laws governing the Sterility of first Crosses and of Hybrids.

We will now consider a little more in detail the laws
governing the sterility of first crosses and of hybrids.
Our chief object will be to see whether or not these
laws indicate that species have been specially endowed
with this quality, in order to prevent their crossing and
blending together in utter confusion. The following
conclusions are drawn up chiefly from Gärtner's admirable work on the hybridisation of plants. I have
taken much pains to ascertain how far they apply to
animals, and, considering how scanty our knowledge is
in regard to hybrid animals, I have been surprised to
find how generally the same rules apply to both
kingdoms.

It has been already remarked, that the degree of
fertility, both of first crosses and of hybrids, graduates
from zero to perfect fertility. It is surprising in how

many curious ways this gradation can be shown; but only the barest outline of the facts can here be given. When pollen from a plant of one family is placed on the stigma of a plant of a distinct family, it exerts no more influence than so much inorganic dust. From this absolute zero of fertility, the pollen of different species applied to the stigma of some one species of the same genus, yields a perfect gradation in the number of seeds produced, up to nearly complete or even quite complete fertility; and, as we have seen, in certain abnormal cases, even to an excess of fertility, beyond that which the plant's own pollen produces. So in hybrids themselves, there are some which never have produced, and probably never would produce, even with the pollen of the pure parents, a single fertile seed: but in some of these cases a first trace of fertility may be detected, by the pollen of one of the pure parent-species causing the flower of the hybrid to wither earlier than it otherwise would have done; and the early withering of the flower is well known to be a sign of incipient fertilisation. From this extreme degree of sterility we have self-fertilised hybrids producing a greater and greater number of seeds up to perfect fertility.

The hybrids raised from two species which are very difficult to cross, and which rarely produce any offspring, are generally very sterile; but the parallelism between the difficulty of making a first cross, and the sterility of the hybrids thus produced—two classes of facts which are generally confounded together—is by no means strict. There are many cases, in which two pure species, as in the genus Verbascum, can be united with unusual facility, and produce numerous hybrid-offspring, yet these hybrids are remarkably sterile. On the other hand, there are species which can be crossed very rarely, or with extreme difficulty, but the hybrids, when at last produced, are very fertile. Even within the limits of the same genus, for instance in Dianthus, these two opposite cases occur.

The fertility, both of first crosses and of hybrids, is more easily affected by unfavourable conditions, than is that of pure species. But the fertility of first crosses is likewise innately variable; for it is not always the same in degree when the same two species are crossed under the same circumstances; it depends in part upon the constitution of the individuals which happen to have been chosen for the experiment. So it is with hybrids, for their degree of fertility is often found to differ greatly in the several individuals raised from seed out of the same capsule and exposed to the same conditions.

By the term systematic affinity is meant, the general resemblance between species in structure and constitution. Now the fertility of first crosses, and of the hybrids produced from them, is largely governed by their systematic affinity. This is clearly shown by hybrids never having been raised between species ranked by systematists in distinct families; and on the other hand, by very closely allied species generally uniting with facility. But the correspondence between systematic affinity and the facility of crossing is by no means strict. A multitude of cases could be given of very closely allied species which will not unite, or only with extreme difficulty; and on the other hand of very distinct species which unite with the utmost facility. In the same family there may be a genus, as Dianthus, in which very many species can most readily be crossed; and another genus, as Silene, in which the most persevering efforts have failed to produce between extremely close species a single hybrid.   Even within the limits of the same genus, we meet with this same difference; for instance, the many species of Nicotiana have been more largely crossed than the species of almost any other genus; but Gärtner found that N. acuminata, which is not a particularly distinct species, obstinately failed to fertilise, or to be fertilised by no less than eight other species of Nicotiana.   Many analogous facts could be given.

No one has been able to point out what kind or what amount of difference, in any recognisable character, is sufficient to prevent two species crossing.   It can be shown that plants most widely different in habit and general appearance, and having strongly marked differences in every part of the flower, even in the pollen, in the fruit, and in the cotyledons, can be crossed.   Annual and perennial plants, deciduous and evergreen trees, plants inhabiting different stations and fitted for extremely different climates, can often be crossed with ease.

By a reciprocal cross between two species, I mean the case, for instance, of a female-ass being first crossed by a stallion, and then a mare by a male-ass; these two species may then be said to have been reciprocally crossed.   There is often the widest possible difference in the facility of making reciprocal crosses.   Such cases are highly important, for they prove that the capacity in any two species to cross is often completely independent of their systematic affinity, that is of any difference in their structure or constitution, excepting in their reproductive systems. The diversity of the result in reciprocal crosses between the same two species was long ago observed by Kölreuter. To give an instance: Mirabilis jalapa can easily be fertilised by the pollen of M. longiflora, and the hybrids thus produced are sufficiently fertile; but Kölreuter tried more than two hundred times, during eight following years, to fertilise reciprocally M. longiflora with the pollen of M. jalapa, and utterly failed. Several other equally striking cases could be given. Thuret has observed the same fact with certain sea-weeds or Fuci. Gärtner, moreover, found that this difference of facility in making reciprocal crosses is extremely common in a lesser degree. He has observed it even between closely related forms (as Matthiola annua and glabra) which many botanists rank only as varieties. It is also a remarkable fact, that hybrids raised from reciprocal crosses, though of course compounded of the very same two species, the one species having first been used as the father and then as the mother, though they rarely differ in external characters, yet generally differ in fertility in a small, and occasionally in a high degree.

Several other singular rules could be given from Gärtner: for instance, some species have a remarkable power of crossing with other species; other species of the same genus have a remarkable power of impressing their likeness on their hybrid offspring; but these two powers do not at all necessarily go together. There are certain hybrids which, instead of having, as is usual, an intermediate character between their two parents, always closely resemble one of them; and such hybrids, though externally so like one of their pure parent-species, are with rare exceptions extremely sterile. So again amongst hybrids which are usually intermediate in structure between their parents, exceptional and abnormal individuals sometimes are born, which closely resemble one of their pure parents; and these hybrids are almost always utterly sterile, even when the other hybrids raised from seed from the same capsule have a considerable degree of fertility. These facts show how completely the fertility of a hybrid may be independent of its external resemblance to either pure parent.

Considering the several rules now given, which govern the fertility of first crosses and of hybrids, we see that when forms, which must be considered as good and distinct species, are united, their fertility graduates from zero to perfect fertility, or even to fertility under certain conditions in excess; that their fertility, besides being eminently susceptible to favourable and unfavourable conditions, is innately variable; that it is by no means always the same in degree in the first cross and in the hybrids produced from this cross; that the fertility of hybrids is not related to the degree in which they resemble in external appearance either parent; and lastly, that the facility of making a first cross between any two species is not always governed by their systematic affinity or degree of resemblance to each other. This latter statement is clearly proved by the difference in the result of reciprocal crosses between the same two species, for, according as the one species or the other is used as the father or the mother, there is generally some difference, and occasionally the widest possible difference, in the facility of effecting an union. The hybrids, moreover, produced from reciprocal crosses often differ in fertility.

Now do these complex and singular rules indicate

that species have been endowed with sterility simply
to prevent their becoming confounded in nature? I
think not. For why should the sterility be so
extremely different in degree, when various species
are crossed, all of which we must suppose it would
be equally important to keep from blending together?
Why should the degree of sterility be innately variable
in the individuals of the same species? Why should
some species cross with facility, and yet produce very
sterile hybrids; and other species cross with extreme
difficulty, and yet produce fairly fertile hybrids? Why
should there often be so great a difference in the result
of a reciprocal cross between the same two species?
Why, it may even be asked, has the production of
hybrids been permitted? To grant to species the
special power of producing hybrids, and then to stop
their further propagation by different degrees of sterility,
not strictly related to the facility of the first union
between their parents, seems a strange arrangement.

The foregoing rules and facts, on the other hand,
appear to me clearly to indicate that the sterility both
of first crosses and of hybrids is simply incidental or
dependent on unknown differences in their reproductive
systems; the differences being of so peculiar and
limited a nature, that, in reciprocal crosses between the
same two species, the male sexual element of the one
will often freely act on the female sexual element of the
other, but not in a reversed direction. It will be
advisable to explain a little more fully by an example
what I mean by sterility being incidental on other
differences, and not a specially endowed quality. As
the capacity of one plant to be grafted or budded on
another is unimportant for their welfare in a state of

nature, I presume that no one will suppose that this
capacity is a *specially* endowed quality, but will admit
that it is incidental on differences in the laws of growth
of the two plants.   We can sometimes see the reason
why one tree will not take on another, from differences
in their rate of growth, in the hardness of their wood,
in the period of the flow or nature of their sap, &c.;
but in a multitude of cases we can assign no reason
whatever.   Great diversity in the size of two plants,
one being woody and the other herbaceous, one being
evergreen and the other deciduous, and adaptation to
widely different climates, do not always prevent the
two grafting together.   As in hybridisation, so with
grafting, the capacity is limited by systematic affinity,
for no one has been able to graft together trees
belonging to quite distinct families; and, on the other
hand, closely allied species, and varieties of the same
species, can usually, but not invariably, be grafted with
ease.   But this capacity, as in hybridisation, is by
no means absolutely governed by systematic affinity.
Although many distinct genera within the same family
have been grafted together, in other cases species of the
same genus will not take on each other.   The pear can
be grafted far more readily on the quince, which is
ranked as a distinct genus, than on the apple, which is
a member of the same genus.   Even different varieties
of the pear take with different degrees of facility on the
quince; so do different varieties of the apricot and
peach on certain varieties of the plum.

As Gärtner found that there was sometimes an
innate difference in different *individuals* of the same
two species in crossing; so Sageret believes this to be
the case with different individuals of the same two

species in being grafted together. As in reciprocal crosses, the facility of effecting an union is often very far from equal, so it sometimes is in grafting; the common gooseberry, for instance, cannot be grafted on the currant, whereas the currant will take, though with difficulty, on the gooseberry.

We have seen that the sterility of hybrids, which have their reproductive organs in an imperfect condition, is a different case from the difficulty of uniting two pure species, which have their reproductive organs perfect; yet these two distinct classes of cases run to a large extent parallel. Something analogous occurs in grafting; for Thouin found that three species of Robinia, which seeded freely on their own roots, and which could be grafted with no great difficulty on a fourth species, when thus grafted were rendered barren. On the other hand, certain species of Sorbus, when grafted on other species yielded twice as much fruit as when on their own roots. We are reminded by this latter fact of the extraordinary cases of Hippeastrum, Passiflora, &c., which seed much more freely when fertilised with the pollen of a distinct species, than when fertilised with pollen from the same plant.

We thus see, that, although there is a clear and great difference between the mere adhesion of grafted stocks, and the union of the male and female elements in the act of reproduction, yet that there is a rude degree of parallelism in the results of grafting and of crossing distinct species. And as we must look at the curious and complex laws governing the facility with which trees can be grafted on each other as incidental on unknown differences in their vegetative systems, so I believe that the still more complex laws governing the facility of first crosses are incidental on unknown differences in their reproductive systems. These differences in both cases, follow to a certain extent, as might have been expected, systematic affinity, by which term every kind of resemblance and dissimilarity between organic beings is attempted to be expressed. The facts by no means seem to indicate that the greater or lesser difficulty of either grafting or crossing various species has been a special endowment; although in the case of crossing, the difficulty is as important for the endurance and stability of specific forms, as in the case of grafting it is unimportant for their welfare.

Origin and Causes of the Sterility of first Crosses and of Hybrids.

At one time it appeared to me probable, as it has to others, that the sterility of first crosses and of hybrids might have been slowly acquired through the natural selection of slightly lessened degrees of fertility, which, like any other variation, spontaneously appeared in certain individuals of one variety when crossed with those of another variety. For it would clearly be advantageous to two varieties or incipient species, if they could be kept from blending, on the same principle that, when man is selecting at the same time two varieties, it is necessary that he should keep them separate. In the first place, it may be remarked that species inhabiting distinct regions are often sterile when crossed; now it could clearly have been of no advantage to such separated species to have been rendered mutually sterile, and consequently this could not have been effected through natural selection; but it may perhaps be argued, that, if a species was rendered sterile with some one compatriot, sterility with other species would follow as a necessary contingency.   In the second place, it is almost as much opposed to the theory of natural selection as to that of special creation, that in reciprocal crosses the male element of one form should have been rendered utterly impotent on a second form, whilst at the same time the male element of this second form is enabled freely to fertilise the first form ; for this peculiar state of the reproductive system could hardly have been advantageous to either species.

In considering the probability of natural selection having come into action, in rendering species mutually sterile, the greatest difficulty will be found to lie in the existence of many graduated steps from slightly lessened fertility to absolute sterility.   It may be admitted that it would profit an incipient species, if it were rendered in some slight degree sterile when crossed with its parent form or with some other variety ; for thus fewer bastardised and deteriorated offspring would be produced to commingle their blood with the new species in process of formation.   But he who will take the trouble to reflect on the steps by which this first degree of sterility could be increased through natural selection to that high degree which is common with so many species, and which is universal with species which have been differentiated to a generic or family rank, will find the subject extraordinarily complex.   After mature reflection it seems to me that this could not have been effected through natural selection.   Take the case of any two species which, when crossed, produced few and sterile offspring ; now, what is there which could favour

2 c

the survival of those individuals which happened to
be endowed in a slightly higher degree with mutual
infertility, and which thus approached by one small
step towards absolute sterility ? Yet an advance of
this kind, if the theory of natural selection be brought to
bear, must have incessantly occurred with many species,
for a multitude are mutually quite barren. With
sterile neuter insects we have reason to believe that
modifications in their structure and fertility have been
slowly accumulated by natural selection, from an
advantage having been thus indirectly given to the
community to which they belonged over other communities of the same species ; but an individual animal
not belonging to a social community, if rendered slightly
sterile when crossed with some other variety, would not
thus itself gain any advantage or indirectly give any
advantage to the other individuals of the same variety,
thus leading to their preservation.

But it would be superfluous to discuss this question in
detail ; for with plants we have conclusive evidence
that the sterility of crossed species must be due to
some principle, quite independent of natural selection.
Both Gärtner and Kölreuter have proved that in genera
including numerous species, a series can be formed from
species which when crossed yield fewer and fewer seeds,
to species which never produce a single seed, but yet
are affected by the pollen of certain other species, for
the germen swells. It is here manifestly impossible to
select the more sterile individuals, which have already
ceased to yield seeds ; so that this acme of sterility,
when the germen alone is affected, cannot have been
gained through selection ; and from the laws governing
the various grades of sterility being so uniform through-

out the animal and vegetable kingdoms, we may infer that the cause, whatever it may be, is the same or nearly the same in all cases.

We will now look a little closer at the probable nature of the differences between species which induce sterility in first crosses and in hybrids. In the case of first crosses, the greater or less difficulty in effecting an union and in obtaining offspring apparently depends on several distinct causes. There must sometimes be a physical impossibility in the male element reaching the ovule, as would be the case with a plant having a pistil too long for the pollen-tubes to reach the ovarium. It has also been observed that when the pollen of one species is placed on the stigma of a distantly allied species, though the pollen-tubes protrude, they do not penetrate the stigmatic surface. Again, the male element may reach the female element but be incapable of causing an embryo to be developed, as seems to have been the case with some of Thuret's experiments on Fuci. No explanation can be given of these facts, any more than why certain trees cannot be grafted on others. Lastly, an embryo may be developed, and then perish at an early period. This latter alternative has not been sufficiently attended to ; but I believe, from observations communicated to me by Mr. Hewitt, who has had great experience in hybridising pheasants and fowls, that the early death of the embryo is a very frequent cause of sterility in first crosses. Mr. Salter has recently given the results of an examination of about 500 eggs produced from various crosses between three species of Gallus and their hybrids; the majority of these eggs had been fertilised ; and in the majority of the fertilised eggs, the embryos had either been partially developed and had then perished, or had become nearly mature, but the young chickens had been unable to break through the shell. Of the chickens which were born, more than four-fifths died within the first few days, or at latest weeks, "without any obvious cause, apparently from mere inability to live;" so that from the 500 eggs only twelve chickens were reared. With plants, hybridised embryos probably often perish in a like manner; at least it is known that hybrids raised from very distinct species are sometimes weak and dwarfed, and perish at an early age; of which fact Max Wichura has recently given some striking cases with hybrid willows. It may be here worth noticing that in some cases of parthenogenesis, the embryos within the eggs of silk moths which had not been fertilised, pass through their early stages of development and then perish like the embryos produced by a cross between distinct species. Until becoming acquainted with these facts, I was unwilling to believe in the frequent early death of hybrid embryos; for hybrids, when once born, are generally healthy and long-lived, as we see in the case of the common mule. Hybrids, however, are differently circumstanced before and after birth: when born and living in a country where their two parents live, they are generally placed under suitable conditions of life. But a hybrid partakes of only half of the nature and constitution of its mother; it may therefore before birth, as long as it is nourished within its mother's womb, or within the egg or seed produced by the mother, be exposed to conditions in some degree unsuitable, and consequently be liable to perish at an early period; more especially as all very young beings are eminently sensitive to injurious or unnatural conditions of life. But after all, the cause more probably lies in some imperfection in the original act of impregnation, causing the embryo to be imperfectly developed, rather than in the conditions to which it is subsequently exposed.

In regard to the sterility of hybrids, in which the sexual elements are imperfectly developed, the case is somewhat different. I have more than once alluded to a large body of facts showing that, when animals and plants are removed from their natural conditions, they are extremely liable to have their reproductive systems seriously affected. This, in fact, is the great bar to the domestication of animals. Between the sterility thus superinduced and that of hybrids, there are many points of similarity. In both cases the sterility is independent of general health, and is often accompanied by excess of size or great luxuriance. In both cases the sterility occurs in various degrees; in both, the male element is the most liable to be affected; but sometimes the female more than the male. In both, the tendency goes to a certain extent with systematic affinity, for whole groups of animals and plants are rendered impotent by the same unnatural conditions; and whole groups of species tend to produce sterile hybrids. On the other hand, one species in a group will sometimes resist great changes of conditions with unimpaired fertility; and certain species in a group will produce unusually fertile hybrids. No one can tell, till he tries, whether any particular animal will breed under confinement, or any exotic plant seed freely under culture; nor can he tell till he tries, whether any two species of a genus will produce more

or less sterile hybrids. Lastly, when organic beings
are placed during several generations under conditions
not natural to them, they are extremely liable to vary,
which seems to be partly due to their reproductive
systems having been specially affected, though in a
lesser degree than when sterility ensues. So it is with
hybrids, for their offspring in successive generations
are eminently liable to vary, as every experimentalist
has observed.

Thus we see that when organic beings are placed
under new and unnatural conditions, and when hybrids
are produced by the unnatural crossing of two species,
the reproductive system, independently of the general
state of health, is affected in a very similar manner.
In the one case, the conditions of life have been disturbed, though often in so slight a degree as to be inappreciable by us; in the other case, or that of hybrids,
the external conditions have remained the same, but
the organisation has been disturbed by two distinct
structures and constitutions, including of course the
reproductive systems, having been blended into one.
For it is scarcely possible that two organisations should
be compounded into one, without some disturbance
occurring in the development, or periodical action, or
mutual relations of the different parts and organs one
to another or to the conditions of life. When hybrids
are able to breed *inter se*, they transmit to their offspring from generation to generation the same compounded organisation, and hence we need not be surprised that their sterility, though in some degree
variable, does not diminish; it is even apt to increase,
this being generally the result, as before explained,
of too close interbreeding. The above view of the

sterility of hybrids being caused by two constitutions being compounded into one has been strongly maintained by Max Wichura.

It must, however, be owned that we cannot understand, on the above or any other view, several facts with respect to the sterility of hybrids; for instance, the unequal fertility of hybrids produced from reciprocal crosses; or the increased sterility in those hybrids which occasionally and exceptionally resemble closely either pure parent. Nor do I pretend that the foregoing remarks go to the root of the matter; no explanation is offered why an organism, when placed under unnatural conditions, is rendered sterile. All that I have attempted to show is, that in two cases, in some respects allied, sterility is the common result,—in the one case from the conditions of life having been disturbed, in the other case from the organisation having been disturbed by two organisations being compounded into one.

A similar parallelism holds good with an allied yet very different class of facts. It is an old and almost universal belief founded on a considerable body of evidence, which I have elsewhere given, that slight changes in the conditions of life are beneficial to all living things. We see this acted on by farmers and gardeners in their frequent exchanges of seed, tubers, &c., from one soil or climate to another, and back again. During the convalescence of animals, great benefit is derived from almost any change in their habits of life. Again, both with plants and animals, there is the clearest evidence that a cross between individuals of the same species, which differ to a certain extent, gives vigour and fertility to the offspring; and that close interbreeding continued during several generations between the nearest relations, if these be kept under the same conditions of life, almost always leads to decreased size, weakness, or sterility.

Hence it seems that, on the one hand, slight changes in the conditions of life benefit all organic beings, and on the other hand, that slight crosses, that is crosses between the males and females of the same species, which have been subjected to slightly different conditions, or which have slightly varied, give vigour and fertility to the offspring. But, as we have seen, organic beings long habituated to certain uniform conditions under a state of nature, when subjected, as under confinement, to a considerable change in their conditions, very frequently are rendered more or less sterile ; and we know that a cross between two forms, that have become widely or specifically different, produce hybrids which are almost always in some degree sterile. I am fully persuaded that this double parallelism is by no means an accident or an illusion. He who is able to explain why the elephant and a multitude of other animals are incapable of breeding when kept under only partial confinement in their native country, will be able to explain the primary cause of hybrids being so generally sterile. He will at the same time be able to explain how it is that the races of some of our domesticated animals, which have often been subjected to new and not uniform conditions, are quite fertile together, although they are descended from distinct species, which would probably have been sterile if aboriginally crossed. The above two parallel series of facts seem to be connected together by some common but unknown bond, which is essentially related to the principle of life; this principle, according to Mr. Herbert Spencer, being that life depends on, or consists in, the incessant action and reaction of various forces, which, as throughout nature, are always tending towards an equilibrium; and when this tendency is slightly disturbed by any change, the vital forces gain in power.

Reciprocal Dimorphism and Trimorphism.

This subject may be here briefly discussed, and will be found to throw some light on hybridism. Several plants belonging to distinct orders present two forms, which exist in about equal numbers and which differ in no respect except in their reproductive organs; one form having a long pistil with short stamens, the other a short pistil with long stamens; the two having differently sized pollen-grains. With trimorphic plants there are three forms likewise differing in the lengths of their pistils and stamens, in the size and colour of the pollen-grains, and in some other respects; and as in each of the three forms there are two sets of stamens, the three forms possess altogether six sets of stamens and three kinds of pistils. These organs are so proportioned in length to each other, that half the stamens in two of the forms stand on a level with the stigma of the third form. Now I have shown, and the result has been confirmed by other observers, that, in order to obtain full fertility with these plants, it is necessary that the stigma of the one form should be fertilised by pollen taken from the stamens of corresponding height in another form. So that with dimorphic species two unions, which may be called legitimate, are fully fertile; and two, which may be called illegitimate, are more or less infertile. With trimorphic species six unions are legitimate, or fully fertile,—and twelve are illegitimate, or more or less infertile.

The infertility which may be observed in various dimorphic and trimorphic plants, when they are illegitimately fertilised, that is by pollen taken from stamens not corresponding in height with the pistil, differs much in degree, up to absolute and utter sterility; just in the same manner as occurs in crossing distinct species. As the degree of sterility in the latter case depends in an eminent degree on the conditions of life being more or less favourable, so I have found it with illegitimate unions. It is well known that if pollen of a distinct species be placed on the stigma of a flower, and its own pollen be afterwards, even after a considerable interval of time, placed on the same stigma, its action is so strongly prepotent that it generally annihilates the effect of the foreign pollen; so it is with the pollen of the several forms of the same species, for legitimate pollen is strongly prepotent over illegitimate pollen, when both are placed on the same stigma. I ascertained this by fertilising several flowers, first illegitimately, and twenty-four hours afterwards legitimately, with pollen taken from a peculiarly coloured variety, and all the seedlings were similarly coloured; this shows that the legitimate pollen, though applied twenty-four hours subsequently, had wholly destroyed or prevented the action of the previously applied illegitimate pollen. Again, as in making reciprocal crosses between the same two species, there is occasionally a great difference in the result, so the same thing occurs with trimorphic plants; for instance, the mid-styled form of Lythrum salicaria was illegitimately fertilised with the greatest ease by pollen from the longer stamens of the short-styled form, and yielded many seeds; but the latter form did not yield a single seed when fertilised by the longer stamens of the mid-styled form.

In all these respects, and in others which might be added, the forms of the same undoubted species when illegitimately united behave in exactly the same manner as do two distinct species when crossed. This led me carefully to observe during four years many seedlings, raised from several illegitimate unions. The chief result is that these illegitimate plants, as they may be called, are not fully fertile. It is possible to raise from dimorphic species, both long-styled and short-styled illegitimate plants, and from trimorphic plants all three illegitimate forms. These can then be properly united in a legitimate manner. When this is done, there is no apparent reason why they should not yield as many seeds as did their parents when legitimately fertilised. But such is not the case. They are all infertile, in various degrees; some being so utterly and incurably sterile that they did not yield during four seasons a single seed or even seed-capsule. The sterility of these illegitimate plants, when united with each other in a legitimate manner, may be strictly compared with that of hybrids when crossed *inter se*. If, on the other hand, a hybrid is crossed with either pure parent-species, the sterility is usually much lessened: and so it is when an illegitimate plant is fertilised by a legitimate plant. In the same manner as the sterility of hybrids does not always run parallel with the difficulty of making the first cross between the two parent-species, so the sterility of certain illegitimate plants was unusually great, whilst the sterility of the union from which they were derived was by no means great. With hybrids raised from the same seed-capsule the degree of sterility is innately variable, so it is in a marked manner with illegitimate plants. Lastly, many hybrids are profuse and persistent flowerers, whilst other and more sterile hybrids produce few flowers, and are weak, miserable dwarfs; exactly similar cases occur with the illegitimate offspring of various dimorphic and trimorphic plants.

Altogether there is the closest identity in character and behaviour between illegitimate plants and hybrids. It is hardly an exaggeration to maintain that illegitimate plants are hybrids, produced within the limits of the same species by the improper union of certain forms, whilst ordinary hybrids are produced from an improper union between so-called distinct species. We have also already seen that there is the closest similarity in all respects between first illegitimate unions and first crosses between distinct species. This will perhaps be made more fully apparent by an illustration; we may suppose that a botanist found two well-marked varieties (and such occur) of the long-styled form of the trimorphic Lythrum salicaria, and that he determined to try by crossing whether they were specifically distinct. He would find that they yielded only about one-fifth of the proper number of seed, and that they behaved in all the other above specified respects as if they had been two distinct species. But to make the case sure, he would raise plants from his supposed hybridised seed, and he would find that the seedlings were miserably dwarfed and utterly sterile, and that they behaved in all other respects like ordinary hybrids. He might then maintain that he had actually proved, in accordance with

the common view, that his two varieties were as good
and as distinct species as any in the world; but he
would be completely mistaken.

The facts now given on dimorphic and trimorphic
plants are important, because they show us, first, that
the physiological test of lessened fertility, both in first
crosses and in hybrids, is no safe criterion of specific
distinction; secondly, because we may conclude that
there is some unknown bond which connects the infertility of illegitimate unions with that of their illegitimate offspring, and we are led to extend the same view
to first crosses and hybrids; thirdly, because we find,
and this seems to me of especial importance, that two
or three forms of the same species may exist and may
differ in no respect whatever, either in structure or in
constitution, relatively to external conditions, and yet
be sterile when united in certain ways.   For we must
remember that it is the union of the sexual elements of
individuals of the same form, for instance, of two long-
styled forms, which results in sterility; whilst it is the
union of the sexual elements proper to two distinct
forms which is fertile.   Hence the case appears at first
sight exactly the reverse of what occurs, in the ordinary
unions of the individuals of the same species and
with crosses between distinct species.   It is, however,
doubtful whether this is really so; but I will not enlarge on this obscure subject.

We may, however, infer as probable from the consideration of dimorphic and trimorphic plants, that the
sterility of distinct species when crossed and of their
hybrid progeny, depends exclusively on the nature of
their sexual elements, and not on any difference in their
structure or general constitution.   We are also led to

this same conclusion by considering reciprocal crosses, in which the male of one species cannot be united, or can be united with great difficulty, with the female of a second species, whilst the converse cross can be effected with perfect facility. That excellent observer, Gärtner, likewise concluded that species when crossed are sterile owing to differences confined to their reproductive systems.

Fertility of Varieties when Crossed, and of their Mongrel Offspring, not universal.

It may be urged, as an overwhelming argument, that there must be some essential distinction between species and varieties, inasmuch as the latter, however much they may differ from each other in external appearance, cross with perfect facility, and yield perfectly fertile offspring. With some exceptions, presently to be given, I fully admit that this is the rule. But the subject is surrounded by difficulties, for, looking to varieties produced under nature, if two forms hitherto reputed to be varieties be found in any degree sterile together, they are at once ranked by most naturalists as species. For instance, the blue and red pimpernel, which are considered by most botanists as varieties, are said by Gärtner to be quite sterile when crossed, and he consequently ranks them as undoubted species. If we thus argue in a circle, the fertility of all varieties produced under nature will assuredly have to be granted.

If we turn to varieties, produced, or supposed to have been produced, under domestication, we are still involved in some doubt. For when it is stated, for

instance, that certain South American indigenous domestic dogs do not readily unite with European dogs,
the explanation which will occur to every one, and
probably the true one, is that they are descended from
aboriginally distinct species.   Nevertheless the perfect
fertility of so many domestic races, differing widely
from each other in appearance, for instance those of the
pigeon, or of the cabbage, is a remarkable fact ; more
especially when we reflect how many species there are,
which, though resembling each other most closely, are
utterly sterile when intercrossed.   Several considerations, however, render the fertility of domestic varieties
less remarkable.   In the first place, it may be observed
that the amount of external difference between two
species is no sure guide to their degree of mutual
sterility, so that similar differences in the case of
varieties would be no sure guide.   It is certain that with
species the cause lies exclusively in differences in their
sexual constitution.   Now the varying conditions to
which domesticated animals and cultivated plants have
been subjected, have had so little tendency towards
modifying the reproductive system in a manner leading
to mutual sterility, that we have good grounds for admitting the directly opposite doctrine of Pallas, namely,
that such conditions generally eliminate this tendency ;
so that the domesticated descendants of species, which
in their natural state probably would have been in
some degree sterile when crossed, become perfectly
fertile together.   With plants, so far is cultivation from
giving a tendency towards sterility between distinct
species, that in several well-authenticated cases already
alluded to, certain plants have been affected in an
opposite manner, for they have become self-impotent

whilst still retaining the capacity of fertilising, and
being fertilised by, other species. If the Pallasian
doctrine of the elimination of sterility through long-
continued domestication be admitted, and it can hardly
be rejected, it becomes in the highest degree improbable
that similar conditions long-continued should likewise
induce this tendency; though in certain cases, with
species having a peculiar constitution, sterility might
occasionally be thus caused. Thus, as I believe, we
can understand why with domesticated animals varieties
have not been produced which are mutually sterile;
and why with plants only a few such cases, immediately
to be given, have been observed.

The real difficulty in our present subject is not, as it
appears to me, why domestic varieties have not become
mutually infertile when crossed, but why this has so
generally occurred with natural varieties, as soon as
they have been permanently modified in a sufficient
degree to take rank as species. We are far from
precisely knowing the cause; nor is this surprising,
seeing how profoundly ignorant we are in regard to the
normal and abnormal action of the reproductive system.
But we can see that species, owing to their struggle for
existence with numerous competitors, will have been
exposed during long periods of time to more uniform
conditions, than have domestic varieties; and this may
well make a wide difference in the result. For we
know how commonly wild animals and plants, when
taken from their natural conditions and subjected to
captivity, are rendered sterile; and the reproductive
functions of organic beings which have always lived
under natural conditions would probably in like
manner be eminently sensitive to the influence of an

unnatural cross. Domesticated productions, on the other hand, which, as shown by the mere fact of their domestication, were not originally highly sensitive to changes in their conditions of life, and which can now generally resist with undiminished fertility repeated changes of conditions, might be expected to produce varieties, which would be little liable to have their reproductive powers injuriously affected by the act of crossing with other varieties which had originated in a like manner.

I have as yet spoken as if the varieties of the same species were invariably fertile when intercrossed. But it is impossible to resist the evidence of the existence of a certain amount of sterility in the few following cases, which I will briefly abstract. The evidence is at least as good as that from which we believe in the sterility of a multitude of species. The evidence is, also, derived from hostile witnesses, who in all other cases consider fertility and sterility as safe criterions of specific distinction. Gärtner kept during several years a dwarf kind of maize with yellow seeds, and a tall variety with red seeds growing near each other in his garden; and although these plants have separated sexes, they never naturally crossed. He then fertilised thirteen flowers of the one kind with pollen of the other; but only a single head produced any seed, and this one head produced only five grains. Manipulation in this case could not have been injurious, as the plants have separated sexes. No one, I believe, has suspected that these varieties of maize are distinct species; and it is important to notice that the hybrid plants thus raised were themselves *perfectly* fertile; so that even Gärtner did not venture to consider the two varieties as specifically distinct.

2 D

Girou de Buzareingues crossed three varieties of gourd, which like the maize has separated sexes, and he asserts that their mutual fertilisation is by so much the less easy as their differences are greater. How far these experiments may be trusted, I know not; but the forms experimented on are ranked by Sageret, who mainly founds his classification by the test of infertility, as varieties, and Naudin has come to the same conclusion.

The following case is far more remarkable, and seems at first incredible; but it is the result of an astonishing number of experiments made during many years on nine species of Verbascum, by so good an observer and so hostile a witness as Gärtner: namely that the yellow and white varieties when crossed produce less seed than the similarly coloured varieties of the same species. Moreover, he asserts that, when yellow and white varieties of one species are crossed with yellow and white varieties of a *distinct* species, more seed is produced by the crosses between the similarly coloured flowers, than between those which are differently coloured. Mr. Scott also has experimented on the species and varieties of Verbascum; and although unable to confirm Gärtner's results on the crossing of the distinct species, he finds that the dissimilarly coloured varieties of the same species yield fewer seeds, in the proportion of 86 to 100, than the similarly coloured varieties. Yet these varieties differ in no respect except in the colour of their flowers; and one variety can sometimes be raised from the seed of another.

Kölreuter, whose accuracy has been confirmed by every subsequent observer, has proved the remarkable fact, that one particular variety of the common tobacco was more fertile than the other varieties, when crossed with a widely distinct species. He experimented on five forms which are commonly reputed to be varieties, and which he tested by the severest trial, namely, by reciprocal crosses, and he found their mongrel offspring perfectly fertile. But one of these five varieties, when used either as the father or mother, and crossed with the Nicotiana glutinosa, always yielded hybrids not so sterile as those which were produced from the four other varieties when crossed with N. glutinosa. Hence the reproductive system of this one variety must have been in some manner and in some degree modified.

From these facts it can no longer be maintained that varieties when crossed are invariably quite fertile. From the great difficulty of ascertaining the infertility of varieties in a state of nature, for a supposed variety, if proved to be infertile in any degree, would almost universally be ranked as a species ;—from man attending only to external characters in his domestic varieties, and from such varieties not having been exposed for very long periods to uniform conditions of life ;—from these several considerations we may conclude that fertility does not constitute a fundamental distinction between varieties and species when crossed. The general sterility of crossed species may safely be looked at, not as a special acquirement or endowment, but as incidental on changes of an unknown nature in their sexual elements.

Hybrids and Mongrels compared, independently of their fertility.

Independently of the question of fertility, the off-spring of species and of varieties when crossed may be compared in several other respects. Gärtner, whose strong wish it was to draw a distinct line between species and varieties, could find very few, and, as it seems to me, quite unimportant differences between the so-called hybrid offspring of species, and the so-called mongrel offspring of varieties. And, on the other hand, they agree most closely in many important respects.

I shall here discuss this subject with extreme brevity. The most important distinction is, that in the first generation mongrels are more variable than hybrids; but Gärtner admits that hybrids from species which have long been cultivated are often variable in the first generation; and I have myself seen striking instances of this fact. Gärtner further admits that hybrids between very closely allied species are more variable than those from very distinct species; and this shows that the difference in the degree of variability graduates away. When mongrels and the more fertile hybrids are propagated for several generations, an extreme amount of variability in the offspring in both cases is notorious; but some few instances of both hybrids and mongrels long retaining a uniform character could be given. The variability, however, in the successive generations of mongrels is, perhaps, greater than in hybrids.

This greater variability in mongrels than in hybrids does not seem at all surprising. For the parents of mongrels are varieties, and mostly domestic varieties (very few experiments having been tried on natural varieties), and this implies that there has been recent variability, which would often continue and would augment that arising from the act of crossing. The slight variability of hybrids in the first generation, in contrast with that in the succeeding generations, is a curious fact and deserves attention. For it bears on the view which I have taken of one of the causes of ordinary variability; namely, that the reproductive system from being eminently sensitive to changed conditions of life, fails under these circumstances to perform its proper function of producing offspring closely similar in all respects to the parent-form. Now hybrids in the first generation are descended from species (excluding those long-cultivated) which have not had their reproductive systems in any way affected, and they are not variable; but hybrids themselves have their reproductive systems seriously affected, and their descendants are highly variable.

But to return to our comparison of mongrels and hybrids: Gärtner states that mongrels are more liable than hybrids to revert to either parent-form; but this, if it be true, is certainly only a difference in degree. Moreover, Gärtner expressly states that hybrids from long cultivated plants are more subject to reversion than hybrids from species in their natural state; and this probably explains the singular difference in the results arrived at by different observers: thus Max Wichura doubts whether hybrids ever revert to their parent-forms, and he experimented on uncultivated species of willows; whilst Naudin, on the other hand, insists in the strongest terms on the almost universal tendency to reversion in hybrids, and he experimented chiefly on cultivated plants. Gärtner further states that when any two species, although most closely allied to each other, are crossed with a third species, the hybrids are widely different from each other; whereas if two very distinct varieties of one species are crossed with another species, the hybrids do not differ much. But this conclusion, as far as I can make out, is founded on a single experiment; and seems directly opposed to the results of several experiments made by Kölreuter.

Such alone are the unimportant differences which Gärtner is able to point out between hybrid and mongrel plants. On the other hand, the degrees and kinds of resemblance in mongrels and in hybrids to their respective parents, more especially in hybrids produced from nearly related species, follow according to Gärtner the same laws. When two species are crossed, one has sometimes a prepotent power of impressing its likeness on the hybrid. So I believe it to be with varieties of plants; and with animals one variety certainly often has this prepotent power over another variety. Hybrid plants produced from a reciprocal cross, generally resemble each other closely; and so it is with mongrel plants from a reciprocal cross. Both hybrids and mongrels can be reduced to either pure parent-form, by repeated crosses in successive generations with either parent.

These several remarks are apparently applicable to animals; but the subject is here much complicated, partly owing to the existence of secondary sexual characters; but more especially owing to prepotency in transmitting likeness running more strongly in one sex than in the other, both when one species is crossed with another, and when one variety is crossed with another variety. For instance, I think those authors are right who maintain that the ass has a prepotent power over the horse, so that both the mule and the hinny resemble more closely the ass than the horse; but that the pre-

potency runs more strongly in the male than in the
female ass, so that the mule, which is the offspring of
the male ass and mare, is more like an ass, than is
the hinny, which is the offspring of the female ass and
stallion.

Much stress has been laid by some authors on the
supposed fact, that it is only with mongrels that the
offspring are not intermediate in character, but closely
resemble one of their parents; but this does sometimes
occur with hybrids, yet I grant much less frequently
than with mongrels. Looking to the cases which I
have collected of cross-bred animals closely resembling
one parent, the resemblances seem chiefly confined to
characters almost monstrous in their nature, and which
have suddenly appeared—such as albinism, melanism,
deficiency of tail or horns, or additional fingers and
toes; and do not relate to characters which have been
slowly acquired through selection. A tendency to
sudden reversions to the perfect character of either
parent would, also, be much more likely to occur with
mongrels, which are descended from varieties often
suddenly produced and semi-monstrous in character,
than with hybrids, which are descended from species
slowly and naturally produced. On the whole, I
entirely agree with Dr. Prosper Lucas, who after
arranging an enormous body of facts with respect to
animals, comes to the conclusion that the laws of
resemblance of the child to its parents are the same,
whether the two parents differ little or much from each
other, namely, in the union of individuals of the same
variety, or of different varieties, or of distinct species.

Independently of the question of fertility and sterility,
in all other respects there seems to be a general and

close similarity in the offspring of crossed species, and
of crossed varieties. If we look at species as having
been specially created, and at varieties as having been
produced by secondary laws, this similarity would be
an astonishing fact. But it harmonises perfectly with
the view that there is no essential distinction between
species and varieties.

Summary of Chapter.

First crosses between forms, sufficiently distinct to
be ranked as species, and their hybrids, are very
generally, but not universally, sterile. The sterility
is of all degrees, and is often so slight that the most
careful experimentalists have arrived at diametrically
opposite conclusions in ranking forms by this test.
The sterility is innately variable in individuals of the
same species, and is eminently susceptible to the action
of favourable and unfavourable conditions. The degree
of sterility does not strictly follow systematic affinity,
but is governed by several curious and complex laws.
It is generally different, and sometimes widely different
in reciprocal crosses between the same two species. It
is not always equal in degree in a first cross and in the
hybrids produced from this cross.

In the same manner as in grafting trees, the capacity
in one species or variety to take on another, is incidental
on differences, generally of an unknown nature, in their
vegetative systems, so in crossing, the greater or less
facility of one species to unite with another is incidental
on unknown differences in their reproductive systems.
There is no more reason to think that species have been
specially endowed with various degrees of sterility to
prevent their crossing and blending in nature, than to

think that trees have been specially endowed with various and somewhat analogous degrees of difficulty in being grafted together in order to prevent their inarching in our forests.

The sterility of first crosses and of their hybrid progeny has not been acquired through natural selection. In the case of first crosses it seems to depend on several circumstances; in some instances in chief part on the early death of the embryo. In the case of hybrids, it apparently depends on their whole organisation having been disturbed by being compounded from two distinct forms; the sterility being closely allied to that which so frequently affects pure species, when exposed to new and unnatural conditions of life. He who will explain these latter cases will be able to explain the sterility of hybrids. This view is strongly supported by a parallelism of another kind: namely, that, firstly, slight changes in the conditions of life add to the vigour and fertility of all organic beings; and secondly, that the crossing of forms, which have been exposed to slightly different conditions of life or which have varied, favours the size, vigour, and fertility of their offspring. The facts given on the sterility of the illegitimate unions of dimorphic and trimorphic plants and of their illegitimate progeny, perhaps render it probable that some unknown bond in all cases connects the degree of fertility of first unions with that of their offspring. The consideration of these facts on dimorphism, as well as of the results of reciprocal crosses, clearly leads to the conclusion that the primary cause of the sterility of crossed species is confined to differences in their sexual elements. But why, in the case of distinct species, the sexual elements should so generally have become more or less modified, leading to their mutual infertility, we do not know; but it seems to stand in some close relation to species having been exposed for long periods of time to nearly uniform conditions of life.

It is not surprising that the difficulty in crossing any two species, and the sterility of their hybrid offspring, should in most cases correspond, even if due to distinct causes: for both depend on the amount of difference between the species which are crossed. Nor is it surprising that the facility of effecting a first cross, and the fertility of the hybrids thus produced, and the capacity of being grafted together—though this latter capacity evidently depends on widely different circumstances—should all run, to a certain extent, parallel with the systematic affinity of the forms subjected to experiment; for systematic affinity includes resemblances of all kinds.

First crosses between forms known to be varieties, or sufficiently alike to be considered as varieties, and their mongrel offspring, are very generally, but not, as is so often stated, invariably fertile. Nor is this almost universal and perfect fertility surprising, when it is remembered how liable we are to argue in a circle with respect to varieties in a state of nature; and when we remember that the greater number of varieties have been produced under domestication by the selection of mere external differences, and that they have not been long exposed to uniform conditions of life. It should also be especially kept in mind, that long-continued domestication tends to eliminate sterility, and is therefore little likely to induce this same quality. Independently of the question of fertility, in all other respects there is the closest general resemblance between hybrids and mongrels,—in their variability, in their power of absorbing each other by repeated crosses, and in their inheritance of characters from both parent-forms. Finally, then, although we are as ignorant of the precise cause of the sterility of first crosses and of hybrids as we are why animals and plants removed from their natural conditions become sterile, yet the facts given in this chapter do not seem to me opposed to the belief that species aboriginally existed as varieties.

CHAPTER X.

On the Imperfection of the Geological Record.

On the absence of intermediate varieties at the present day—On the nature of extinct intermediate varieties ; on their number— On the lapse of time, as inferred from the rate of denudation and of deposition—On the lapse of time as estimated by years —On the poorness of our palæontological collections—On the intermittence of geological formations—On the denudation of granitic areas—On the absence of intermediate varieties in any one formation—On the sudden appearance of groups of species— On their sudden appearance in the lowest known fossiliferous strata—Antiquity of the habitable earth.

In the sixth chapter I enumerated the chief objections which might be justly urged against the views maintained in this volume. Most of them have now been discussed. One, namely the distinctness of specific forms, and their not being blended together by innumerable transitional links, is a very obvious difficulty. I assigned reasons why such links do not commonly occur at the present day under the circumstances apparently most favourable for their presence, namely on an extensive and continuous area with graduated physical conditions. I endeavoured to show, that the life of each species depends in a more important manner on the presence of other already defined organic forms, than on climate, and, therefore, that the really governing conditions of life do not graduate away quite insensibly like heat or moisture. I endeavoured, also, to show that intermediate varieties, from existing in lesser numbers than the forms which they connect, will generally be beaten out and exterminated during the course of further modification and improvement. The main cause, however, of innumerable intermediate links not now occurring everywhere throughout nature, depends on the very process of natural selection, through which new varieties continually take the places of and supplant their parent-forms. But just in proportion as this process of extermination has acted on an enormous scale, so must the number of intermediate varieties, which have formerly existed, be truly enormous. Why then is not every geological formation and every stratum full of such intermediate links ? Geology assuredly does not reveal any such finely-graduated organic chain ; and this, perhaps, is the most obvious and serious objection which can be urged against the theory. The explanation lies, as I believe, in the extreme imperfection of the geological record.

In the first place, it should always be borne in mind what sort of intermediate forms must, on the theory, have formerly existed. I have found it difficult, when looking at any two species, to avoid picturing to myself forms *directly* intermediate between them. But this is a wholly false view ; we should always look for forms intermediate between each species and a common but unknown progenitor ; and the progenitor will generally have differed in some respects from all its modified descendants. To give a simple illustration : the fantail and pouter pigeons are both descended from the rock-pigeon ; if we possessed all the intermediate varieties which have ever existed, we should have an extremely close series between both and the rock-pigeon; but we should have no varieties directly intermediate between the fantail and pouter; none, for instance, combining a tail somewhat expanded with a crop somewhat enlarged, the characteristic features of these two breeds. These two breeds, moreover, have become so much modified, that, if we had no historical or indirect evidence regarding their origin, it would not have been possible to have determined, from a mere comparison of their structure with that of the rock-pigeon, C. livia, whether they had descended from this species or from some other allied form, such as C. oenas.

So, with natural species, if we look to forms very distinct, for instance to the horse and tapir, we have no reason to suppose that links directly intermediate between them ever existed, but between each and an unknown common parent. The common parent will have had in its whole organisation much general resemblance to the tapir and to the horse; but in some points of structure may have differed considerably from both, even perhaps more than they differ from each other. Hence, in all such cases, we should be unable to recognise the parent-form of any two or more species, even if we closely compared the structure of the parent with that of its modified descendants, unless at the same time we had a nearly perfect chain of the intermediate links.

It is just possible by the theory, that one of two living forms might have descended from the other; for instance, a horse from a tapir; and in this case *direct* intermediate links will have existed between them. But such a case would imply that one form had remained for a very long period unaltered, whilst its descendants had undergone a vast amount of change;

and the principle of competition between organism and organism, between child and parent, will render this a very rare event; for in all cases the new and improved forms of life tend to supplant the old and unimproved forms.

By the theory of natural selection all living species have been connected with the parent-species of each genus, by differences not greater than we see between the natural and domestic varieties of the same species at the present day; and these parent-species, now generally extinct, have in their turn been similarly connected with more ancient forms; and so on backwards, always converging to the common ancestor of each great class. So that the number of intermediate and transitional links, between all living and extinct species, must have been inconceivably great. But assuredly, if this theory be true, such have lived upon the earth.

On the Lapse of Time, as inferred from the rate of Deposition and extent of Denudation.

Independently of our not finding fossil remains of such infinitely numerous connecting links, it may be objected that time cannot have sufficed for so great an amount of organic change, all changes having been effected slowly. It is hardly possible for me to recall to the reader who is not a practical geologist, the facts leading the mind feebly to comprehend the lapse of time. He who can read Sir Charles Lyell's grand work on the Principles of Geology, which the future historian will recognise as having produced a revolution in natural science, and yet does not admit how vast have been the past periods of time, may at once close this volume. Not that it suffices to study the Principles of Geology, or to read special treatises by different observers on separate formations, and to mark how each author attempts to give an inadequate idea of the duration of each formation, or even of each stratum. We can best gain some idea of past time by knowing the agencies at work, and learning how deeply the surface of the land has been denuded, and how much sediment has been deposited. As Lyell has well remarked, the extent and thickness of our sedimentary formations are the result and the measure of the denudation which the earth's crust has elsewhere undergone. Therefore a man should examine for himself the great piles of super-imposed strata, and watch the rivulets bringing down mud, and the waves wearing away the sea-cliffs, in order to comprehend something about the duration of past time, the monuments of which we see all around us.

It is good to wander along the coast, when formed of moderately hard rocks, and mark the process of degrad-ation. The tides in most cases reach the cliffs only for a short time twice a day, and the waves eat into them only when they are charged with sand or pebbles; for there is good evidence that pure water effects nothing in wearing away rock. At last the base of the cliff is undermined, huge fragments fall down, and these, remaining fixed, have to be worn away atom by atom, until after being reduced in size they can be rolled about by the waves, and then they are more quickly ground into pebbles, sand, or mud. But how often do we see along the bases of retreating cliffs rounded boulders, all thickly clothed by marine productions, showing how little they are abraded and how seldom they are rolled about! Moreover, if we follow for a few miles any line of rocky cliff, which is undergoing degradation, we find that it is only here and there, along a short length or round a promontory, that the cliffs are at the present time suffering. The appearance of the surface and the vegetation show that elsewhere years have elapsed since the waters washed their base.

We have, however, recently learnt from the observations of Ramsay, in the van of many excellent observers—of Jukes, Geikie, Croll, and others, that subaerial degradation is a much more important agency than coast-action, or the power of the waves. The whole surface of the land is exposed to the chemical action of the air and of the rain-water with its dissolved carbonic acid, and in colder countries to frost; the disintegrated matter is carried down even gentle slopes during heavy rain, and to a greater extent than might be supposed, especially in arid districts, by the wind; it is then transported by the streams and rivers, which when rapid deepen their channels, and triturate the fragments. On a rainy day, even in a gently undulating country, we see the effects of subaerial degradation in the muddy rills which flow down every slope. Messrs. Ramsay and Whitaker have shown, and the observation is a most striking one, that the great lines of escarpment in the Wealden district and those ranging across England, which formerly were looked at as ancient sea-coasts, cannot have been thus formed, for each line is composed of one and the same formation, whilst our sea-cliffs are everywhere formed by the intersection of various formations. This being the case, we are compelled to admit that the escarpments owe

2 E

their origin in chief part to the rocks of which they are composed having resisted subaerial denudation better than the surrounding surface ; this surface consequently has been gradually lowered, with the lines of harder rock left projecting.   Nothing impresses the mind with the vast duration of time, according to our ideas of time, more forcibly than the conviction thus gained that subaerial agencies which apparently have so little power, and which seem to work so slowly, have produced great results.

When thus impressed with the slow rate at which the land is worn away through subaerial and littoral action, it is good, in order to appreciate the past duration of time, to consider, on the one hand, the masses of rock which have been removed over many extensive areas, and on the other hand the thickness of our sedimentary formations.   I remember having been much struck when viewing volcanic islands, which have been worn by the waves and pared all round into perpendicular cliffs of one or two thousand feet in height; for the gentle slope of the lava-streams, due to their formerly liquid state, showed at a glance how far the hard, rocky beds had once extended into the open ocean.   The same story is told still more plainly by faults,—those great cracks along which the strata have been upheaved on one side, or thrown down on the other, to the height or depth of thousands of feet; for since the crust cracked, and it makes no great difference whether the upheaval was sudden, or, as most geologists now believe, was slow and effected by many starts, the surface of the land has been so completely planed down that no trace of these vast dislocations is externally visible.   The Craven fault, for instance extends for upwards of 30 miles, and along this line the vertical displacement of the strata varies from 600 to 3000 feet. Professor Ramsay has published an account of a down-throw in Anglesea of 2300 feet; and he informs me that he fully believes that there is one in Merioneth-shire of 12,000 feet; yet in these cases there is nothing on the surface of the land to show such prodigious movements; the pile of rocks on either side of the crack having been smoothly swept away.

On the other hand, in all parts of the world the piles of sedimentary strata are of wonderful thickness. In the Cordillera I estimated one mass of conglomerate at ten thousand feet; and although conglomerates have probably been accumulated at a quicker rate than finer sediments, yet from being formed of worn and rounded pebbles, each of which bears the stamp of time, they are good to show how slowly the mass must have been heaped together. Professor Ramsay has given me the maximum thickness, from actual measurement in most cases, of the successive formations in *different* parts of Great Britain; and this is the result :—

	Feet.
Palæozoic strata (not including igneous beds)	57,154
Secondary strata	13,190
Tertiary strata	2,240

—making altogether 72,584 feet; that is, very nearly thirteen and three-quarters British miles. Some of the formations, which are represented in England by thin beds, are thousands of feet in thickness on the Continent. Moreover, between each successive formation, we have, in the opinion of most geologists, blank periods of enormous length. So that the lofty pile of sedimentary rocks in Britain gives but an inadequate idea of the time which has elapsed during their accumulation. The consideration of these various facts impresses the mind almost in the same manner as does the vain endeavour to grapple with the idea of eternity.

Nevertheless this impression is partly false. Mr. Croll, in an interesting paper, remarks that we do not err " in forming too great a conception of the length of "geological periods," but in estimating them by years. When geologists look at large and complicated phenomena, and then at the figures representing several million years, the two produce a totally different effect on the mind, and the figures are at once pronounced too small. In regard to subaerial denudation, Mr. Croll shows, by calculating the known amount of sediment annually brought down by certain rivers, relatively to their areas of drainage, that 1000 feet of solid rock, as it became gradually disintegrated, would thus be removed from the mean level of the whole area in the course of six million years. This seems an astonishing result, and some considerations lead to the suspicion that it may be too large, but even if halved or quartered it is still very surprising. Few of us, however, know what a million really means: Mr. Croll gives the following illustration: take a narrow strip of paper, 83 feet 4 inches in length, and stretch it along the wall of a large hall; then mark off at one end the tenth of an inch. This tenth of an inch will represent one hundred years, and the entire strip a million years. But let it be borne in mind, in relation to the subject of this work, what a hundred years implies, represented as it is by a measure utterly insignificant in a hall of the above dimensions. Several eminent breeders, during a single lifetime, have so largely modified some of the higher animals, which propagate their kind much more slowly than most of the lower animals, that they have formed what well deserves to be called a new sub-breed. Few men have attended with due care to any one strain for more than half a century, so that a hundred years represents the work of two breeders in succession. It is not to be supposed that species in a state of nature ever change so quickly as domestic animals under the guidance of methodical selection. The comparison would be in every way fairer with the effects which follow from unconscious selection, that is the preservation of the most useful or beautiful animals, with no intention of modifying the breed; but by this process of unconscious selection, various breeds have been sensibly changed in the course of two or three centuries.

Species, however, probably change much more slowly, and within the same country only a few change at the same time. This slowness follows from all the inhabitants of the same country being already so well adapted to each other, that new places in the polity of nature do not occur until after long intervals, due to the occurrence of physical changes of some kind, or through the immigration of new forms. Moreover variations or individual differences of the right nature, by which some of the inhabitants might be better fitted to their new places under the altered circumstances, would not always occur at once. Unfortunately we have no means of determining, according to the standard of years, how long a period it takes to modify a species; but to the subject of time we must return.

On the Poorness of Palæontological Collections.

Now let us turn to our richest geological museums, and what a paltry display we behold! That our collections are imperfect is admitted by every one. The remark of that admirable palæontologist, Edward Forbes, should never be forgotten, namely, that very many fossil species are known and named from single and often broken specimens, or from a few specimens collected on some one spot. Only a small portion of the surface of the earth has been geologically explored, and no part with sufficient care, as the important discoveries made every year in Europe prove. No organism wholly soft can be preserved. Shells and bones decay and disappear when left on the bottom of the sea, where sediment is not accumulating. We probably take a quite erroneous view, when we assume that sediment is being deposited over nearly the whole bed of the sea, at a rate sufficiently quick to embed and preserve fossil remains. Throughout an enormously large proportion of the ocean, the bright blue tint of the water bespeaks its purity. The many cases on record of a formation conformably covered, after an immense interval of time, by another and later formation, without the underlying bed having suffered in the interval any wear and tear, seem explicable only on the view of the bottom of the sea not rarely lying for ages in an unaltered condition. The remains which do become embedded, if in sand or gravel, will, when the beds are upraised, generally be dissolved by the percolation of rain-water charged with carbolic acid. Some of the many kinds of animals which live on the beach between high and low water mark seem to be rarely preserved. For instance, the

several species of the Chthamalinæ (a sub-family of
sessile cirripedes) coat the rocks all over the world in
infinite numbers : they are all strictly littoral, with the
exception of a single Mediterranean species, which
inhabits deep water, and this has been found fossil in
Sicily, whereas not one other species has hitherto been
found in any tertiary formation : yet it is known that
the genus Chthamalus existed during the Chalk period.
Lastly, many great deposits requiring a vast length of
time for their accumulation, are entirely destitute of
organic remains, without our being able to assign any
reason : one of the most striking instances is that of the
Flysch formation, which consists of shale and sandstone,
several thousand, occasionally even six thousand feet in
thickness, and extending for at least 300 miles from
Vienna to Switzerland; and although this great mass
has been most carefully searched, no fossils, except a
few vegetable remains, have been found.

With respect to the terrestrial productions which lived
during the Secondary and Palæozoic periods, it is superfluous to state that our evidence is fragmentary in an
extreme degree.   For instance, until recently not a land-
shell was known belonging to either of these vast
periods, with the exception of one species discovered by
Sir C. Lyell and Dr. Dawson in the carboniferous strata
of North America; but now land-shells have been found
in the lias.   In regard to mammiferous remains, a
glance at the historical table published in Lyell's
Manual will bring home the truth, how accidental and
rare is their preservation, far better than pages of detail.
Nor is their rarity surprising, when we remember how
large a proportion of the bones of tertiary mammals
have been discovered either in caves or in lacustrine

deposits; and that not a cave or true lacustrine bed is known belonging to the age of our secondary or palæozoic formations.

But the imperfection in the geological record largely results from another and more important cause than any of the foregoing; namely, from the several formations being separated from each other by wide intervals of time. This doctrine has been emphatically admitted by many geologists and palæontologists, who, like E. Forbes, entirely disbelieve in the change of species. When we see the formations tabulated in written works, or when we follow them in nature, it is difficult to avoid believing that they are closely consecutive. But we know, for instance, from Sir R. Murchison's great work on Russia, what wide gaps there are in that country between the superimposed formations; so it is in North America, and in many other parts of the world. The most skilful geologist, if his attention had been confined exclusively to these large territories, would never have suspected that, during the periods which were blank and barren in his own country, great piles of sediment, charged with new and peculiar forms of life, had elsewhere been accumulated. And if, in each separate territory, hardly any idea can be formed of the length of time which has elapsed between the consecutive formations, we may infer that this could nowhere be ascertained. The frequent and great changes in the mineralogical composition of consecutive formations, generally implying great changes in the geography of the surrounding lands, whence the sediment was derived, accord with the belief of vast intervals of time having elapsed between each formation.

We can, I think, see why the geological formations

of each region are almost invariably intermittent; that is, have not followed each other in close sequence. Scarcely any fact struck me more when examining many hundred miles of the South American coasts, which have been upraised several hundred feet within the recent period, than the absence of any recent deposits sufficiently extensive to last for even a short geological period. Along the whole west coast, which is inhabited by a peculiar marine fauna, tertiary beds are so poorly developed, that no record of several successive and peculiar marine faunas will probably be preserved to a distant age. A little reflection will explain why, along the rising coast of the western side of South America, no extensive formations with recent or tertiary remains can anywhere be found, though the supply of sediment must for ages have been great, from the enormous degradation of the coast-rocks and from muddy streams entering the sea. The explanation, no doubt, is, that the littoral and sub-littoral deposits are continually worn away, as soon as they are brought up by the slow and gradual rising of the land within the grinding action of the coast-waves.

We may, I think, conclude that sediment must be accumulated in extremely thick, solid, or extensive masses, in order to withstand the incessant action of the waves, when first upraised and during successive oscillations of level, as well as the subsequent subaerial degradation. Such thick and extensive accumulations of sediment may be formed in two ways; either in profound depths of the sea, in which case the bottom will not be inhabited by so many and such varied forms of life, as the more shallow seas; and the mass when upraised will give an imperfect record of the organisms

which existed in the neighbourhood during the period
of its accumulation.   Or, sediment may be deposited to
any thickness and extent over a shallow bottom, if it
continue slowly to subside.   In this latter case, as long
as the rate of subsidence and the supply of sediment
nearly balance each other, the sea will remain shallow
and favourable for many and varied forms, and thus
a rich fossiliferous formation, thick enough, when upraised, to resist a large amount of denudation, may be
formed.

I am convinced that nearly all our ancient formations,
which are throughout the greater part of their thickness
rich in fossils, have thus been formed during subsidence.
Since publishing my views on this subject in 1845, I
have watched the progress of Geology, and have been
surprised to note how author after author, in treating
of this or that great formation, has come to the conclusion that it was accumulated during subsidence.   I
may add, that the only ancient tertiary formation on
the west coast of South America, which has been bulky
enough to resist such degradation as it has as yet
suffered, but which will hardly last to a distant
geological age, was deposited during a downward
oscillation of level, and thus gained considerable
thickness.

All geological facts tell us plainly that each area
has undergone numerous slow oscillations of level,
and apparently these oscillations have affected wide
spaces.   Consequently, formations rich in fossils and
sufficiently thick and extensive to resist subsequent
degradation, will have been formed over wide spaces
during periods of subsidence, but only where the
supply of sediment was sufficient to keep the sea

shallow and to embed and preserve the remains before they had time to decay. On the other hand, as long as the bed of the sea remains stationary, *thick* deposits cannot have been accumulated in the shallow parts, which are the most favourable to life. Still less can this have happened during the alternate periods of elevation; or, to speak more accurately, the beds which were then accumulated will generally have been destroyed by being upraised and brought within the limits of the coast-action.

These remarks apply chiefly to littoral and sublittoral deposits. In the case of an extensive and shallow sea, such as that within a large part of the Malay Archipelago, where the depth varies from 30 or 40 to 60 fathoms, a widely extended formation might be formed during a period of elevation, and yet not suffer excessively from denudation during its slow upheaval; but the thickness of the formation could not be great, for owing to the elevatory movement it would be less than the depth in which it was formed; nor would the deposit be much consolidated, nor be capped by overlying formations, so that it would run a good chance of being worn away by atmospheric degradation and by the action of the sea during subsequent oscillations of level. It has, however, been suggested by Mr. Hopkins, that if one part of the area, after rising and before being denuded, subsided, the deposit formed during the rising movement, though not thick, might afterwards become protected by fresh accumulations, and thus be preserved for a long period.

Mr. Hopkins also expresses his belief that sedimentary beds of considerable horizontal extent have rarely been completely destroyed. But all geologists, excepting

the few who believe that our present metamorphic
schists and plutonic rocks once formed the primordial
nucleus of the globe, will admit that these latter rocks
have been stript of their covering to an enormous
extent. For it is scarcely possible that such rocks
could have been solidified and crystallized whilst
uncovered; but if the metamorphic action occurred at
profound depths of the ocean, the former protecting
mantle of rock may not have been very thick. Admitting then that gneiss, mica-schist, granite, diorite,
&c., were once necessarily covered up, how can we
account for the naked and extensive areas of such rocks
in many parts of the world, except on the belief that
they have subsequently been completely denuded of all
overlying strata? That such extensive areas do exist
cannot be doubted: the granitic region of Parime is
described by Humboldt as being at least nineteen times
as large as Switzerland. South of the Amazon, Boué
colours an area composed of rocks of this nature as
equal to that of Spain, France, Italy, part of Germany,
and the British Islands, all conjoined. This region has
not been carefully explored, but from the concurrent
testimony of travellers, the granitic area is very large:
thus, Von Eschwege gives a detailed section of these
rocks, stretching from Rio de Janeiro for 260 geographical miles inland in a straight line; and I
travelled for 150 miles in another direction, and saw
nothing but granitic rocks. Numerous specimens,
collected along the whole coast from near Rio Janeiro to
the mouth of the Plata, a distance of 1100 geographical
miles, were examined by me, and they all belonged to
this class. Inland, along the whole northern bank of
the Plata I saw, besides modern tertiary beds, only one

small patch of slightly metamorphosed rock, which alone could have formed a part of the original capping of the granitic series. Turning to a well-known region, namely, to the United States and Canada, as shown in Professor H. D. Rogers's beautiful map, I have estimated the areas by cutting out and weighing the paper, and I find that the metamorphic (excluding " the semi-meta- "morphic ") and granitic rocks exceed, in the proportion of 19 to 12·5, the whole of the newer Palæozoic formations. In many regions the metamorphic and granitic rocks would be found much more widely extended than they appear to be, if all the sedimentary beds were removed which rest unconformably on them, and which could not have formed part of the original mantle under which they were crystallized. Hence it is probable that in some parts of the world whole formations have been completely denuded, with not a wreck left behind.

One remark is here worth a passing notice. During periods of elevation the area of the land and of the adjoining shoal parts of the sea will be increased, and new stations will often be formed :—all circumstances favourable, as previously explained, for the formation of new varieties and species; but during such periods there will generally be a blank in the geological record. On the other hand, during subsidence, the inhabited area and number of inhabitants will decrease (excepting on the shores of a continent when first broken up into an archipelago), and consequently during subsidence, though there will be much extinction, few new varieties or species will be formed ; and it is during these very periods of subsidence, that the deposits which are richest in fossils have been accumulated.

On the Absence of Numerous Intermediate Varieties in any Single Formation.

From these several considerations, it cannot be doubted that the geological record, viewed as a whole, is extremely imperfect; but if we confine our attention to any one formation, it becomes much more difficult to understand why we do not therein find closely graduated varieties between the allied species which lived at its commencement and at its close. Several cases are on record of the same species presenting varieties in the upper and lower parts of the same formation; thus, Trautschold gives a number of instances with Ammonites; and Hilgendorf has described a most curious case of ten graduated forms of Planorbis multiformis in the successive beds of a fresh-water formation in Switzerland. Although each formation has indisputably required a vast number of years for its deposition, several reasons can be given why each should not commonly include a graduated series of links between the species which lived at its commencement and close; but I cannot assign due proportional weight to the following considerations.

Although each formation may mark a very long lapse of years, each probably is short compared with the period requisite to change one species into another. I am aware that two palæontologists, whose opinions are worthy of much deference, namely Bronn and Woodward, have concluded that the average duration of each formation is twice or thrice as long as the average duration of specific forms. But insuperable difficulties, as it seems to me, prevent us from coming to any just conclusion on this head. When we see a species first

appearing in the middle of any formation, it would be
rash in the extreme to infer that it had not elsewhere
previously existed. So again when we find a species
disappearing before the last layers have been deposited,
it would be equally rash to suppose that it then became
extinct. We forget how small the area of Europe is
compared with the rest of the world ; nor have the
several stages of the same formation throughout Europe
been correlated with perfect accuracy.

We may safely infer that with marine animals of all
kinds there has been a large amount of migration due
to climatal and other changes ; and when we see a
species first appearing in any formation, the probability
is that it only then first immigrated into that area. It
is well-known, for instance, that several species appear
somewhat earlier in the palæozoic beds of North
America than in those of Europe ; time having apparently been required for their migration from the
American to the European seas. In examining the
latest deposits in various quarters of the world, it has
everywhere been noted, that some few still existing
species are common in the deposit, but have become
extinct in the immediately surrounding sea ; or, conversely, that some are now abundant in the neighbouring
sea, but are rare or absent in this particular deposit.
It is an excellent lesson to reflect on the ascertained
amount of migration of the inhabitants of Europe
during the glacial epoch, which forms only a part of
one whole geological period ; and likewise to reflect on
the changes of level, on the extreme change of climate,
and on the great lapse of time, all included within this
same glacial period. Yet it may be doubted whether,
in any quarter of the world, sedimentary deposits,

including fossil remains, have gone on accumulating within the same area during the whole of this period. It is not, for instance, probable that sediment was deposited during the whole of the glacial period near the mouth of the Mississippi, within that limit of depth at which marine animals can best flourish: for we know that great geographical changes occurred in other parts of America during this space of time. When such beds as were deposited in shallow water near the mouth of the Mississippi during some part of the glacial period shall have been upraised, organic remains will probably first appear and disappear at different levels, owing to the migrations of species and to geographical changes. And in the distant future, a geologist, examining these beds, would be tempted to conclude that the average duration of life of the embedded fossils had been less than that of the glacial period, instead of having been really far greater, that is, extending from before the glacial epoch to the present day.

In order to get a perfect gradation between two forms in the upper and lower parts of the same formation, the deposit must have gone on continuously accumulating during a long period, sufficient for the slow process of modification; hence the deposit must be a very thick one; and the species undergoing change must have lived in the same district throughout the whole time. But we have seen that a thick formation, fossiliferous throughout its entire thickness, can accumulate only during a period of subsidence; and to keep the depth approximately the same, which is necessary that the same marine species may live on the same space, the supply of sediment must nearly counterbalance the amount of subsidence. But this same movement of subsidence will tend to submerge the area whence the sediment is derived, and thus diminish the supply, whilst the downward movement continues. In fact, this nearly exact balancing between the supply of sediment and the amount of subsidence is probably a rare contingency; for it has been observed by more than one palæontologist, that very thick deposits are usually barren of organic remains, except near their upper or lower limits.

It would seem that each separate formation, like the whole pile of formations in any country, has generally been intermittent in its accumulation. When we see, as is so often the case, a formation composed of beds of widely different mineralogical composition, we may reasonably suspect that the process of deposition has been more or less interrupted. Nor will the closest inspection of a formation give us any idea of the length of time which its deposition may have consumed. Many instances could be given of beds only a few feet in thickness, representing formations, which are elsewhere thousands of feet in thickness, and which must have required an enormous period for their accumulation; yet no one ignorant of this fact would have even suspected the vast lapse of time represented by the thinner formation. Many cases could be given of the lower beds of a formation having been upraised, denuded, submerged, and then re-covered by the upper beds of the same formation,—facts, showing what wide, yet easily overlooked, intervals have occurred in its accumulation. In other cases we have the plainest evidence in great fossilised trees, still standing upright as they grew, of many long intervals of time and changes of level during the process of deposition, which

2 F

would not have been suspected, had not the trees been preserved: thus Sir C. Lyell and Dr. Dawson found carboniferous beds 1400 feet thick in Nova Scotia, with ancient root-bearing strata, one above the other at no less than sixty-eight different levels. Hence, when the same species occurs at the bottom, middle, and top of a formation, the probability is that it has not lived on the same spot during the whole period of deposition, but has disappeared and reappeared, perhaps many times, during the same geological period. Consequently if it were to undergo a considerable amount of modification during the deposition of any one geological formation, a section would not include all the fine intermediate gradations which must on our theory have existed, but abrupt, though perhaps slight, changes of form.

It is all-important to remember that naturalists have no golden rule by which to distinguish species and varieties; they grant some little variability to each species, but when they meet with a somewhat greater amount of difference between any two forms, they rank both as species, unless they are enabled to connect them together by the closest intermediate gradations; and this, from the reasons just assigned, we can seldom hope to effect in any one geological section. Supposing B and C to be two species, and a third, A, to be found in an older and underlying bed; even if A were strictly intermediate between B and C, it would simply be ranked as a third and distinct species, unless at the same time it could be closely connected by intermediate varieties with either one or both forms. Nor should it be forgotten, as before explained, that A might be the actual progenitor of B and C, and yet would not necessarily be strictly intermediate between

them in all respects. So that we might obtain the
parent-species and its several modified descendants
from the lower and upper beds of the same formation,
and unless we obtained numerous transitional gradations, we should not recognise their blood-relationship,
and should consequently rank them as distinct species.

It is notorious on what excessively slight differences
many palæontologists have founded their species; and
they do this the more readily if the specimens come
from different sub-stages of the same formation. Some
experienced conchologists are now sinking many of the
very fine species of D'Orbigny and others into the rank
of varieties; and on this view we do find the kind of
evidence of change which on the theory we ought to
find. Look again at the later tertiary deposits, which
include many shells believed by the majority of naturalists to be identical with existing species; but some
excellent naturalists, as Agassiz and Pictet, maintain
that all these tertiary species are specifically distinct,
though the distinction is admitted to be very slight;
so that here, unless we believe that these eminent
naturalists have been misled by their imaginations,
and that these late tertiary species really present no
difference whatever from their living representatives, or
unless we admit, in opposition to the judgment of
most naturalists, that these tertiary species are all
truly distinct from the recent, we have evidence of the
frequent occurrence of slight modifications of the kind
required. If we look to rather wider intervals of time,
namely, to distinct but consecutive stages of the same
great formation, we find that the embedded fossils,
though universally ranked as specifically different, yet
are far more closely related to each other than are the

species found in more widely separated formations; so that here again we have undoubted evidence of change in the direction required by the theory; but to this latter subject I shall return in the following chapter.

With animals and plants that propagate rapidly and do not wander much, there is reason to suspect, as we have formerly seen, that their varieties are generally at first local; and that such local varieties do not spread widely and supplant their parent-forms until they have been modified and perfected in some considerable degree. According to this view, the chance of discovering in a formation in any one country all the early stages of transition between any two forms, is small, for the successive changes are supposed to have been local or confined to some one spot. Most marine animals have a wide range; and we have seen that with plants it is those which have the widest range, that oftenest present varieties; so that, with shells and other marine animals, it is probable that those which had the widest range, far exceeding the limits of the known geological formations in Europe, have oftenest given rise, first to local varieties and ultimately to new species; and this again would greatly lessen the chance of our being able to trace the stages of transition in any one geological formation.

It is a more important consideration, leading to the same result, as lately insisted on by Dr. Falconer, namely, that the period during which each species underwent modification, though long as measured by years, was probably short in comparison with that during which it remained without undergoing any change.

It should not be forgotten, that at the present day, with perfect specimens for examination, two forms

can seldom be connected by intermediate varieties,
and thus proved to be the same species, until many
specimens are collected from many places; and with
fossil species this can rarely be done. We shall,
perhaps, best perceive the improbability of our being
enabled to connect species by numerous, fine, intermediate, fossil links, by asking ourselves whether, for
instance, geologists at some future period will be able
to prove that our different breeds of cattle, sheep,
horses, and dogs are descended from a single stock or
from several aboriginal stocks; or, again, whether
certain sea-shells inhabiting the shores of North
America, which are ranked by some conchologists as
distinct species from their European representatives,
and by other conchologists as only varieties, are really
varieties, or are, as it is called, specifically distinct.
This could be effected by the future geologist only by
his discovering in a fossil state numerous intermediate
gradations; and such success is improbable in the
highest degree.

It has been asserted over and over again, by writers
who believe in the immutability of species, that geology
yields no linking forms. This assertion, as we shall
see in the next chapter, is certainly erroneous. As Sir
J. Lubbock has remarked, "Every species is a link
"between other allied forms." If we take a genus
having a score of species, recent and extinct, and
destroy four-fifths of them, no one doubts that the
remainder will stand much more distinct from each
other. If the extreme forms in the genus happen to
have been thus destroyed, the genus itself will stand
more distinct from other allied genera. What geological research has not revealed, is the former existence

of infinitely numerous gradations, as fine as existing varieties, connecting together nearly all existing and extinct species. But this ought not to be expected; yet this has been repeatedly advanced as a most serious objection against my views.

It may be worth while to sum up the foregoing remarks on the causes of the imperfection of the geological record under an imaginary illustration. The Malay Archipelago is about the size of Europe from the North Cape to the Mediterranean, and from Britain to Russia; and therefore equals all the geological formations which have been examined with any accuracy, excepting those of the United States of America. I fully agree with Mr. Godwin-Austen, that the present condition of the Malay Archipelago, with its numerous large islands separated by wide and shallow seas, probably represents the former state of Europe, whilst most of our formations were accumulating. The Malay Archipelago is one of the richest regions in organic beings; yet if all the species were to be collected which have ever lived there, how imperfectly would they represent the natural history of the world!

But we have every reason to believe that the terrestrial productions of the archipelago would be preserved in an extremely imperfect manner in the formations which we suppose to be there accumulating. Not many of the strictly littoral animals, or of those which lived on naked submarine rocks, would be embedded; and those embedded in gravel or sand would not endure to a distant epoch. Wherever sediment did not accumulate on the bed of the sea, or where it did not accumulate at a sufficient rate to protect organic bodies from decay, no remains could be preserved.

Formations rich in fossils of many kinds, and of thickness sufficient to last to an age as distant in futurity as the secondary formations lie in the past, would generally be formed in the archipelago only during periods of subsidence. These periods of subsidence would be separated from each other by immense intervals of time, during which the area would be either stationary or rising; whilst rising, the fossiliferous formations on the steeper shores would be destroyed, almost as soon as accumulated, by the incessant coast-action, as we now see on the shores of South America. Even throughout the extensive and shallow seas within the archipelago, sedimentary beds could hardly be accumulated of great thickness during the periods of elevation, or become capped and protected by subsequent deposits, so as to have a good chance of enduring to a very distant future. During the periods of subsidence, there would probably be much extinction of life; during the periods of elevation, there would be much variation, but the geological record would then be less perfect.

It may be doubted whether the duration of any one great period of subsidence over the whole or part of the archipelago, together with a contemporaneous accumulation of sediment, would *exceed* the average duration of the same specific forms; and these contingencies are indispensable for the preservation of all the transitional gradations between any two or more species. If such gradations were not all fully preserved, transitional varieties would merely appear as so many new, though closely allied species. It is also probable that each great period of subsidence would be interrupted by oscillations of level, and that slight climatal changes would intervene during such lengthy periods; and in these cases the inhabitants of the archipelago would migrate, and no closely consecutive record of their modifications could be preserved in any one formation.

Very many of the marine inhabitants of the archipelago now range thousands of miles beyond its confines; and analogy plainly leads to the belief that it would be chiefly these far-ranging species, though only some of them, which would oftenest produce new varieties; and the varieties would at first be local or confined to one place, but if possessed of any decided advantage, or when further modified and improved, they would slowly spread and supplant their parent-forms. When such varieties returned to their ancient homes, as they would differ from their former state in a nearly uniform, though perhaps extremely slight degree, and as they would be found embedded in slightly different sub-stages of the same formation, they would, according to the principles followed by many palæontologists, be ranked as new and distinct species.

If then there be some degree of truth in these remarks, we have no right to expect to find, in our geological formations, an infinite number of those fine transitional forms which, on our theory, have connected all the past and present species of the same group into one long and branching chain of life. We ought only to look for a few links, and such assuredly we do find —some more distantly, some more closely, related to each other; and these links, let them be ever so close, if found in different stages of the same formation, would, by many palæontologists, be ranked as distinct species. But I do not pretend that I should ever have suspected how poor was the record in the best preserved geological sections, had not the absence of innumerable transitional links between the species which lived at the commencement and close of each formation, pressed so hardly on my theory.

On the sudden Appearance of whole Groups of allied Species.

The abrupt manner in which whole groups of species suddenly appear in certain formations, has been urged by several palæontologists—for instance, by Agassiz, Pictet, and Sedgwick—as a fatal objection to the belief in the transmutation of species. If numerous species, belonging to the same genera or families, have really started into life at once, the fact would be fatal to the theory of evolution through natural selection. For the development by this means of a group of forms, all of which are descended from some one progenitor, must have been an extremely slow process; and the progenitors must have lived long before their modified descendants. But we continually overrate the perfection of the geological record, and falsely infer, because certain genera or families have not been found beneath a certain stage, that they did not exist before that stage. In all cases positive palæontological evidence may be implicitly trusted; negative evidence is worthless, as experience has so often shown. We continually forget how large the world is, compared with the area over which our geological formations have been carefully examined; we forget that groups of species may else-where have long existed, and have slowly multiplied, before they invaded the ancient archipelagoes of Europe and the United States. We do not make due allowance for the intervals of time which have elapsed between our consecutive formations,—longer perhaps in many cases than the time required for the accumulation of each formation. These intervals will have given time for the multiplication of species from some one parent-form: and in the succeeding formation, such groups or species will appear as if suddenly created.

I may here recall a remark formerly made, namely, that it might require a long succession of ages to adapt an organism to some new and peculiar line of life, for instance, to fly through the air; and consequently that the transitional forms would often long remain confined to some one region; but that, when this adaptation had once been effected, and a few species had thus acquired a great advantage over other organisms, a comparatively short time would be necessary to produce many divergent forms, which would spread rapidly and widely, throughout the world. Professor Pictet, in his excellent Review of this work, in commenting on early transitional forms, and taking birds as an illustration, cannot see how the successive modifications of the anterior limbs of a supposed prototype could possibly have been of any advantage. But look at the penguins of the Southern Ocean; have not these birds their front limbs in this precise intermediate state of "neither true "arms nor true wings"? Yet these birds hold their place victoriously in the battle for life; for they exist in infinite numbers and of many kinds. I do not suppose that we here see the real transitional grades through which the wings of birds have passed; but what special difficulty is there in believing that it might profit the modified descendants of the penguin, first to become enabled to flap along the surface of the

sea like the logger-headed duck, and ultimately to rise
from its surface and glide through the air ?

I will now give a few examples to illustrate the
foregoing remarks, and to show how liable we are to
error in supposing that whole groups of species have
suddenly been produced. Even in so short an interval
as that between the first and second editions of Pictet's
great work on Palæontology, published in 1844–46 and
in 1853–57, the conclusions on the first appearance and
disappearance of several groups of animals have been
considerably modified; and a third edition would
require still further changes. I may recall the well-
known fact that in geological treatises, published not
many years ago, mammals were always spoken of as
having abruptly come in at the commencement of the
tertiary series. And now one of the richest known accumulations of fossil mammals belongs to the middle of
the secondary series; and true mammals have been
discovered in the new red sandstone at nearly the commencement of this great series. Cuvier used to urge that
no monkey occurred in any tertiary stratum; but now
extinct species have been discovered in India, South
America and in Europe, as far back as the miocene stage.
Had it not been for the rare accident of the preservation
of footsteps in the new red sandstone of the United States,
who would have ventured to suppose that no less than
at least thirty different bird-like animals, some of
gigantic size, existed during that period ? Not a fragment of bone has been discovered in these beds. Not
long ago, palæontologists maintained that the whole
class of birds came suddenly into existence during the
eocene period; but now we know, on the authority of
Professor Owen, that a bird certainly lived during the

deposition of the upper greensand; and still more recently, that strange bird, the Archeopteryx, with a long lizard-like tail, bearing a pair of feathers on each joint, and with its wings furnished with two free claws, has been discovered in the oolitic slates of Solenhofen. Hardly any recent discovery shows more forcibly than this, how little we as yet know of the former inhabitants of the world.

I may give another instance, which, from having passed under my own eyes, has much struck me. In a memoir on Fossil Sessile Cirripedes, I stated that, from the large number of existing and extinct tertiary species; from the extraordinary abundance of the individuals of many species all over the world, from the Arctic regions to the equator, inhabiting various zones of depths from the upper tidal limits to 50 fathoms; from the perfect manner in which specimens are preserved in the oldest tertiary beds; from the ease with which even a fragment of a valve can be recognised; from all these circumstances, I inferred that, had sessile cirripedes existed during the secondary periods, they would certainly have been preserved and discovered; and as not one species had then been discovered in beds of this age, I concluded that this great group had been suddenly developed at the commencement of the tertiary series. This was a sore trouble to me, adding as I then thought one more instance of the abrupt appearance of a great group of species. But my work had hardly been published, when a skilful palæontologist, M. Bosquet, sent me a drawing of a perfect specimen of an unmistakeable sessile cirripede, which he had himself extracted from the chalk of Belgium. And, as if to make the case as striking as possible, this cirripede was a Chthamalus, a very common, large, and ubiquitous genus, of which not one species has as yet been found even in any tertiary stratum. Still more recently, a Pyrgoma, a member of a distinct sub-family of sessile cirripedes, has been discovered by Mr. Woodward in the upper chalk; so that we now have abundant evidence of the existence of this group of animals during the secondary period.

The case most frequently insisted on by palæontologists of the apparently sudden appearance of a whole group of species, is that of the teleostean fishes, low down, according to Agassiz, in the Chalk period. This group includes the large majority of existing species. But certain Jurassic and Triassic forms are now commonly admitted to be teleostean; and even some palæozoic forms have thus been classed by one high authority. If the teleosteans had really appeared suddenly in the northern hemisphere at the commencement of the chalk formation the fact would have been highly remarkable; but it would not have formed an insuperable difficulty, unless it could likewise have been shown that at the same period the species were suddenly and simultaneously developed in other quarters of the world. It is almost superfluous to remark that hardly any fossil-fish are known from south of the equator; and by running through Pictet's Palæontology it will be seen that very few species are known from several formations in Europe. Some few families of fish now have a confined range; the teleostean fishes might formerly have had a similarly confined range, and after having been largely developed in some one sea, have spread widely. Nor have we any right to suppose that the seas of the world have always been so freely open from south to north as they are at present. Even at this day, if the Malay Archipelago were converted into land, the tropical parts of the Indian Ocean would form a large and perfectly enclosed basin, in which any great group of marine animals might be multiplied; and here they would remain confined, until some of the species became adapted to a cooler climate, and were enabled to double the Southern capes of Africa or Australia, and thus reach other and distant seas.

From these considerations, from our ignorance of the geology of other countries beyond the confines of Europe and the United States, and from the revolution in our palæontological knowledge effected by the discoveries of the last dozen years, it seems to me to be about as rash to dogmatize on the succession of organic forms throughout the world, as it would be for a naturalist to land for five minutes on a barren point in Australia, and then to discuss the number and range of its productions.

On the sudden Appearance of Groups of allied Species in the lowest known Fossiliferous Strata.

There is another and allied difficulty, which is much more serious. I allude to the manner in which species belonging to several of the main divisions of the animal kingdom suddenly appear in the lowest known fossiliferous rocks. Most of the arguments which have convinced me that all the existing species of the same group are descended from a single progenitor, apply with equal force to the earliest known species. For instance, it cannot be doubted that all the Cambrian and Silurian trilobites are descended from some one crustacean, which must have lived long before the Cambrian age, and which probably differed greatly from any known animal.   Some of the most ancient animals, as the Nautilus, Lingula, &c., do not differ much from living species ; and it cannot on our theory be supposed, that these old species were the progenitors of all the species belonging to the same groups which have subsequently appeared, for they are not in any degree intermediate in character.

Consequently, if the theory be true, it is indisputable that before the lowest Cambrian stratum was deposited long periods elapsed, as long as, or probably far longer than, the whole interval from the Cambrian age to the present day ; and that during these vast periods the world swarmed with living creatures. Here we encounter a formidable objection ; for it seems doubtful whether the earth, in a fit state for the habitation of living creatures, has lasted long enough.   Sir W. Thompson concludes that the consolidation of the crust can hardly have occurred less than 20 or more than 400 million years ago, but probably not less than 98 or more than 200 million years.   These very wide limits show how doubtful the data are ; and other elements may have hereafter to be introduced into the problem. Mr. Croll estimates that about 60 million years have elapsed since the Cambrian period, but this, judging from the small amount of organic change since the commencement of the Glacial epoch, appears a very short time for the many and great mutations of life, which have certainly occurred since the Cambrian formation ; and the previous 140 million years can hardly be considered as sufficient for the development of the varied forms of life which already existed during the Cambrian period. It is, however, probable, as Sir William Thompson insists, that the world at a very early period was subjected to more rapid and violent changes in its physical conditions than those now occurring; and such changes would have tended to induce changes at a corresponding rate in the organisms which then existed.

To the question why we do not find rich fossiliferous deposits belonging to these assumed earliest periods prior to the Cambrian system, I can give no satisfactory answer. Several eminent geologists, with Sir R. Murchison at their head, were until recently convinced that we beheld in the organic remains of the lowest Silurian stratum the first dawn of life. Other highly competent judges, as Lyell and E. Forbes, have disputed this conclusion. We should not forget that only a small portion of the world is known with accuracy. Not very long ago M. Barrande added another and lower stage, abounding with new and peculiar species, beneath the then known Silurian system; and now, still lower down in the Lower Cambrian formation, Mr. Hicks has found in South Wales beds rich in trilobites, and containing various molluscs and annelids. The presence of phosphatic nodules and bituminous matter, even in some of the lowest azoic rocks, probably indicates life at these periods; and the existence of the Eozoon in the Laurentian formation of Canada is generally admitted. There are three great series of strata beneath the Silurian system in Canada, in the lowest of which the Eozoon is found. Sir W. Logan states that their "united thickness "may possibly far surpass that of all the succeeding "rocks, from the base of the palæozoic series to the

" present time. We are thus carried back to a period
" so remote, that the appearance of the so-called
" Primordial fauna (of Barrande) may by some be
" considered as a comparatively modern event." The
Eozoon belongs to the most lowly organised of all
classes of animals, but is highly organised for its class ;
it existed in countless numbers, and, as Dr. Dawson has
remarked, certainly preyed on other minute organic
beings, which must have lived in great numbers.   Thus
the words, which I wrote in 1859, about the existence
of living beings long before the Cambrian period, and
which are almost the same with those since used by Sir
W. Logan, have proved true.   Nevertheless, the difficulty
of assigning any good reason for the absence of vast
piles of strata rich in fossils beneath the Cambrian
system is very great.   It does not seem probable that
the most ancient beds have been quite worn away by
denudation, or that their fossils have been wholly
obliterated by metamorphic action, for if this had been
the case we should have found only small remnants of
the formations next succeeding them in age, and these
would always have existed in a partially metamorphosed
condition.   But the descriptions which we possess of
the Silurian deposits over immense territories in Russia
and in North America, do not support the view, that
the older a formation is, the more invariably it has
suffered extreme denudation and metamorphism.

The case at present must remain inexplicable ; and
may be truly urged as a valid argument against the
views here entertained.   To show that it may hereafter
receive some explanation, I will give the following
hypothesis.   From the nature of the organic remains
which do not appear to have inhabited profound depths,

2 G

in the several formations of Europe and of the United States; and from the amount of sediment, miles in thickness, of which the formations are composed, we may infer that from first to last large islands or tracts of land, whence the sediment was derived, occurred in the neighbourhood of the now existing continents of Europe and North America. This same view has since been maintained by Agassiz and others. But we do not know what was the state of things in the intervals between the several successive formations; whether Europe and the United States during these intervals existed as dry land, or as a submarine surface near land, on which sediment was not deposited, or as the bed of an open and unfathomable sea.

Looking to the existing oceans, which are thrice as extensive as the land, we see them studded with many islands; but hardly one truly oceanic island (with the exception of New Zealand, if this can be called a truly oceanic island) is as yet known to afford even a remnant of any palæozoic or secondary formation. Hence we may perhaps infer, that during the palæozoic and secondary periods, neither continents nor continental islands existed where our oceans now extend; for had they existed, palæozoic and secondary formations would in all probability have been accumulated from sediment derived from their wear and tear; and these would have been at least partially upheaved by the oscillations of level, which must have intervened during these enormously long periods. If then we may infer anything from these facts, we may infer that, where our oceans now extend, oceans have extended from the remotest period of which we have any record; and on the other hand, that where continents now exist, large tracts of

land have existed, subjected no doubt to great oscillations of level, since the Cambrian period. The coloured
map appended to my volume on Coral Reefs, led me to
conclude that the great oceans are still mainly areas of
subsidence, the great archipelagoes still areas of oscillations of level, and the continents areas of elevation.
But we have no reason to assume that things have thus
remained from the beginning of the world. Our continents seem to have been formed by a preponderance, during
many oscillations of level, of the force of elevation ; but
may not the areas of preponderant movement have
changed in the lapse of ages ? At a period long antecedent
to the Cambrian epoch, continents may have existed where
oceans are now spread out; and clear and open oceans
may have existed where our continents now stand. Nor
should we be justified in assuming that if, for instance,
the bed of the Pacific Ocean were now converted into a
continent we should there find sedimentary formations in
a recognisable condition older than the Cambrian strata,
supposing such to have been formerly deposited; for it
might well happen that strata which had subsided some
miles nearer to the centre of the earth, and which had
been pressed on by an enormous weight of superincum-
bent water, might have undergone far more metamorphic
action than strata which have always remained nearer
to the surface. The immense areas in some parts of the
world, for instance in South America, of naked metamorphic rocks, which must have been heated under
great pressure, have always seemed to me to require
some special explanation; and we may perhaps believe
that we see in these large areas, the many formations
long anterior to the Cambrian epoch in a completely
metamorphosed and denuded condition.

The several difficulties here discussed, namely—that, though we find in our geological formations many links between the species which now exist and which formerly existed, we do not find infinitely numerous fine transitional forms closely joining them all together;—the sudden manner in which several groups of species first appear in our European formations;—the almost entire absence, as at present known, of formations rich in fossils beneath the Cambrian strata,—are all undoubtedly of the most serious nature. We see this in the fact that the most eminent palæontologists, namely, Cuvier, Agassiz Barrande, Pictet, Falconer, E. Forbes, &c., and all our greatest geologists, as Lyell, Murchison, Sedgwick, &c., have unanimously, often vehemently, maintained the immutability of species. But Sir Charles Lyell now gives the support of his high authority to the opposite side; and most geologists and palæontologists are much shaken in their former belief. Those who believe that the geological record is in any degree perfect, will undoubtedly at once reject the theory. For my part, following out Lyell's metaphor, I look at the geological record as a history of the world imperfectly kept, and written in a changing dialect; of this history we possess the last volume alone, relating only to two or three countries. Of this volume, only here and there a short chapter has been preserved; and of each page, only here and there a few lines. Each word of the slowly-changing language, more or less different in the successive chapters, may represent the forms of life, which are entombed in our consecutive formations, and which falsely appear to have been abruptly introduced. On this view, the difficulties above discussed are greatly diminished, or even disappear.

CHAPTER XI.

On the Geological Succession of Organic Beings.

On the slow and successive appearance of new species—On their different rates of change—Species once lost do not reappear—Groups of species follow the same general rules in their appearance and disappearance as do single species—On extinction—On simultaneous changes in the forms of life throughout the world—On the affinities of extinct species to each other and to living species—On the state of development of ancient forms—On the succession of the same types within the same areas—Summary of preceding and present chapter.

LET us now see whether the several facts and laws relating to the geological succession of organic beings accord best with the common view of the immutability of species, or with that of their slow and gradual modification, through variation and natural selection.

New species have appeared very slowly, one after another, both on the land and in the waters. Lyell has shown that it is hardly possible to resist the evidence on this head in the case of the several tertiary stages; and every year tends to fill up the blanks between the stages, and to make the proportion between the lost and existing forms more gradual. In some of the most recent beds, though undoubtedly of high antiquity if measured by years, only one or two species are extinct, and only one or two are new, having appeared there for the first time, either locally, or, as far as we know, on the face of the earth. The secondary formations are more broken;

but, as Bronn has remarked, neither the appearance nor disappearance of the many species embedded in each formation has been simultaneous.

Species belonging to different genera and classes have not changed at the same rate, or in the same degree. In the older tertiary beds a few living shells may still be found in the midst of a multitude of extinct forms. Falconer has given a striking instance of a similar fact, for an existing crocodile is associated with many lost mammals and reptiles in the sub-Himalayan deposits. The Silurian Lingula differs but little from the living species of this genus ; whereas most of the other Silurian Molluscs and all the Crustaceans have changed greatly. The productions of the land seem to have changed at a quicker rate than those of the sea, of which a striking instance has been observed in Switzerland.   There is some reason to believe that organisms high in the scale, change more quickly than those that are low : though there are exceptions to this rule.   The amount of organic change, as Pictet has remarked, is not the same in each successive so-called formation.   Yet if we compare any but the most closely related formations, all the species will be found to have undergone some change.   When a species has once disappeared from the face of the earth, we have no reason to believe that the same identical form ever reappears.   The strongest apparent exception to this latter rule is that of the so-called " colonies " of M. Barrande, which intrude for a period in the midst of an older formation, and then allow the pre-existing fauna to reappear ; but Lyell's explanation, namely, that it is a case of temporary migration from a distinct geographical province, seems satisfactory.

These several facts accord well with our theory, which

includes no fixed law of development, causing all the
inhabitants of an area to change abruptly, or simultaneously, or to an equal degree. The process of modification must be slow, and will generally affect only a
few species at the same time ; for the variability of each
species is independent of that of all others. Whether
such variations or individual differences as may arise will
be accumulated through natural selection in a greater
or less degree, thus causing a greater or less amount of
permanent modification, will depend on many complex
contingencies—on the variations being of a beneficial
nature, on the freedom of intercrossing, on the slowly
changing physical conditions of the country, on the
immigration of new colonists, and on the nature of the
other inhabitants with which the varying species come
into competition. Hence it is by no means surprising
that one species should retain the same identical form
much longer than others ; or, if changing, should change
in a less degree. We find similar relations between the
existing inhabitants of distinct countries ; for instance,
the land-shells and coleopterous insects of Madeira have
come to differ considerably from their nearest allies on
the continent of Europe, whereas the marine shells and
birds have remained unaltered. We can perhaps understand the apparently quicker rate of change in terrestrial
and in more highly organised productions compared
with marine and lower productions, by the more complex
relations of the higher beings to their organic and inorganic conditions of life, as explained in a former
chapter. When many of the inhabitants of any area
have become modified and improved, we can understand,
on the principle of competition, and from the all-important relations of organism to organism in the struggle for

life, that any form which did not become in some degree modified and improved, would be liable to extermination. Hence we see why all the species in the same region do at last, if we look to long enough intervals of time, become modified, for otherwise they would become extinct.

In members of the same class the average amount of change, during long and equal periods of time, may, perhaps, be nearly the same; but as the accumulation of enduring formations, rich in fossils, depends on great masses of sediment being deposited on subsiding areas, our formations have been almost necessarily accumulated at wide and irregularly intermittent intervals of time; consequently the amount of organic change exhibited by the fossils embedded in consecutive formations is not equal. Each formation, on this view, does not mark a new and complete act of creation, but only an occasional scene, taken almost at hazard, in an ever slowly changing drama.

We can clearly understand why a species when once lost should never reappear, even if the very same conditions of life, organic and inorganic, should recur. For though the offspring of one species might be adapted (and no doubt this has occurred in innumerable instances) to fill the place of another species in the economy of nature, and thus supplant it; yet the two forms—the old and the new—would not be identically the same; for both would almost certainly inherit different characters from their distinct progenitors; and organisms already differing would vary in a different manner. For instance, it is possible, if all our fantail pigeons were destroyed, that fanciers might make a new breed hardly distinguishable from the present breed; but if the parent rock-pigeon were likewise destroyed, and under nature we have every reason to believe that parent-forms are generally supplanted and exterminated by their improved offspring, it is incredible that a fantail, identical with the existing breed, could be raised from any other species of pigeon, or even from any other well-established race of the domestic pigeon, for the successive variations would almost certainly be in some degree different, and the newly-formed variety would probably inherit from its progenitor some characteristic differences.

Groups of species, that is, genera and families, follow the same general rules in their appearance and disappearance as do single species, changing more or less quickly, and in a greater or lesser degree. A group, when it has once disappeared, never reappears; that is, its existence, as long as it lasts, is continuous. I am aware that there are some apparent exceptions to this rule, but the exceptions are surprisingly few, so few that E. Forbes, Pictet, and Woodward (though all strongly opposed to such views as I maintain) admit its truth; and the rule strictly accords with the theory. For all the species of the same group, however long it may have lasted, are the modified descendants one from the other, and all from a common progenitor. In the genus Lingula, for instance, the species which have successively appeared at all ages must have been connected by an unbroken series of generations, from the lowest Silurian stratum to the present day.

We have seen in the last chapter that whole groups of species sometimes falsely appear to have been abruptly developed; and I have attempted to give an explanation of this fact, which if true would be fatal to my views. But such cases are certainly exceptional; the general rule being a gradual increase in number, until the group reaches its maximum, and then, sooner or later, a gradual decrease. If the number of the species included within a genus, or the number of the genera within a family, be represented by a vertical line of varying thickness, ascending through the successive geological formations, in which the species are found, the line will sometimes falsely appear to begin at its lower end, not in a sharp point, but abruptly; it then gradually thickens upwards, often keeping of equal thickness for a space, and ultimately thins out in the upper beds, marking the decrease and final extinction of the species. This gradual increase in number of the species of a group is strictly conformable with the theory, for the species of the same genus, and the genera of the same family, can increase only slowly and progressively; the process of modification and the production of a number of allied forms necessarily being a slow and gradual process,—one species first giving rise to two or three varieties, these being slowly converted into species, which in their turn produce by equally slow steps other varieties and species, and so on, like the branching of a great tree from a single stem, till the group becomes large.

On Extinction.

We have as yet only spoken incidentally of the disappearance of species and of groups of species. On the theory of natural selection, the extinction of old forms and the production of new and improved forms are intimately connected together. The old notion of all the inhabitants of the earth having been swept away by catastrophes at successive periods is very generally given up, even by those geologists, as Elie de Beaumont, Murchison, Barrande, &c., whose general views would naturally lead them to this conclusion. On the contrary, we have every reason to believe, from the study of the tertiary formations, that species and groups of species gradually disappear, one after another, first from one spot, then from another, and finally from the world. In some few cases however, as by the breaking of an isthmus and the consequent irruption of a multitude of new inhabitants into an adjoining sea, or by the final subsidence of an island, the process of extinction may have been rapid. Both single species and whole groups of species last for very unequal periods; some groups, as we have seen, have endured from the earliest known dawn of life to the present day; some have disappeared before the close of the palæozoic period. No fixed law seems to determine the length of time during which any single species or any single genus endures. There is reason to believe that the extinction of a whole group of species is generally a slower process than their production: if their appearance and disappearance be represented, as before, by a vertical line of varying thickness the line is found to taper more gradually at its upper end, which marks the progress of extermination, than at its lower end, which marks the first appearance and the early increase in number of the species. In some cases, however, the extermination of whole groups, as of ammonites, towards the close of the secondary period, has been wonderfully sudden.

The extinction of species has been involved in the most gratuitous mystery. Some authors have even supposed that, as the individual has a definite length of life, so have species a definite duration. No one can have marvelled more than I have done at the extinction of species. When I found in La Plata the tooth of a horse embedded with the remains of Mastodon, Megatherium, Toxodon, and other extinct monsters, which all co-existed with still living shells at a very late geological period, I was filled with astonishment; for, seeing that the horse, since its introduction by the Spaniards into South America, has run wild over the whole country and has increased in numbers at an unparalleled rate, I asked myself what could so recently have exterminated the former horse under conditions of life apparently so favourable. But my astonishment was groundless. Professor Owen soon perceived that the tooth, though so like that of the existing horse, belonged to an extinct species. Had this horse been still living, but in some degree rare, no naturalist would have felt the least surprise at its rarity; for rarity is the attribute of a vast number of species of all classes, in all countries. If we ask ourselves why this or that species is rare, we answer that something is unfavourable in its conditions of life; but what that something is we can hardly ever tell. On the supposition of the fossil horse still existing as a rare species, we might have felt certain, from the analogy of all other mammals, even of the slow-breeding elephant, and from the history of the naturalisation of the domestic horse in South America, that under more favourable conditions it would in a very few years have stocked the whole continent. But we could not have told what the unfavourable conditions were which checked its increase, whether some one or several contingencies, and at what period of the horse's life, and in what degree they severally acted. If the conditions had gone on, however slowly, becoming less and less favourable, we assuredly should not have perceived the fact, yet the fossil horse would certainly have become rarer and rarer, and finally extinct;—its place being seized on by some more successful competitor.

It is most difficult always to remember that the increase of every creature is constantly being checked by unperceived hostile agencies; and that these same unperceived agencies are amply sufficient to cause rarity, and finally extinction. So little is this subject understood, that I have heard surprise repeatedly expressed at such great monsters as the Mastodon and the more ancient Dinosaurians having become extinct; as if mere bodily strength gave victory in the battle of life. Mere size, on the contrary, would in some cases determine, as has been remarked by Owen, quicker extermination from the greater amount of requisite food. Before man inhabited India or Africa, some cause must have checked the continued increase of the existing elephant. A highly capable judge, Dr. Falconer, believes that it is chiefly insects which, from incessantly harassing and weakening the elephant in India, check its increase; and this was Bruce's conclusion with respect to the African elephant in Abyssinia. It is certain that insects and blood-sucking bats determine the existence of the larger naturalized quadrupeds in several parts of S. America.

We see in many cases in the more recent tertiary formations, that rarity precedes extinction; and we know that this has been the progress of events with those animals which have been exterminated, either locally or wholly, through man's agency. I may repeat what I published in 1845, namely, that to admit that species generally become rare before they become extinct—to feel no surprise at the rarity of a species, and yet to marvel greatly when the species ceases to exist, is much the same as to admit that sickness in the individual is the forerunner of death—to feel no surprise at sickness, but, when the sick man dies, to wonder and to suspect that he died by some deed of violence.

The theory of natural selection is grounded on the belief that each new variety and ultimately each new species, is produced and maintained by having some advantage over those with which it comes into competition; and the consequent extinction of the less-favoured forms almost inevitably follows. It is the same with our domestic productions; when a new and slightly improved variety has been raised, it at first supplants the less improved varieties in the same neighbourhood; when much improved it is transported far and near, like our short-horn cattle, and takes the place of other breeds in other countries. Thus the appearance of new forms and the disappearance of old forms, both those naturally and those artificially produced, are bound together. In flourishing groups, the number of new specific forms which have been produced within a given time has at some periods probably been greater than the number of the old specific forms which have been exterminated; but we know that species have not gone on indefinitely increasing, at least during the later geological epochs, so that, looking to later times, we may believe that the production of new forms has caused the extinction of about the same number of old forms.

The competition will generally be most severe, as formerly explained and illustrated by examples, between the forms which are most like each other in all respects. Hence the improved and modified descendants of a species will generally cause the extermination of the parent-species; and if many new forms have been developed from any one species, the nearest allies of that species,

i.e. the species of the same genus, will be the most liable
to extermination. Thus, as I believe, a number of new
species descended from one species, that is a new genus,
comes to supplant an old genus, belonging to the same
family. But it must often have happened that a new
species belonging to some one group has seized on the
place occupied by a species belonging to a distinct group,
and thus have caused its extermination. If many allied
forms be developed from the successful intruder, many
will have to yield their places; and it will generally be
the allied forms, which will suffer from some inherited
inferiority in common. But whether it be species
belonging to the same or to a distinct class, which have
yielded their places to other modified and improved
species, a few of the sufferers may often be preserved
for a long time, from being fitted to some peculiar line
of life, or from inhabiting some distant and isolated
station, where they will have escaped severe competition. For instance, some species of Trigonia, a great
genus of shells in the secondary formations, survive in
the Australian seas; and a few members of the great and
almost extinct group of Ganoid fishes still inhabit our
fresh waters. Therefore the utter extinction of a group
is generally, as we have seen, a slower process than its
production.

With respect to the apparently sudden extermination
of whole families or orders, as of Trilobites at the close
of the palæozoic period and of Ammonites at the close of
the secondary period, we must remember what has been
already said on the probable wide intervals of time
between our consecutive formations; and in these
intervals there may have been much slow extermination. Moreover, when, by sudden immigration or by

unusually rapid development, many species of a new group have taken possession of an area, many of the older species will have been exterminated in a correspondingly rapid manner; and the forms which thus yield their places will commonly be allied, for they will partake of the same inferiority in common.

Thus, as it seems to me, the manner in which single species and whole groups of species become extinct accords well with the theory of natural selection. We need not marvel at extinction; if we must marvel, let it be at our own presumption in imagining for a moment that we understand the many complex contingencies on which the existence of each species depends. If we forget for an instant that each species tends to increase inordinately, and that some check is always in action, yet seldom perceived by us, the whole economy of nature will be utterly obscured. Whenever we can precisely say why this species is more abundant in individuals than that; why this species and not another can be naturalised in a given country; then, and not until then, we may justly feel surprise why we cannot account for the extinction of any particular species or group of species.

On the Forms of Life changing almost simultaneously throughout the World.

Scarcely any palæontological discovery is more striking than the fact that the forms of life change almost simultaneously throughout the world. Thus our European Chalk formation can be recognised in many distant regions, under the most different climates, where not a fragment of the mineral chalk itself can be found; namely in North America, in equatorial South America,

in Tierra del Fuego, at the Cape of Good Hope, and in
the peninsula of India. For at these distant points,
the organic remains in certain beds present an unmistakeable resemblance to those of the Chalk. It is not
that the same species are met with; for in some cases
not one species is identically the same, but they belong to the same families, genera, and sections of
genera, and sometimes are similarly characterised in
such trifling points as mere superficial sculpture. Moreover, other forms, which are not found in the Chalk of
Europe, but which occur in the formations either above
or below, occur in the same order at these distant
points of the world. In the several successive palæozoic formations of Russia, Western Europe, and North
America, a similar parallelism in the forms of life has
been observed by several authors; so it is, according
to Lyell, with the European and North American
tertiary deposits. Even if the few fossil species which
are common to the Old and New Worlds were kept
wholly out of view, the general parallelism in the
successive forms of life, in the palæozoic and tertiary
stages, would still be manifest, and the several formations could be easily correlated.

These observations, however, relate to the marine
inhabitants of the world: we have not sufficient data
to judge whether the productions of the land and of
fresh water at distant points change in the same parallel
manner. We may doubt whether they have thus
changed: if the Megatherium, Mylodon, Macrauchenia,
and Toxodon had been brought to Europe from La
Plata, without any information in regard to their geological position, no one would have suspected that they
had co-existed with sea-shells all still living; but as

2 H

these anomalous monsters co-existed with the Mastodon and Horse, it might at least have been inferred that they had lived during one of the later tertiary stages.

When the marine forms of life are spoken of as having changed simultaneously throughout the world, it must not be supposed that this expression relates to the same year, or to the same century, or even that it has a very strict geological sense; for if all the marine animals now living in Europe, and all those that lived in Europe during the pleistocene period (a very remote period as measured by years, including the whole glacial epoch) were compared with those now existing in South America or in Australia, the most skilful naturalist would hardly be able to say whether the present or the pleistocene inhabitants of Europe resembled most closely those of the southern hemisphere. So, again, several highly competent observers maintain that the existing productions of the United States are more closely related to those which lived in Europe during certain late tertiary stages, than to the present inhabitants of Europe; and if this be so, it is evident that fossiliferous beds now deposited on the shores of North America would hereafter be liable to be classed with somewhat older European beds. Nevertheless, looking to a remotely future epoch, there can be little doubt that all the more modern *marine* formations, namely, the upper pliocene, the pleistocene and strictly modern beds of Europe, North and South America, and Australia, from containing fossil remains in some degree allied, and from not including those forms which are found only in the older underlying deposits, would be correctly ranked as simultaneous in a geological sense.

The fact of the forms of life changing simultaneously,

in the above large sense, at distant parts of the world, has greatly struck those admirable observers, MM. de Verneuil and d'Archiac.   After referring to the parallelism of the palæozoic forms of life in various parts of Europe, they add, " If, struck by this strange sequence, we turn our attention to North America, and there discover a series of analogous phenomena, it will appear certain that all these modifications of species, their extinction, and the introduction of new ones, cannot be owing to mere changes in marine currents or other causes more or less local and temporary, but depend on general laws which govern the whole animal kingdom." M. Barrande has made forcible remarks to precisely the same effect.   It is, indeed, quite futile to look to changes of currents, climate, or other physical conditions, as the cause of these great mutations in the forms of life throughout the world, under the most different climates.   We must, as Barrande has remarked, look to some special law.   We shall see this more clearly when we treat of the present distribution of organic beings, and find how slight is the relation between the physical conditions of various countries and the nature of their inhabitants.

This great fact of the parallel succession of the forms of life throughout the world, is explicable on the theory of natural selection.   New species are formed by having some advantage over older forms; and the forms, which are already dominant, or have some advantage over the other forms in their own country, give birth to the greatest number of new varieties or incipient species. We have distinct evidence on this head, in the plants which are dominant, that is, which are commonest and most widely diffused, producing the greatest number of new varieties. It is also natural that the dominant, varying, and far-spreading species, which have already invaded to a certain extent the territories of other species, should be those which would have the best chance of spreading still further, and of giving rise in new countries to other new varieties and species. The process of diffusion would often be very slow, depending on climatal and geographical changes, on strange accidents, and on the gradual acclimatisation of new species to the various climates through which they might have to pass, but in the course of time the dominant forms would generally succeed in spreading and would ultimately prevail. The diffusion would, it is probable, be slower with the terrestrial inhabitants of the distinct continents than with the marine inhabitants of the continuous sea. We might therefore expect to find, as we do find, a less strict degree of parallelism in the succession of the productions of the land than with those of the sea.

Thus, as it seems to me, the parallel, and, taken in a large sense, simultaneous, succession of the same forms of life throughout the world, accords well with the principle of new species having been formed by dominant species spreading widely and varying; the new species thus produced being themselves dominant, owing to their having had some advantage over their already dominant parents, as well as over other species, and again spreading, varying, and producing new forms. The old forms which are beaten and which yield their places to the new and victorious forms, will generally be allied in groups, from inheriting some inferiority in common; and therefore, as new and improved groups spread throughout the world, old groups disappear from the world; and the succession of forms everywhere tends to correspond both in their first appearance and final disappearance.

There is one other remark connected with this subject worth making. I have given my reasons for believing that most of our great formations, rich in fossils, were deposited during periods of subsidence; and that blank intervals of vast duration, as far as fossils are concerned, occurred during the periods when the bed of the sea was either stationary or rising, and likewise when sediment was not thrown down quickly enough to embed and preserve organic remains. During these long and blank intervals I suppose that the inhabitants of each region underwent a considerable amount of modification and extinction, and that there was much migration from other parts of the world. As we have reason to believe that large areas are affected by the same movement, it is probable that strictly contemporaneous formations have often been accumulated over very wide spaces in the same quarter of the world; but we are very far from having any right to conclude that this has invariably been the case, and that large areas have invariably been affected by the same movements. When two formations have been deposited in two regions during nearly, but not exactly, the same period, we should find in both, from the causes explained in the foregoing paragraphs, the same general succession in the forms of life; but the species would not exactly correspond; for there will have been a little more time in the one region than in the other for modification, extinction, and immigration.

I suspect that cases of this nature occur in Europe. Mr. Prestwich, in his admirable Memoirs on the eocene deposits of England and France, is able to draw a close

general parallelism between the successive stages in
the two countries ; but when he compares certain stages
in England with those in France, although he finds in
both a curious accordance in the numbers of the species
belonging to the same genera, yet the species themselves differ in a manner very difficult to account for
considering the proximity of the two areas,—unless,
indeed, it be assumed that an isthmus separated two
seas inhabited by distinct, but contemporaneous, faunas.
Lyell has made similar observations on some of the
later tertiary formations.   Barrande, also, shows that
there is a striking general parallelism in the successive
Silurian deposits of Bohemia and Scandinavia ; nevertheless he finds a surprising amount of difference in
the species.   If the several formations in these regions
have not been deposited during the same exact periods,
—a formation in one region often corresponding with a
blank interval in the other,—and if in both regions the
species have gone on slowly changing during the accumulation of the several formations and during the long
intervals of time between them ; in this case the several
formations in the two regions could be arranged in the
same order, in accordance with the general succession
of the forms of life, and the order would falsely appear
to be strictly parallel ; nevertheless the species would
not be all the same in the apparently corresponding
stages in the two regions

*On the Affinities of Extinct Species to each other, and
to Living Forms.*

Let us now look to the mutual affinities of extinct
and living species.   All fall into a few grand classes ;
and this fact is at once explained on the principle of

descent. The more ancient any form is, the more, as a general rule, it differs from living forms. But, as Buckland long ago remarked, extinct species can all be classed either in still existing groups, or between them. That the extinct forms of life help to fill up the intervals between existing genera, families, and orders, is certainly true; but as this statement has often been ignored or even denied, it may be well to make some remarks on this subject, and to give some instances. If we confine our attention either to the living or to the extinct species of the same class, the series is far less perfect than if we combine both into one general system. In the writings of Professor Owen we continually meet with the expression of generalised forms, as applied to extinct animals; and in the writings of Agassiz, of prophetic or synthetic types; and these terms imply that such forms are in fact intermediate or connecting links. Another distinguished palæontologist, M. Gaudry, has shown in the most striking manner that many of the fossil mammals discovered by him in Attica serve to break down the intervals between existing genera. Cuvier ranked the Ruminants and Pachyderms, as two of the most distinct orders of mammals: but so many fossil links have been disentombed that Owen has had to alter the whole classification, and has placed certain pachyderms in the same sub-order with ruminants; for example, he dissolves by gradations the apparently wide interval between the pig and the camel. The Ungulata or hoofed quadrupeds are now divided into the even-toed or odd-toed divisions; but the Macrauchenia of S. America connects to a certain extent these two grand divisions. No one will deny that the Hipparion is intermediate between the existing horse and certain older ungulate forms. What a wonderful connecting link in the chain of mammals is the Typotherium from S. America, as the name given to it by Professor Gervais expresses, and which cannot be placed in any existing order. The Sirenia form a very distinct group of mammals, and one of the most remarkable peculiarities in the existing dugong and lamentin is the entire absence of hind limbs without even a rudiment being left; but the extinct Halitherium had, according to Professor Flower, an ossified thigh-bone "articulated to a well-defined acetabulum in the pelvis," and it thus makes some approach to ordinary hoofed quadrupeds, to which the Sirenia are in other respects allied. The cetaceans or whales are widely different from all other mammals, but the tertiary Zeuglodon and Squalodon, which have been placed by some naturalists in an order by themselves, are considered by Professor Huxley to be undoubtedly cetaceans, "and to constitute connecting links with the aquatic carnivora."

Even the wide interval between birds and reptiles has been shown by the naturalist just quoted to be partially bridged over in the most unexpected manner, on the one hand, by the ostrich and extinct Archeopteryx, and on the other hand, by the Compsognathus, one of the Dinosaurians—that group which includes the most gigantic of all terrestrial reptiles. Turning to the Invertebrata, Barrande asserts, a higher authority could not be named, that he is every day taught that, although palæozoic animals can certainly be classed under existing groups, yet that at this ancient period the groups were not so distinctly separated from each other as they now are.

Some writers have objected to any extinct species,

or group of species, being considered as intermediate
between any two living species, or groups of species.
If by this term it is meant that an extinct form is
directly intermediate in all its characters between two
living forms or groups, the objection is probably valid.
But in a natural classification many fossil species
certainly stand between living species, and some extinct
genera between living genera, even between genera
belonging to distinct families. The most common case,
especially with respect to very distinct groups, such as
fish and reptiles, seems to be, that, supposing them to
be distinguished at the present day by a score of characters, the ancient members are separated by a somewhat lesser number of characters ; so that the two groups
formerly made a somewhat nearer approach to each
other than they now do.

It is a common belief that the more ancient a form
is, by so much the more it tends to connect by some of
its characters groups now widely separated from each
other. This remark no doubt must be restricted to
those groups which have undergone much change in
the course of geological ages ; and it would be difficult
to prove the truth of the proposition, for every now and
then even a living animal, as the Lepidosiren, is discovered having affinities directed towards very distinct
groups. Yet if we compare the older Reptiles and
Batrachians, the older Fish, the older Cephalopods, and
the eocene Mammals, with the more recent members of
the same classes, we must admit that there is truth in
the remark.

Let us see how far these several facts and inferences
accord with the theory of descent with modification.
As the subject is somewhat complex, I must request

the reader to turn to the diagram in the fourth chapter. We may suppose that the numbered letters in italics represent genera, and the dotted lines diverging from them the species in each genus. The diagram is much too simple, too few genera and too few species being given, but this is unimportant for us. The horizontal lines may represent successive geological formations, and all the forms beneath the uppermost line may be considered as extinct. The three existing genera a^{14}, q^{14}, p^{14}, will form a small family; b^{14} and f^{14} a closely allied family or sub-family; and o^{14}, e^{14}, m^{14}, a third family. These three families, together with the many extinct genera on the several lines of descent diverging from the parent-form (A) will form an order, for all will have inherited something in common from their ancient progenitor. On the principle of the continued tendency to divergence of character, which was formerly illustrated by this diagram, the more recent any form is, the more it will generally differ from its ancient progenitor. Hence we can understand the rule that the most ancient fossils differ most from existing forms. We must not, however, assume that divergence of character is a necessary contingency; it depends solely on the descendants from a species being thus enabled to seize on many and different places in the economy of nature. Therefore it is quite possible, as we have seen in the case of some Silurian forms, that a species might go on being slightly modified in relation to its slightly altered conditions of life, and yet retain throughout a vast period the same general characteristics. This is represented in the diagram by the letter F^{14}.

All the many forms, extinct and recent, descended from (A), make, as before remarked, one order; and this order, from the continued effects of extinction and divergence of character, has become divided into several sub-families and families, some of which are supposed to have perished at different periods, and some to have endured to the present day.

By looking at the diagram we can see that if many of the extinct forms supposed to be imbedded in the successive formations, were discovered at several points low down in the series, the three existing families on the uppermost line would be rendered less distinct from each other. If, for instance, the genera a^1, a^5, a^{10}, f^8, m^3, m^6, m^9, were disinterred, these three families would be so closely linked together that they probably would have to be united into one great family, in nearly the same manner as has occurred with ruminants and certain pachyderms. Yet he who objected to consider as intermediate the extinct genera, which thus link together the living genera of three families, would be partly justified, for they are intermediate, not directly, but only by a long and circuitous course through many widely different forms. If many extinct forms were to be discovered above one of the middle horizontal lines or geological formations—for instance, above No. VI.—but none from beneath this line, then only two of the families (those on the left hand, a^{14}, &c., and b^{14}, &c.) would have to be united into one; and there would remain two families, which would be less distinct from each other than they were before the discovery of the fossils. So again if the three families formed of eight genera (a^{14} to m^{14}), on the uppermost line, be supposed to differ from each other by half-a-dozen important characters, then the families which existed at the period marked VI. would certainly have differed from each other by a less number of characters; for they would at this early stage of descent have diverged in a less degree from their common progenitor. Thus it comes that ancient and extinct genera are often in a greater or less degree intermediate in character between their modified descendants, or between their collateral relations.

Under nature the process will be far more complicated than is represented in the diagram; for the groups will have been more numerous; they will have endured for extremely unequal lengths of time, and will have been modified in various degrees. As we possess only the last volume of the geological record, and that in a very broken condition, we have no right to expect, except in rare cases, to fill up the wide intervals in the natural system, and thus to unite distinct families or orders. All that we have a right to expect is, that those groups which have, within known geological periods, undergone much modification, should in the older formations make some slight approach to each other; so that the older members should differ less from each other in some of their characters than do the existing members of the same groups; and this by the concurrent evidence of our best palæontologists is frequently the case.

Thus, on the theory of descent with modification, the main facts with respect to the mutual affinities of the extinct forms of life to each other and to living forms, are explained in a satisfactory manner. And they are wholly inexplicable on any other view.

On this same theory, it is evident that the fauna during any one great period in the earth's history will be intermediate in general character between that which preceded and that which succeeded it. Thus the species which lived at the sixth great stage of descent in the diagram are the modified offspring of those which lived at the fifth stage, and are the parents of those which became still more modified at the seventh stage; hence they could hardly fail to be nearly intermediate in character between the forms of life above and below. We must, however, allow for the entire extinction of some preceding forms, and in any one region for the immigration of new forms from other regions, and for a large amount of modification during the long and blank intervals between the successive formations. Subject to these allowances, the fauna of each geological period undoubtedly is intermediate in character, between the preceding and succeeding faunas. I need give only one instance, namely, the manner in which the fossils of the Devonian system, when this system was first discovered, were at once recognised by palæontologists as intermediate in character between those of the overlying carboniferous, and underlying Silurian systems. But each fauna is not necessarily exactly intermediate, as unequal intervals of time have elapsed between consecutive formations.

It is no real objection to the truth of the statement that the fauna of each period as a whole is nearly intermediate in character between the preceding and succeeding faunas, that certain genera offer exceptions to the rule. For instance, the species of mastodons and elephants, when arranged by Dr. Falconer in two series,—in the first place according to their mutual affinities, and in the second place according to their periods of existence, —do not accord in arrangement. The species extreme in character are not the oldest or the most recent; nor are those which are intermediate in character, inter-

mediate in age.   But supposing for an instant, in this
and other such cases, that the record of the first appearance and disappearance of the species was complete,
which is far from the case, we have no reason to believe
that forms successively produced necessarily endure for
corresponding lengths of time.   A very ancient form
may occasionally have lasted much longer than a form
elsewhere subsequently produced, especially in the case
of terrestrial productions inhabiting separated districts.
To compare small things with great; if the principal
living and extinct races of the domestic pigeon were
arranged in serial affinity, this arrangement would not
closely accord with the order in time of their production, and even less with the order of their disappearance ;
for the parent rock-pigeon still lives ; and many varieties
between the rock-pigeon and the carrier have become
extinct; and carriers which are extreme in the important character of length of beak originated earlier than
short-beaked tumblers, which are at the opposite end of
the series in this respect.

Closely connected with the statement, that the organic
remains from an intermediate formation are in some
degree intermediate in character, is the fact, insisted on
by all palæontologists, that fossils from two consecutive
formations are far more closely related to each other,
than are the fossils from two remote formations.   Pictet
gives as a well-known instance, the general resemblance of the organic remains from the several stages
of the Chalk formation, though the species are distinct in each stage.   This fact alone, from its generality,
seems to have shaken Professor Pictet in his belief
in the immutability of species.   He who is acquainted
with the distribution of existing species over the globe,

will not attempt to account for the close resemblance of distinct species in closely consecutive formations, by the physical conditions of the ancient areas having remained nearly the same. Let it be remembered that the forms of life, at least those inhabiting the sea, have changed almost simultaneously throughout the world, and therefore under the most different climates and conditions. Consider the prodigious vicissitudes of climate during the pleistocene period, which includes the whole glacial epoch, and note how little the specific forms of the inhabitants of the sea have been affected.

On the theory of descent, the full meaning of the fossil remains from closely consecutive formations being closely related, though ranked as distinct species, is obvious. As the accumulation of each formation has often been interrupted, and as long blank intervals have intervened between successive formations, we ought not to expect to find, as I attempted to show in the last chapter, in any one or in any two formations, all the intermediate varieties between the species which appeared at the commencement and close of these periods : but we ought to find after intervals, very long as measured by years, but only moderately long as measured geologically, closely allied forms, or, as they have been called by some authors, representative species; and these assuredly we do find. We find, in short, such evidence of the slow and scarcely sensible mutations of specific forms, as we have the right to expect.

On the State of Development of Ancient compared with Living Forms.

We have seen in the fourth chapter that the degree of differentiation and specialisation of the parts in organic beings, when arrived at maturity, is the best standard, as yet suggested, of their degree of perfection or highness. We have also seen that, as the specialisation of parts is an advantage to each being, so natural selection will tend to render the organisation of each being more specialised and perfect, and in this sense higher; not but that it may leave many creatures with simple and unimproved structures fitted for simple conditions of life, and in some cases will even degrade or simplify the organisation, yet leaving such degraded beings better fitted for their new walks of life. In another and more general manner, new species become superior to their predecessors; for they have to beat in the struggle for life all the older forms, with which they come into close competition. We may therefore conclude that if under a nearly similar climate the eocene inhabitants of the world could be put into competition with the existing inhabitants, the former would be beaten and exterminated by the latter, as would the secondary by the eocene, and the palæozoic by the secondary forms. So that by this fundamental test of victory in the battle for life, as well as by the standard of the specialisation of organs, modern forms ought, on the theory of natural selection, to stand higher than ancient forms. Is this the case? A large majority of palæontologists would answer in the affirmative; and it seems that this answer must be admitted as true, though difficult of proof.

It is no valid objection to this conclusion, that certain
Brachiopods have been but slightly modified from an
extremely remote geological epoch; and that certain
land and fresh-water shells have remained nearly the
same, from the time when, as far as is known, they
first appeared. It is not an insuperable difficulty that
Foraminifera have not, as insisted on by Dr. Carpenter,
progressed in organisation since even the Laurentian
epoch; for some organisms would have to remain fitted
for simple conditions of life, and what could be better
fitted for this end than these lowly organised Protozoa?
Such objections as the above would be fatal to my view,
if it included advance in organisation as a necessary
contingent. They would likewise be fatal, if the above
Foraminifera, for instance, could be proved to have
first come into existence during the Laurentian epoch,
or the above Brachiopods during the Cambrian formation; for in this case, there would not have been time
sufficient for the development of these organisms up to
the standard which they had then reached. When
advanced up to any given point, there is no necessity,
on the theory of natural selection, for their further
continued progress; though they will, during each
successive age, have to be slightly modified, so as to
hold their places in relation to slight changes in their
conditions. The foregoing objections hinge on the
question whether we really know how old the world
is, and at what period the various forms of life first
appeared; and this may well be disputed.

The problem whether organisation on the whole has
advanced is in many ways excessively intricate. The
geological record, at all times imperfect, does not
extend far enough back, to shew with unmistakeable

2 I

clearness that within the known history of the world organisation has largely advanced. Even at the present day, looking to members of the same class, naturalists are not unanimous which forms ought to be ranked as highest: thus, some look at the selaceans or sharks, from their approach in some important points of structure to reptiles, as the highest fish; others look at the teleosteans as the highest. The ganoids stand intermediate between the selaceans and teleosteans; the latter at the present day are largely preponderant in number; but formerly selaceans and ganoids alone existed; and in this case, according to the standard of highness chosen, so will it be said that fishes have advanced or retrograded in organisation. To attempt to compare members of distinct types in the scale of highness seems hopeless; who will decide whether a cuttle-fish be higher than a bee—that insect which the great Von Baer believed to be "in fact more highly organised than a fish, although upon another type"? In the complex struggle for life it is quite credible that crustaceans, not very high in their own class, might beat cephalopods, the highest molluscs; and such crustaceans, though not highly developed, would stand very high in the scale of invertebrate animals, if judged by the most decisive of all trials—the law of battle. Beside these inherent difficulties in deciding which forms are the most advanced in organisation, we ought not solely to compare the highest members of a class at any two periods—though undoubtedly this is one and perhaps the most important element in striking a balance—but we ought to compare all the members, high and low, at the two periods. At an ancient epoch the highest and lowest molluscoidal animals, namely, cephalopods and brachiopods, swarmed in numbers; at the present time both groups are greatly reduced, whilst others, intermediate in organisation, have largely increased; consequently some naturalists maintain that molluscs were formerly more highly developed than at present; but a stronger case can be made out on the opposite side, by considering the vast reduction of brachiopods, and the fact that our existing cephalopods, though few in number, are more highly organised than their ancient representatives. We ought also to compare the relative proportional numbers at any two periods of the high and low classes throughout the world: if, for instance, at the present day fifty thousand kinds of vertebrate animals exist, and if we knew that at some former period only ten thousand kinds existed, we ought to look at this increase in number in the highest class, which implies a great displacement of lower forms, as a decided advance in the organisation of the world. We thus see how hopelessly difficult it is to compare with perfect fairness under such extremely complex relations, the standard of organisation of the imperfectly-known faunas of successive periods.

We shall appreciate this difficulty more clearly, by looking to certain existing faunas and floras. From the extraordinary manner in which European productions have recently spread over New Zealand, and have seized on places which must have been previously occupied by the indigenes, we must believe, that if all the animals and plants of Great Britain were set free in New Zealand, a multitude of British forms would in the course of time become thoroughly naturalised there, and would exterminate many of the natives. On the other hand, from the fact that hardly a single inhabitant of the southern hemisphere has become wild in any part of Europe, we may well doubt whether, if all the productions of New Zealand were set free in Great Britain, any considerable number would be enabled to seize on places now occupied by our native plants and animals. Under this point of view, the productions of Great Britain stand much higher in the scale than those of New Zealand. Yet the most skilful naturalist, from an examination of the species of the two countries, could not have foreseen this result.

Agassiz and several other highly competent judges insist that ancient animals resemble to a certain extent the embryos of recent animals belonging to the same classes; and that the geological succession of extinct forms is nearly parallel with the embryological development of existing forms. This view accords admirably well with our theory. In a future chapter I shall attempt to show that the adult differs from its embryo, owing to variations having supervened at a not early age, and having been inherited at a corresponding age. This process, whilst it leaves the embryo almost unaltered, continually adds, in the course of successive generations, more and more difference to the adult. Thus the embryo comes to be left as a sort of picture, preserved by nature, of the former and less modified condition of the species. This view may be true, and yet may never be capable of proof. Seeing, for instance, that the oldest known mammals, reptiles, and fishes strictly belong to their proper classes, though some of these old forms are in a slight degree less distinct from each other than are the typical members of the same groups at the present day, it would be vain to look for animals having the common embryological character of the Vertebrata, until beds rich in fossils are discovered far beneath the lowest Cambrian strata—a discovery of which the chance is small.

On the Succession of the same Types within the same Areas, during the later Tertiary periods.

Mr. Clift many years ago showed that the fossil mammals from the Australian caves were closely allied to the living marsupials of that continent. In South America, a similar relationship is manifest, even to an uneducated eye, in the gigantic pieces of armour, like those of the armadillo, found in several parts of La Plata; and Professor Owen has shown in the most striking manner that most of the fossil mammals, buried there in such numbers, are related to South American types. This relationship is even more clearly seen in the wonderful collection of fossil bones made by MM. Lund and Clausen in the caves of Brazil. I was so much impressed with these facts that I strongly insisted, in 1839 and 1845, on this "law of the succession of types,"—on "this wonderful relationship in the same continent between the dead and the living." Professor Owen has subsequently extended the same generalisation to the mammals of the Old World. We see the same law in this author's restorations of the extinct and gigantic birds of New Zealand. We see it also in the birds of the caves of Brazil. Mr. Woodward has shown that the same law holds good with sea-shells, but, from the wide distribution of most molluscs, it is not well displayed by them. Other cases could be added, as the relation between the extinct and living land-shells of Madeira; and between the extinct and living brackish water-shells of the Aralo-Caspian Sea.

Now what does this remarkable law of the succession of the same types within the same areas mean ? He would be a bold man who, after comparing the present climate of Australia and of parts of South America, under the same latitude, would attempt to account, on the one hand through dissimilar physical conditions, for the dissimilarity of the inhabitants of these two continents; and, on the other hand through similarity of conditions, for the uniformity of the same types in each continent during the later tertiary periods. Nor can it be pretended that it is an immutable law that marsupials should have been chiefly or solely produced in Australia; or that Edentata and other American types should have been solely produced in South America. For we know that Europe in ancient times was peopled by numerous marsupials; and I have shown in the publications above alluded to, that in America the law of distribution of terrestrial mammals was formerly different from what it now is. North America formerly partook strongly of the present character of the southern half of the continent; and the southern half was formerly more closely allied, than it is at present, to the northern half. In a similar manner we know, from Falconer and Cautley's discoveries, that Northern India was formerly more closely related in its mammals to Africa than it is at the present time. Analogous facts could be given in relation to the distribution of marine animals.

On the theory of descent with modification, the great law of the long enduring, but not immutable, succession of the same types within the same areas, is at once explained; for the inhabitants of each quarter of the world will obviously tend to leave in that quarter, during the next succeeding period of time, closely allied though in

some degree modified descendants. If the inhabitants
of one continent formerly differed greatly from those of
another continent, so will their modified descendants still
differ in nearly the same manner and degree. But after
very long intervals of time, and after great geographical
changes, permitting much intermigration, the feebler
will yield to the more dominant forms, and there will
be nothing immutable in the distribution of organic
beings.

It may be asked in ridicule, whether I suppose that
the megatherium and other allied huge monsters, which
formerly lived in South America, have left behind them
the sloth, armadillo, and anteater, as their degenerate
descendants. This cannot for an instant be admitted.
These huge animals have become wholly extinct, and
have left no progeny. But in the caves of Brazil, there
are many extinct species which are closely allied in size
and in all other characters to the species still living in
South America; and some of these fossils may have been
the actual progenitors of the living species. It must not
be forgotten that, on our theory, all the species of the
same genus are the descendants of some one species; so
that, if six genera, each having eight species, be found in
one geological formation, and in a succeeding formation
there be six other allied or representative genera each
with the same number of species, then we may conclude
that generally only one species of each of the older
genera has left modified descendants, which constitute
the new genera containing the several species; the other
seven species of each old genus having died out and left
no progeny. Or, and this will be a far commoner case,
two or three species in two or three alone of the six older
genera will be the parents of the new genera: the other

species and the other old genera having become utterly
extinct. In failing orders, with the genera and species
decreasing in numbers as is the case with the Edentata
of South America, still fewer genera and species will
leave modified blood-descendants.

Summary of the preceding and present Chapters.

I have attempted to show that the geological record is
extremely imperfect; that only a small portion of the
globe has been geologically explored with care; that
only certain classes of organic beings have been largely
preserved in a fossil state; that the number both of
specimens and of species, preserved in our museums, is
absolutely as nothing compared with the number of
generations which must have passed away even during a
single formation; that, owing to subsidence being almost
necessary for the accumulation of deposits rich in fossil
species of many kinds, and thick enough to outlast future
degradation, great intervals of time must have elapsed
between most of our successive formations; that there
has probably been more extinction during the periods of
subsidence, and more variation during the periods of
elevation, and during the latter the record will have been
least perfectly kept; that each single formation has not
been continuously deposited; that the duration of each
formation is probably short compared with the average
duration of specific forms; that migration has played an
important part in the first appearance of new forms in
any one area and formation; that widely ranging species
are those which have varied most frequently, and have
oftenest given rise to new species; that varieties have
at first been local; and lastly, although each species

must have passed through numerous transitional stages, it is probable that the periods, during which each underwent modification, though many and long as measured by years, have been short in comparison with the periods during which each remained in an unchanged condition.   These causes, taken conjointly, will to a large extent explain why—though we do find many links—we do not find interminable varieties, connecting together all extinct and existing forms by the finest graduated steps. It should also be constantly borne in mind that any linking variety between two forms, which might be found, would be ranked, unless the whole chain could be perfectly restored, as a new and distinct species; for it is not pretended that we have any sure criterion by which species and varieties can be discriminated.

He who rejects this view of the imperfection of the geological record, will rightly reject the whole theory. For he may ask in vain where are the numberless transitional links which must formerly have connected the closely allied or representative species, found in the successive stages of the same great formation ?   He may disbelieve in the immense intervals of time which must have elapsed between our consecutive formations ; he may overlook how important a part migration has played, when the formations of any one great region, as those of Europe, are considered ; he may urge the apparent, but often falsely apparent, sudden coming in of whole groups of species.   He may ask where are the remains of those infinitely numerous organisms which must have existed long before the Cambrian system was deposited ?   We now know that at least one animal did then exist ; but I can answer this last question only by supposing that where our oceans now extend

they have extended for an enormous period, and where
our oscillating continents now stand they have stood
since the commencement of the Cambrian system; but
that, long before that epoch, the world presented a
widely different aspect; and that the older continents,
formed of formations older than any known to us, exist
now only as remnants in a metamorphosed condition,
or lie still buried under the ocean.

Passing from these difficulties, the other great leading
facts in palæontology agree admirably with the theory
of descent with modification through variation and
natural selection. We can thus understand how it is
that new species come in slowly and successively; how
species of different classes do not necessarily change
together, or at the same rate, or in the same degree;
yet in the long run that all undergo modification to
some extent. The extinction of old forms is the almost
inevitable consequence of the production of new forms.
We can understand why, when a species has once disappeared, it never reappears. Groups of species increase
in numbers slowly, and endure for unequal periods of
time; for the process of modification is necessarily
slow, and depends on many complex contingencies.
The dominant species belonging to large and dominant
groups tend to leave many modified descendants, which
form new sub-groups and groups. As these are formed,
the species of the less vigorous groups, from their inferiority inherited from a common progenitor, tend to
become extinct together, and to leave no modified offspring on the face of the earth. But the utter extinction of a whole group of species has sometimes been a
slow process, from the survival of a few descendants,
lingering in protected and isolated situations. When a

group has once wholly disappeared, it does not reappear; for the link of generation has been broken.

We can understand how it is that dominant forms which spread widely and yield the greatest number of varieties tend to people the world with allied, but modified, descendants; and these will generally succeed in displacing the groups which are their inferiors in the struggle for existence. Hence, after long intervals of time, the productions of the world appear to have changed simultaneously.

We can understand how it is that all the forms of life, ancient and recent, make together a few grand classes. We can understand, from the continued tendency to divergence of character, why the more ancient a form is, the more it generally differs from those now living; why ancient and extinct forms often tend to fill up gaps between existing forms, sometimes blending two groups, previously classed as distinct, into one; but more commonly bringing them only a little closer together. The more ancient a form is, the more often it stands in some degree intermediate between groups now distinct; for the more ancient a form is, the more nearly it will be related to, and consequently resemble, the common progenitor of groups, since become widely divergent. Extinct forms are seldom directly intermediate between existing forms; but are intermediate only by a long and circuitous course through other extinct and different forms. We can clearly see why the organic remains of closely consecutive formations are closely allied; for they are closely linked together by generation. We can clearly see why the remains of an intermediate formation are intermediate in character.

The inhabitants of the world at each successive

period in its history have beaten their predecessors
in the race for life, and are, in so far, higher in the
scale, and their structure has generally become more
specialised; and this may account for the common
belief held by so many palæontologists, that organisation on the whole has progressed.   Extinct and ancient
animals resemble to a certain extent the embryos of
the more recent animals belonging to the same classes,
and this wonderful fact receives a simple explanation
according to our views.   The succession of the same
types of structure within the same areas during the
later geological periods ceases to be mysterious, and is
intelligible on the principle of inheritance.

If then the geological record be as imperfect as many
believe, and it may at least be asserted that the record
cannot be proved to be much more perfect, the main
objections to the theory of natural selection are greatly
diminished or disappear.   On the other hand, all the
chief laws of palæontology plainly proclaim, as it seems
to me, that species have been produced by ordinary
generation : old forms having been supplanted by new
and improved forms of life, the products of Variation
and the Survival of the Fittest.

CHAPTER XII.

GEOGRAPHICAL DISTRIBUTION.

Present distribution cannot be accounted for by differences in physical conditions—Importance of barriers—Affinity of the productions of the same continent—Centres of creation—Means of dispersal by changes of climate and of the level of the land, and by occasional means—Dispersal during the Glacial period—Alternate Glacial periods in the North and South.

IN considering the distribution of organic beings over the face of the globe, the first great fact which strikes us is, that neither the similarity nor the dissimilarity of the inhabitants of various regions can be wholly accounted for by climatal and other physical conditions. Of late, almost every author who has studied the subject has come to this conclusion. The case of America alone would almost suffice to prove its truth; for if we exclude the arctic and northern temperate parts, all authors agree that one of the most fundamental divisions in geographical distribution is that between the New and Old Worlds; yet if we travel over the vast American continent, from the central parts of the United States to its extreme southern point, we meet with the most diversified conditions; humid districts, arid deserts, lofty mountains, grassy plains, forests, marshes, lakes, and great rivers, under almost every temperature. There is hardly a climate or condition in the Old World which cannot be paralleled in the New—at

least as closely as the same species generally require.
No doubt small areas can be pointed out in the Old
World hotter than any in the New World; but these
are not inhabited by a fauna different from that of the
surrounding districts; for it is rare to find a group of
organisms confined to a small area, of which the conditions are peculiar in only a slight degree. Notwithstanding this general parallelism in the conditions of
the Old and New Worlds, how widely different are
their living productions !

In the southern hemisphere, if we compare large
tracts of land in Australia, South Africa, and western
South America, between latitudes 25° and 35°, we shall
find parts extremely similar in all their conditions, yet
it would not be possible to point out three faunas and
floras more utterly dissimilar. Or, again, we may compare the productions of South America south of lat.
35° with those north of 25°, which consequently are
separated by a space of ten degrees of latitude, and are
exposed to considerably different conditions; yet they
are incomparably more closely related to each other
than they are to the productions of Australia or Africa
under nearly the same climate. Analogous facts could
be given with respect to the inhabitants of the sea.

A second great fact which strikes us in our general
review is, that barriers of any kind, or obstacles to
free migration, are related in a close and important
manner to the differences between the productions of
various regions. We see this in the great difference
in nearly all the terrestrial productions of the New
and Old Worlds, excepting in the northern parts, where
the land almost joins, and where, under a slightly
different climate, there might have been free migration

for the northern temperate forms, as there now is for the strictly arctic productions. We see the same fact in the great difference between the inhabitants of Australia, Africa, and South America under the same latitude ; for these countries are almost as much isolated from each other as is possible. On each continent, also, we see the same fact ; for on the opposite sides of lofty and continuous mountain-ranges, of great deserts and even of large rivers, we find different productions ; though as mountain-chains, deserts, &c., are not as impassable, or likely to have endured so long, as the oceans separating continents, the differences are very inferior in degree to those characteristic of distinct continents.

Turning to the sea, we find the same law. The marine inhabitants of the eastern and western shores of South America are very distinct, with extremely few shells, crustacea, or echinodermata in common ; but Dr. Günther has recently shown that about thirty per cent. of the fishes are the same on the opposite sides of the isthmus of Panama ; and this fact has led naturalists to believe that the isthmus was formerly open. Westward of the shores of America, a wide space of open ocean extends, with not an island as a halting-place for emigrants ; here we have a barrier of another kind, and as soon as this is passed we meet in the eastern islands of the Pacific with another and totally distinct fauna. So that three marine faunas range far northward and southward in parallel lines not far from each other, under corresponding climates ; but from being separated from each other by impass-able barriers, either of land or open sea, they are almost wholly distinct. On the other hand, proceeding still

farther westward from the eastern islands of the tropical
parts of the Pacific, we encounter no impassable barriers,
and we have innumerable islands as halting-places, or
continuous coasts, until, after travelling over a hemisphere, we come to the shores of Africa; and over this
vast space we meet with no well-defined and distinct
marine faunas. Although so few marine animals are
common to the above-named three approximate faunas
of Eastern and Western America and the eastern
Pacific islands, yet many fishes range from the Pacific
into the Indian Ocean, and many shells are common
to the eastern islands of the Pacific and the eastern
shores of Africa on almost exactly opposite meridians
of longitude.

A third great fact, partly included in the foregoing
statement, is the affinity of the productions of the same
continent or of the same sea, though the species themselves are distinct at different points and stations. It
is a law of the widest generality, and every continent
offers innumerable instances. Nevertheless the naturalist, in travelling, for instance, from north to south,
never fails to be struck by the manner in which successive groups of beings, specifically distinct, though
nearly related, replace each other. He hears from
closely allied, yet distinct kinds of birds, notes nearly
similar, and sees their nests similarly constructed, but
not quite alike, with eggs coloured in nearly the same
manner. The plains near the Straits of Magellan are
inhabited by one species of Rhea (American ostrich),
and northward the plains of La Plata by another species
of the same genus; and not by a true ostrich or emu,
like those inhabiting Africa and Australia under the
same latitude. On these same plains of La Plata we

see the agouti and bizcacha, animals having nearly the same habits as our hares and rabbits, and belonging to the same order of Rodents, but they plainly display an American type of structure. We ascend the lofty peaks of the Cordillera, and we find an alpine species of bizcacha; we look to the waters, and we do not find the beaver or musk-rat, but the coypu and capybara, rodents of the S. American type. Innumerable other instances could be given. If we look to the islands off the American shore, however much they may differ in geological structure, the inhabitants are essentially American, though they may be all peculiar species. We may look back to past ages, as shown in the last chapter, and we find American types then prevailing on the American continent and in the American seas. We see in these facts some deep organic bond, throughout space and time, over the same areas of land and water, independently of physical conditions. The naturalist must be dull who is not led to inquire what this bond is.

The bond is simply inheritance, that cause which alone, as far as we positively know, produces organisms quite like each other, or, as we see in the case of varieties, nearly alike. The dissimilarity of the inhabitants of different regions may be attributed to modification through variation and natural selection, and probably in a subordinate degree to the definite influence of different physical conditions. The degrees of dissimilarity will depend on the migration of the more dominant forms of life from one region into another having been more or less effectually prevented, at periods more or less remote;—on the nature and number of the former immigrants;—and on the action

2 K

of the inhabitants on each other in leading to the preservation of different modifications; the relation of organism to organism in the struggle for life being, as I have already often remarked, the most important of all relations. Thus the high importance of barriers comes into play by checking migration; as does time for the slow process of modification through natural selection. Widely-ranging species, abounding in individuals, which have already triumphed over many competitors in their own widely-extended homes, will have the best chance of seizing on new places, when they spread into new countries. In their new homes they will be exposed to new conditions, and will frequently undergo further modification and improvement; and thus they will become still further victorious, and will produce groups of modified descendants. On this principle of inheritance with modification we can understand how it is that sections of genera, whole genera, and even families, are confined to the same areas, as is so commonly and notoriously the case.

There is no evidence, as was remarked in the last chapter, of the existence of any law of necessary development. As the variability of each species is an independent property, and will be taken advantage of by natural selection, only so far as it profits each individual in its complex struggle for life, so the amount of modification in different species will be no uniform quantity. If a number of species, after having long competed with each other in their old home, were to migrate in a body into a new and afterwards isolated country, they would be little liable to modification; for neither migration nor isolation in themselves effect anything. These principles come into play only by bringing organisms into new relations with each other and in a lesser degree with the surrounding physical conditions. As we have seen in the last chapter that some forms have retained nearly the same character from an enormously remote geological period, so certain species have migrated over vast spaces, and have not become greatly or at all modified.

According to these views, it is obvious that the several species of the same genus, though inhabiting the most distant quarters of the world, must originally have proceeded from the same source, as they are descended from the same progenitor. In the case of those species which have undergone during whole geological periods little modification, there is not much difficulty in believing that they have migrated from the same region; for during the vast geographical and climatal changes which have supervened since ancient times, almost any amount of migration is possible. But in many other cases, in which we have reason to believe that the species of a genus have been produced within comparatively recent times, there is great difficulty on this head. It is also obvious that the individuals of the same species, though now inhabiting distant and isolated regions, must have proceeded from one spot, where their parents were first produced: for, as has been explained, it is incredible that individuals identically the same should have been produced from parents specifically distinct.

Single Centres of supposed Creation.—We are thus brought to the question which has been largely discussed by naturalists, namely, whether species have been created at one or more points of the earth's surface. Undoubtedly there are many cases of extreme difficulty in understanding how the same species could possibly have migrated from some one point to the several distant and isolated points, where now found. Nevertheless the simplicity of the view that each species was first produced within a single region captivates the mind. He who rejects it, rejects the *vera causa* of ordinary generation with subsequent migration, and calls in the agency of a miracle. It is universally admitted, that in most cases the area inhabited by a species is continuous; and that when a plant or animal inhabits two points so distant from each other, or with an interval of such a nature, that the space could not have been easily passed over by migration, the fact is given as something remarkable and exceptional. The incapacity of migrating across a wide sea is more clear in the case of terrestrial mammals than perhaps with any other organic beings; and, accordingly, we find no inexplicable instances of the same mammals inhabiting distant points of the world. No geologist feels any difficulty in Great Britain possessing the same quadrupeds with the rest of Europe, for they were no doubt once united. But if the same species can be produced at two separate points, why do we not find a single mammal common to Europe and Australia or South America? The conditions of life are nearly the same, so that a multitude of European animals and plants have become naturalised in America and Australia; and some of the aboriginal plants are identically the same at these distant points of the northern and southern hemispheres? The answer, as I believe, is, that mammals have not been able to migrate, whereas some plants, from their varied means of dispersal, have migrated across the wide and broken

interspaces.   The great and striking influence of barriers
of all kinds, is intelligible only on the view that the
great majority of species have been produced on one
side, and have not been able to migrate to the opposite
side.   Some few families, many sub-families, very many
genera, and a still greater number of sections of genera,
are confined to a single region ; and it has been observed
by several naturalists that the most natural genera, or
those genera in which the species are most closely
related to each other, are generally confined to the
same country, or if they have a wide range that their
range is continuous.   What a strange anomaly it would
be, if a directly opposite rule were to prevail, when we
go down one step lower in the series, namely, to the
individuals of the same species, and these had not
been, at least at first, confined to some one region !

Hence it seems to me, as it has to many other
naturalists, that the view of each species having been
produced in one area alone, and having subsequently
migrated from that area as far as its powers of migration
and subsistence under past and present conditions
permitted, is the most probable.   Undoubtedly many
cases occur, in which we cannot explain how the same
species could have passed from one point to the other.
But the geographical and climatal changes which have
certainly occurred within recent geological times, must
have rendered discontinuous the formerly continuous
range of many species.   So that we are reduced to consider
whether the exceptions to continuity of range are so
numerous and of so grave a nature, that we ought to give
up the belief, rendered probable by general considerations,
that each species has been produced within one area, and
has migrated thence as far as it could.   It would be

hopelessly tedious to discuss all the exceptional cases of the same species, now living at distant and separated points, nor do I for a moment pretend that any explanation could be offered of many instances. But, after some preliminary remarks, I will discuss a few of the most striking classes of facts ; namely, the existence of the same species on the summits of distant mountain ranges, and at distant points in the arctic and antarctic regions ; and secondly (in the following chapter), the wide distribution of freshwater productions ; and thirdly, the occurrence of the same terrestrial species on islands and on the nearest mainland, though separated by hundreds of miles of open sea. If the existence of the same species at distant and isolated points of the earth's surface, can in many instances be explained on the view of each species having migrated from a single birthplace ; then, considering our ignorance with respect to former climatal and geographical changes and to the various occasional means of transport, the belief that a single birthplace is the law, seems to me incomparably the safest.

In discussing this subject, we shall be enabled at the same time to consider a point equally important for us, namely, whether the several species of a genus which must on our theory all be descended from a common progenitor, can have migrated, undergoing modification during their migration, from some one area. If, when most of the species inhabiting one region are different from those of another region, though closely allied to them, it can be shown that migration from the one region to the other has probably occurred at some former period, our general view will be much strengthened ; for the explanation is obvious on the principle of descent with

modification.   A volcanic island, for instance, upheaved
and formed at the distance of a few hundreds of miles
from a continent, would probably receive from it in the
course of time a few colonists, and their descendants,
though modified, would still be related by inheritance to
the inhabitants of that continent.   Cases of this nature
are common, and are, as we shall hereafter see, inexplicable on the theory of independent creation.   This view
of the relation of the species of one region to those of
another, does not differ much from that advanced by Mr.
Wallace, who concludes that " every species has come
into existence coincident both in space and time with a
pre-existing closely allied species."   And it is now well
known that he attributes this coincidence to descent with
modification.

The question of single or multiple centres of creation
differs from another though allied question,—namely,
whether all the individuals of the same species are
descended from a single pair, or single hermaphrodite, or
whether, as some authors suppose, from many individuals
simultaneously created.   With organic beings which
never intercross, if such exist, each species must be descended from a succession of modified varieties, that
have supplanted each other, but have never blended with
other individuals or varieties of the same species ; so that,
at each successive stage of modification, all the individuals
of the same form will be descended from a single parent.
But in the great majority of cases, namely, with all
organisms which habitually unite for each birth, or which
occasionally intercross, the individuals of the same
species inhabiting the same area will be kept nearly
uniform by intercrossing; so that many individuals will
go on simultaneously changing, and the whole amount

of modification at each stage will not be due to descent
from a single parent.   To illustrate what I mean : our
English race-horses differ from the horses of every
other breed; but they do not owe their difference and
superiority to descent from any single pair, but to
continued care in the selecting and training of many
individuals during each generation.

Before discussing the three classes of facts, which I
have selected as presenting the greatest amount of
difficulty on the theory of "single centres of creation,"
I must say a few words on the means of dispersal.

Means of Dispersal.

Sir C. Lyell and other authors have ably treated
this subject.   I can give here only the briefest abstract
of the more important facts.   Change of climate must
have had a powerful influence on migration.   A region
now impassable to certain organisms from the nature of
its climate, might have been a high road for migration,
when the climate was different.   I shall, however,
presently have to discuss this branch of the subject in
some detail.   Changes of level in the land must also
have been highly influential: a narrow isthmus now
separates two marine faunas; submerge it, or let it
formerly have been submerged, and the two faunas will
now blend together, or may formerly have blended.
Where the sea now extends, land may at a former period
have connected islands or possibly even continents together, and thus have allowed terrestrial productions to
pass from one to the other.   No geologist disputes that
great mutations of level have occurred within the period
of existing organisms.   Edward Forbes insisted that all
the islands in the Atlantic must have been recently

connected with Europe or Africa, and Europe likewise
with America.   Other authors have thus hypothetically
bridged over every ocean, and united almost every island
with some mainland.   If indeed the arguments used by
Forbes are to be trusted, it must be admitted that
scarcely a single island exists which has not recently
been united to some continent.   This view cuts the
Gordian knot of the dispersal of the same species to the
most distant points, and removes many a difficulty ; but
to the best of my judgment we are not authorised in
admitting such enormous geographical changes within
the period of existing species.   It seems to me that we
have abundant evidence of great oscillations in the level
of the land or sea ; but not of such vast changes in the
position and extension of our continents, as to have united
them within the recent period to each other and to the
several intervening oceanic islands.   I freely admit the
former existence of many islands, now buried beneath
the sea, which may have served as halting-places for
plants and for many animals during their migration.   In
the coral-producing oceans such sunken islands are now
marked by rings of coral or atolls standing over them.
Whenever it is fully admitted, as it will some day be, that
each species has proceeded from a single birthplace, and
when in the course of time we know something definite
about the means of distribution, we shall be enabled
to speculate with security on the former extension of the
land.   But I do not believe that it will ever be proved
that within the recent period most of our continents
which now stand quite separate, have been continuously,
or almost continuously united with each other, and with
the many existing oceanic islands.   Several facts in
distribution—such as the great difference in the marine

faunas on the opposite sides of almost every continent, —the close relation of the tertiary inhabitants of several lands and even seas to their present inhabitants,—the degree of affinity between the mammals inhabiting islands with those of the nearest continent, being in part determined (as we shall hereafter see) by the depth of the intervening ocean,—these and other such facts are opposed to the admission of such prodigious geographical revolutions within the recent period, as are necessary on the view advanced by Forbes and admitted by his followers. The nature and relative proportions of the inhabitants of oceanic islands are likewise opposed to the belief of their former continuity with continents. Nor does the almost universally volcanic composition of such islands favour the admission that they are the wrecks of sunken continents;—if they had originally existed as continental mountain ranges, some at least of the islands would have been formed, like other mountain summits, of granite, metamorphic schists, old fossiliferous and other rocks, instead of consisting of mere piles of volcanic matter.

I must now say a few words on what are called accidental means, but which more properly should be called occasional means of distribution. I shall here confine myself to plants. In botanical works, this or that plant is often stated to be ill adapted for wide dissemination; but the greater or less facilities for transport across the sea may be said to be almost wholly unknown. Until I tried, with Mr. Berkeley's aid, a few experiments, it was not even known how far seeds could resist the injurious action of sea-water. To my surprise I found that out of 87 kinds, 64 germinated after an immersion of 28 days, and a few survived an immersion

of 137 days. It deserves notice that certain orders were
far more injured than others: nine Leguminosæ were
tried, and, with one exception, they resisted the salt-water
badly; seven species of the allied orders, Hydrophyllaceæ
and Polemoniaceæ, were all killed by a month's immersion. For convenience' sake I chiefly tried small
seeds without the capsule or fruit; and as all of these
sank in a few days they could not have been floated
across wide spaces of the sea, whether or not they were
injured by the salt-water. Afterwards I tried some larger
fruits, capsules, &c., and some of these floated for a long
time. It is well known what a difference there is in the
buoyancy of green and seasoned timber; and it occurred
to me that floods would often wash into the sea dried
plants or branches with seed-capsules or fruit attached
to them. Hence I was led to dry the stems and branches
of 94 plants with ripe fruit, and to place them on sea-
water. The majority sank quickly, but some which,
whilst green, floated for a very short time, when dried
floated much longer; for instance, ripe hazel-nuts sank
immediately, but when dried they floated for 90 days,
and afterwards when planted germinated; an asparagus-
plant with ripe berries floated for 23 days, when dried it
floated for 85 days, and the seeds afterwards germinated;
the ripe seeds of Helosciadium sank in two days, when
dried they floated for above 90 days, and afterwards
germinated. Altogether, out of the 94 dried plants, 18
floated for above 28 days; and some of the 18 floated
for a very much longer period. So that as $\frac{64}{87}$ kinds of
seeds germinated after an immersion of 28 days; and as
$\frac{18}{94}$ distinct species with ripe fruit (but not all the same
species as in the foregoing experiment) floated, after
being dried, for above 28 days, we may conclude, as far

as anything can be inferred from these scanty facts, that the seeds of $\frac{14}{100}$ kinds of plants of any country might be floated by sea-currents during 28 days and would retain their power of germination. In Johnston's Physical Atlas, the average rate of the several Atlantic currents is 33 miles per diem (some currents running at the rate of 60 miles per diem); on this average, the seeds of $\frac{14}{100}$ plants belonging to one country might be floated across 924 miles of sea to another country, and when stranded, if blown by an inland gale to a favourable spot, would germinate.

Subsequently to my experiments, M. Martens tried similar ones, but in a much better manner, for he placed the seeds in a box in the actual sea, so that they were alternately wet and exposed to the air like really floating plants. He tried 98 seeds, mostly different from mine; but he chose many large fruits and likewise seeds from plants which live near the sea; and this would have favoured both the average length of their flotation and their resistance to the injurious action of the salt-water. On the other hand, he did not previously dry the plants or branches with the fruit; and this, as we have seen, would have caused some of them to have floated much longer. The result was that $\frac{18}{98}$ of his seeds of different kinds floated for 42 days, and were then capable of germination. But I do not doubt that plants exposed to the waves would float for a less time than those protected from violent movement as in our experiments. Therefore it would perhaps be safer to assume that the seeds of about $\frac{10}{100}$ plants of a flora, after having been dried, could be floated across a space of sea 900 miles in width, and would then germinate. The fact of the larger fruits often floating longer than the small, is interesting; as plants with large seeds or fruit which, as Alph. de Candolle has shown, generally have restricted ranges, could hardly be transported by any other means.

Seeds may be occasionally transported in another manner. Drift timber is thrown up on most islands, even on those in the midst of the widest oceans; and the natives of the coral-islands in the Pacific procure stones for their tools, solely from the roots of drifted trees, these stones being a valuable royal tax. I find that when irregularly shaped stones are embedded in the roots of trees, small parcels of earth are frequently enclosed in their interstices and behind them,—so perfectly that not a particle could be washed away during the longest transport: out of one small portion of earth thus *completely* enclosed by the roots of an oak about 50 years old, three dicotyledonous plants germinated: I am certain of the accuracy of this observation. Again, I can show that the carcases of birds, when floating on the sea, sometimes escape being immediately devoured: and many kinds of seeds in the crops of floating birds long retain their vitality: peas and vetches, for instance, are killed by even a few days' immersion in sea-water; but some taken out of the crop of a pigeon, which had floated on artificial sea-water for 30 days, to my surprise nearly all germinated.

Living birds can hardly fail to be highly effective agents in the transportation of seeds. I could give many facts showing how frequently birds of many kinds are blown by gales to vast distances across the ocean. We may safely assume that under such circumstances their rate of flight would often be 35 miles an hour; and some authors have given a far higher estimate. I

have never seen an instance of nutritious seeds passing
through the intestines of a bird; but hard seeds of fruit
pass uninjured through even the digestive organs of a
turkey. In the course of two months, I picked up in
my garden 12 kinds of seeds, out of the excrement of
small birds, and these seemed perfect, and some of
them, which were tried, germinated. But the following
fact is more important: the crops of birds do not secrete
gastric juice, and do not, as I know by trial, injure in
the least the germination of seeds; now, after a bird
has found and devoured a large supply of food, it is
positively asserted that all the grains do not pass into
the gizzard for twelve or even eighteen hours. A bird
in this interval might easily be blown to the distance
of 500 miles, and hawks are known to look out for tired
birds, and the contents of their torn crops might thus
readily get scattered. Some hawks and owls bolt their
prey whole, and, after an interval of from twelve to
twenty hours, disgorge pellets, which, as I know from
experiments made in the Zoological Gardens, include
seeds capable of germination. Some seeds of the oat
wheat, millet, canary, hemp, clover, and beet germinated
after having been from twelve to twenty-one hours
in the stomachs of different birds of prey; and two
seeds of beet grew after having been thus retained
for two days and fourteen hours. Fresh-water fish, I
find, eat seeds of many land and water plants; fish are
frequently devoured by birds, and thus the seeds might
be transported from place to place. I forced many
kinds of seeds into the stomachs of dead fish, and then
gave their bodies to fishing-eagles, storks, and pelicans;
these birds, after an interval of many hours, either
rejected the seeds in pellets or passed them in their

excrement; and several of these seeds retained the power of germination. Certain seeds, however, were always killed by this process.

Locusts are sometimes blown to great distances from the land; I myself caught one 370 miles from the coast of Africa, and have heard of others caught at greater distances. The Rev. R. T. Lowe informed Sir C. Lyell that in November 1844 swarms of locusts visited the island of Madeira. They were in countless numbers, as thick as the flakes of snow in the heaviest snowstorm, and extended upwards as far as could be seen with a telescope. During two or three days they slowly careered round and round in an immense ellipse, at least five or six miles in diameter, and at night alighted on the taller trees, which were completely coated with them. They then disappeared over the sea, as suddenly as they had appeared, and have not since visited the island. Now, in parts of Natal it is believed by some farmers, though on insufficient evidence, that injurious seeds are introduced into their grass-land in the dung left by the great flights of locusts which often visit that country. In consequence of this belief Mr. Weale sent me in a letter a small packet of the dried pellets, out of which I extracted under the microscope several seeds, and raised from them seven grass plants, belonging to two species, of two genera. Hence a swarm of locusts, such as that which visited Madeira, might readily be the means of introducing several kinds of plants into an island lying far from the mainland.

Although the beaks and feet of birds are generally clean, earth sometimes adheres to them: in one case I removed sixty-one grains, and in another case twenty-two grains of dry argillaceous earth from the foot of a partridge, and in the earth there was a pebble as large as the seed of a vetch.   Here is a better case: the leg of a woodcock was sent to me by a friend, with a little cake of dry earth attached to the shank, weighing only nine grains; and this contained a seed of the toad-rush (Juncus bufonius) which germinated and flowered.   Mr. Swaysland, of Brighton, who during the last forty years has paid close attention to our migratory birds, informs me that he has often shot wagtails (Motacillæ), wheat-ears, and whinchats (Saxicolæ), on their first arrival on our shores, before they had alighted; and he has several times noticed little cakes of earth attached to their feet. Many facts could be given showing how generally soil is charged with seeds.   For instance, Prof. Newton sent me the leg of a red-legged partridge (Caccabis rufa) which had been wounded and could not fly, with a ball of hard earth adhering to it, and weighing six and a half ounces.   The earth had been kept for three years, but when broken, watered and placed under a bell glass, no less than 82 plants sprung from it: these consisted of 12 monocotyledons, including the common oat, and at least one kind of grass, and of 70 dicotyledons, which consisted, judging from the young leaves, of at least three distinct species.   With such facts before us, can we doubt that the many birds which are annually blown by gales across great spaces of ocean, and which annually migrate—for instance, the millions of quails across the Mediterranean—must occasionally transport a few seeds embedded in dirt adhering to their feet or beaks?   But I shall have to recur to this subject.

As icebergs are known to be sometimes loaded with earth and stones, and have even carried brushwood, bones, and the nest of a land-bird, it can hardly be doubted that they must occasionally, as suggested by Lyell, have transported seeds from one part to another of the arctic and antarctic regions; and during the Glacial period from one part of the now temperate regions to another. In the Azores, from the large number of plants common to Europe, in comparison with the species on the other islands of the Atlantic, which stand nearer to the mainland, and (as remarked by Mr. H. C. Watson) from their somewhat northern character in comparison with the latitude, I suspected that these islands had been partly stocked by ice-borne seeds, during the Glacial epoch. At my request Sir C. Lyell wrote to M. Hartung to inquire whether he had observed erratic boulders on these islands, and he answered that he had found large fragments of granite and other rocks, which do not occur in the archipelago. Hence we may safely infer that icebergs formerly landed their rocky burthens on the shores of these mid-ocean islands, and it is at least possible that they may have brought thither some few seeds of northern plants.

Considering that these several means of transport, and that other means, which without doubt remain to be discovered, have been in action year after year for tens of thousands of years, it would, I think, be a marvellous fact if many plants had not thus become widely transported. These means of transport are sometimes called accidental, but this is not strictly correct: the currents of the sea are not accidental, nor is the direction of prevalent gales of wind. It should be observed that scarcely any means of transport would carry seeds for very great distances: for seeds do not

2 L

retain their vitality when exposed for a great length of
time to the action of sea-water; nor could they be long
carried in the crops or intestines of birds. These
means, however, would suffice for occasional transport
across tracts of sea some hundred miles in breadth, or
from island to island, or from a continent to a neighbouring island, but not from one distant continent to
another. The floras of distant continents would not by
such means become mingled; but would remain as
distinct as they now are. The currents, from their
course, would never bring seeds from North America to
Britain, though they might and do bring seeds from
the West Indies to our western shores, where, if not
killed by their very long immersion in salt water, they
could not endure our climate. Almost every year, one
or two land-birds are blown across the whole Atlantic
Ocean, from North America to the western shores of
Ireland and England; but seeds could be transported
by these rare wanderers only by one means, namely, by
dirt adhering to their feet or beaks, which is in itself a
rare accident. Even in this case, how small would be
the chance of a seed falling on favourable soil, and
coming to maturity! But it would be a great error to
argue that because a well-stocked island, like Great
Britain, has not, as far as is known (and it would be
very difficult to prove this), received within the last
few centuries, through occasional means of transport,
immigrants from Europe or any other continent, that a
poorly-stocked island, though standing more remote
from the mainland, would not receive colonists by similar means. Out of a hundred kinds of seeds or animals
transported to an island, even if far less well-stocked
than Britain, perhaps not more than one would be so

well fitted to its new home, as to become naturalised.
But this is no valid argument against what would be
effected by occasional means of transport, during the
long lapse of geological time, whilst the island was
being upheaved, and before it had become fully stocked
with inhabitants.   On almost bare land, with few or no
destructive insects or birds living there, nearly every
seed which chanced to arrive, if fitted for the climate,
would germinate and survive.

Dispersal during the Glacial Period.

The identity of many plants and animals, on mountain-summits, separated from each other by hundreds
of miles of lowlands, where Alpine species could not
possibly exist, is one of the most striking cases known
of the same species living at distant points, without
the apparent possibility of their having migrated from
one point to the other.   It is indeed a remarkable fact
to see so many plants of the same species living on the
snowy regions of the Alps or Pyrenees, and in the
extreme northern parts of Europe; but it is far more
remarkable, that the plants on the White Mountains, in
the United States of America, are all the same with
those of Labrador, and nearly all the same, as we hear
from Asa Gray, with those on the loftiest mountains
of Europe.   Even as long ago as 1747, such facts led
Gmelin to conclude that the same species must have
been independently created at many distinct points ;
and we might have remained in this same belief, had
not Agassiz and others called vivid attention to the
Glacial period, which, as we shall immediately see,
affords a simple explanation of these facts.   We have
evidence of almost every conceivable kind, organic and

inorganic, that, within a very recent geological period, central Europe and North America suffered under an arctic climate. The ruins of a house burnt by fire do not tell their tale more plainly than do the mountains of Scotland and Wales, with their scored flanks, polished surfaces, and perched boulders, of the icy streams with which their valleys were lately filled. So greatly has the climate of Europe changed, that in Northern Italy, gigantic moraines, left by old glaciers, are now clothed by the vine and maize. Throughout a large part of the United States, erratic boulders and scored rocks plainly reveal a former cold period.

The former influence of the glacial climate on the distribution of the inhabitants of Europe, as explained by Edward Forbes, is substantially as follows. But we shall follow the changes more readily, by supposing a new glacial period slowly to come on, and then pass away, as formerly occurred. As the cold came on, and as each more southern zone became fitted for the inhabitants of the north, these would take the places of the former inhabitants of the temperate regions. The latter, at the same time, would travel further and further southward, unless they were stopped by barriers, in which case they would perish. The mountains would become covered with snow and ice, and their former Alpine inhabitants would descend to the plains. By the time that the cold had reached its maximum, we should have an arctic fauna and flora, covering the central parts of Europe, as far south as the Alps and Pyrenees, and even stretching into Spain. The now temperate regions of the United States would likewise be covered by arctic plants and animals and these would be nearly the same with those of Europe; for the present circumpolar inhabitants, which we suppose to have everywhere travelled southward, are remarkably uniform round the world.

As the warmth returned, the arctic forms would retreat northward, closely followed up in their retreat by the productions of the more temperate regions. And as the snow melted from the bases of the mountains, the arctic forms would seize on the cleared and thawed ground, always ascending, as the warmth increased and the snow still further disappeared, higher and higher, whilst their brethren were pursuing their northern journey. Hence, when the warmth had fully returned, the same species, which had lately lived together on the European and North American lowlands, would again be found in the arctic regions of the Old and New Worlds, and on many isolated mountain-summits far distant from each other.

Thus we can understand the identity of many plants at points so immensely remote as the mountains of the United States and those of Europe. We can thus also understand the fact that the Alpine plants of each mountain-range are more especially related to the arctic forms living due north or nearly due north of them: for the first migration when the cold came on, and the re-migration on the returning warmth, would generally have been due south and north. The Alpine plants, for example, of Scotland, as remarked by Mr. H. C. Watson, and those of the Pyrenees, as remarked by Ramond, are more especially allied to the plants of northern Scandinavia; those of the United States to Labrador; those of the mountains of Siberia to the arctic regions of that country. These views, grounded as they are on the perfectly well-ascertained occurrence

of a former Glacial period, seem to me to explain in so
satisfactory a manner the present distribution of the
Alpine and Arctic productions of Europe and America,
that when in other regions we find the same species
on distant mountain-summits, we may almost conclude,
without other evidence, that a colder climate formerly
permitted their migration across the intervening lowlands, now become too warm for their existence.

As the arctic forms moved first southward and afterwards backwards to the north, in unison with the
changing climate, they will not have been exposed
during their long migrations to any great diversity of
temperature; and as they all migrated in a body together, their mutual relations will not have been much
disturbed. Hence, in accordance with the principles
inculcated in this volume, these forms will not have
been liable to much modification. But with the Alpine
productions, left isolated from the moment of the returning warmth, first at the bases and ultimately on
the summits of the mountains, the case will have been
somewhat different; for it is not likely that all the same
arctic species will have been left on mountain-ranges far
distant from each other, and have survived there ever
since; they will also in all probability, have become
mingled with ancient Alpine species, which must have
existed on the mountains before the commencement of the
Glacial epoch, and which during the coldest period will
have been temporarily driven down to the plains; they
will, also, have been subsequently exposed to somewhat
different climatal influences. Their mutual relations
will thus have been in some degree disturbed; consequently they will have been liable to modification; and
they have been modified; for if we compare the present

Alpine plants and animals of the several great European mountain-ranges one with another, though many of the species remain identically the same, some exist as varieties, some as doubtful forms or sub-species, and some as distinct yet closely allied species representing each other on the several ranges.

In the foregoing illustration I have assumed that at the commencement of our imaginary Glacial period, the arctic productions were as uniform round the polar regions as they are at the present day. But it is also necessary to assume that many sub-arctic and some few temperate forms were the same round the world, for some of the species which now exist on the lower mountain-slopes and on the plains of North America and Europe are the same; and it may be asked how I account for this degree of uniformity in the sub-arctic and temperate forms round the world, at the commencement of the real Glacial period. At the present day, the sub-arctic and northern temperate productions of the Old and New Worlds are separated from each other by the whole Atlantic Ocean and by the northern part of the Pacific. During the Glacial period, when the inhabitants of the Old and New Worlds lived farther southwards than they do at present, they must have been still more completely separated from each other by wider spaces of ocean; so that it may well be asked how the same species could then or previously have entered the two continents. The explanation, I believe, lies in the nature of the climate before the commencement of the Glacial period. At this, the newer Pliocene period, the majority of the inhabitants of the world were specifically the same as now, and we have good reason to believe that the climate was warmer than at the present

day. Hence we may suppose that the organisms which
now live under latitude 60°, lived during the Pliocene
period farther north under the Polar Circle, in latitude
66°-67°; and that the present arctic productions then
lived on the broken land still nearer to the pole. Now,
if we look at a terrestrial globe, we see under the Polar
Circle that there is almost continuous land from western Europe, through Siberia, to eastern America. And
this continuity of the circumpolar land, with the consequent freedom under a more favourable climate for intermigration, will account for the supposed uniformity of
the sub-arctic and temperate productions of the Old and
New Worlds, at a period anterior to the Glacial epoch.

Believing, from reasons before alluded to, that our
continents have long remained in nearly the same
relative position, though subjected to great oscillations
of level, I am strongly inclined to extend the above
view, and to infer that during some still earlier and
still warmer period, such as the older Pliocene period,
a large number of the same plants and animals inhabited the almost continuous circumpolar land; and
that these plants and animals, both in the Old and
New Worlds, began slowly to migrate southwards as
the climate became less warm, long before the commencement of the Glacial period. We now see, as
I believe, their descendants, mostly in a modified condition, in the central parts of Europe and the United
States. On this view we can understand the relationship with very little identity, between the productions
of North America and Europe,—a relationship which is
highly remarkable, considering the distance of the two
areas, and their separation by the whole Atlantic Ocean.
We can further understand the singular fact remarked

on by several observers that the productions of Europe and America during the later tertiary stages were more closely related to each other than they are at the present time; for during these warmer periods the northern parts of the Old and New Worlds will have been almost continuously united by land, serving as a bridge, since rendered impassable by cold, for the intermigration of their inhabitants.

During the slowly decreasing warmth of the Pliocene period, as soon as the species in common, which inhabited the New and Old Worlds, migrated south of the Polar Circle, they will have been completely cut off from each other. This separation, as far as the more temperate productions are concerned, must have taken place long ages ago. As the plants and animals migrated southward, they will have become mingled in the one great region with the native American productions, and would have had to compete with them; and in the other great region, with those of the Old World. Consequently we have here everything favourable for much modification,—for far more modification than with the Alpine productions, left isolated, within a much more recent period, on the several mountain-ranges and on the arctic lands of Europe and N. America. Hence it has come, that when we compare the now living productions of the temperate regions of the New and Old Worlds, we find very few identical species (though Asa Gray has lately shown that more plants are identical than was formerly supposed), but we find in every great class many forms, which some naturalists rank as geographical races, and others as distinct species; and a host of closely allied or representative forms which are ranked by all naturalists as specifically distinct.

As on the land, so in the waters of the sea, a slow southern migration of a marine fauna, which, during the Pliocene or even a somewhat earlier period, was nearly uniform along the continuous shores of the Polar Circle, will account, on the theory of modification, for many closely allied forms now living in marine areas completely sundered. Thus, I think, we can understand the presence of some closely allied, still existing and extinct tertiary forms, on the eastern and western shores of temperate North America; and the still more striking fact of many closely allied crustaceans (as described in Dana's admirable work), some fish and other marine animals, inhabiting the Mediterranean and the seas of Japan,—these two areas being now completely separated by the breadth of a whole continent and by wide spaces of ocean.

These cases of close relationship in species either now or formerly inhabiting the seas on the eastern and western shores of North America, the Mediterranean and Japan, and the temperate lands of North America and Europe, are inexplicable on the theory of creation. We cannot maintain that such species have been created alike, in correspondence with the nearly similar physical conditions of the areas; for if we compare, for instance, certain parts of South America with parts of South Africa or Australia, we see countries closely similar in all their physical conditions, with their inhabitants utterly dissimilar.

Alternate Glacial Periods in the North and South.

But we must return to our more immediate subject. I am convinced that Forbes's view may be largely extended. In Europe we meet with the plainest evidence of the Glacial period, from the western shores of Britain to the Oural range, and southward to the Pyrenees. We may infer from the frozen mammals and nature of the mountain vegetation, that Siberia was similarly affected. In the Lebanon, according to Dr. Hooker, perpetual snow formerly covered the central axis, and fed glaciers which rolled 4000 feet down the valleys. The same observer has recently found great moraines at a low level on the Atlas range in N. Africa. Along the Himalaya, at points 900 miles apart, glaciers have left the marks of their former low descent; and in Sikkim, Dr. Hooker saw maize growing on ancient and gigantic moraines. Southward of the Asiatic continent, on the opposite side of the equator, we know, from the excellent researches of Dr. J. Haast and Dr. Hector, that in New Zealand immense glaciers formerly descended to a low level; and the same plants found by Dr. Hooker on widely separated mountains in this island tell the same story of a former cold period. From facts communicated to me by the Rev. W. B. Clarke, it appears also that there are traces of former glacial action on the mountains of the south-eastern corner of Australia.

Looking to America; in the northern half, ice-borne fragments of rock have been observed on the eastern side of the continent, as far south as lat. 36°-37°, and on the shores of the Pacific, where the climate is now so different, as far south as lat. 46°. Erratic boulders have, also, been noticed on the Rocky Mountains. In the Cordillera of South America, nearly under the equator, glaciers once extended far below their present level. In Central Chile I examined a vast mound of detritus with great boulders, crossing the Portillo valley, which there can hardly be a doubt once formed a huge moraine; and Mr. D. Forbes informs me that he found in various parts of the Cordillera, from lat. 13° to 30° S., at about the height of 12,000 feet, deeply-furrowed rocks, resembling those with which he was familiar in Norway, and likewise great masses of detritus, including grooved pebbles.   Along this whole space of the Cordillera true glaciers do not now exist even at much more considerable heights.   Farther south on both sides of the continent, from lat. 41° to the southernmost extremity, we have the clearest evidence of former glacial action, in numerous immense boulders transported far from their parent source.

From these several facts, namely from the glacial action having extended all round the northern and southern hemispheres—from the period having been in a geological sense recent in both hemispheres—from its having lasted in both during a great length of time, as may be inferred from the amount of work effected—and lastly from glaciers having recently descended to a low level along the whole line of the Cordillera, it at one time appeared to me that we could not avoid the conclusion that the temperature of the whole world had been simultaneously lowered during the Glacial period.   But now Mr. Croll, in a series of admirable memoirs, has attempted to show that a glacial condition of climate is the result of various physical causes, brought into operation by an increase in the eccentricity of the earth's orbit.   All these causes tend towards the same end; but the most powerful appears to be the indirect influence of the eccentricity of the orbit upon oceanic currents.   According to Mr. Croll, cold periods

regularly recur every ten or fifteen thousand years; and
these at long intervals are extremely severe, owing to
certain contingencies, of which the most important, as Sir
C. Lyell has shown, is the relative position of the land
and water. Mr. Croll believes that the last great Glacial
period occurred about 240,000 years ago, and endured
with slight alterations of climate for about 160,000
years. With respect to more ancient Glacial periods,
several geologists are convinced from direct evidence
that such occurred during the Miocene and Eocene
formations, not to mention still more ancient formations.
But the most important result for us, arrived at by Mr.
Croll, is that whenever the northern hemisphere passes
through a cold period the temperature of the southern
hemisphere is actually raised, with the winters rendered
much milder, chiefly through changes in the direction of
the ocean-currents. So conversely it will be with the
northern hemisphere, whilst the southern passes though
a Glacial period. This conclusion throws so much light
on geographical distribution that I am strongly inclined
to trust in it; but I will first give the facts, which
demand an explanation.

In South America, Dr. Hooker has shown that
besides many closely allied species, between forty and
fifty of the flowering plants of Tierra del Fuego, forming
no inconsiderable part of its scanty flora, are common
to North America and Europe, enormously remote as
these areas in opposite hemispheres are from each
other. On the lofty mountains of equatorial America
a host of peculiar species belonging to European
genera occur. On the Organ mountains of Brazil,
some few temperate European, some Antarctic, and
some Andean genera were found by Gardner, which do

not exist in the low intervening hot countries. On the Silla of Caraccas, the illustrious Humboldt long ago found species belonging to genera characteristic of the Cordillera.

In Africa, several forms characteristic of Europe and some few representatives of the flora of the Cape of Good Hope occur on the mountains of Abyssinia. At the Cape of Good Hope a very few European species, believed not to have been introduced by man, and on the mountains several representative European forms are found, which have not been discovered in the intertropical parts of Africa. Dr. Hooker has also lately shown that several of the plants living on the upper parts of the lofty island of Fernando Po and on the neighbouring Cameroon mountains, in the Gulf of Guinea, are closely related to those on the mountains of Abyssinia, and likewise to those of temperate Europe. It now also appears, as I hear from Dr. Hooker, that some of these same temperate plants have been discovered by the Rev. R. T. Lowe on the mountains of the Cape Verde islands. This extension of the same temperate forms, almost under the equator, across the whole continent of Africa and to the mountains of the Cape Verde archipelago, is one of the most astonishing facts ever recorded in the distribution of plants.

On the Himalaya, and on the isolated mountain-ranges of the peninsula of India, on the heights of Ceylon, and on the volcanic cones of Java, many plants occur, either identically the same or representing each other, and at the same time representing plants of Europe, not found in the intervening hot lowlands. A list of the genera of plants collected on the loftier peaks of Java, raises a picture of a collection made on

a hillock in Europe! Still more striking is the fact
that peculiar Australian forms are represented by
certain plants growing on the summits of the mountains of Borneo. Some of these Australian forms, as I
hear from Dr. Hooker, extend along the heights of the
peninsula of Malacca, and are thinly scattered on the
one hand over India, and on the other hand as far
north as Japan.

On the southern mountains of Australia, Dr.
F. Müller has discovered several European species;
other species, not introduced by man, occur on the
lowlands; and a long list can be given, as I am
informed by Dr. Hooker, of European genera, found in
Australia, but not in the intermediate torrid regions.
In the admirable 'Introduction to the Flora of New
Zealand,' by Dr. Hooker, analogous and striking facts
are given in regard to the plants of that large island.
Hence we see that certain plants growing on the more
lofty mountains of the tropics in all parts of the world,
and on the temperate plains of the north and south, are
either the same species or varieties of the same species.
It should, however, be observed that these plants are
not strictly arctic forms; for, as Mr. H. C. Watson
has remarked, " in receding from polar towards equatorial latitudes, the Alpine or mountain floras really
become less and less Arctic." Besides these identical
and closely allied forms, many species inhabiting the
same widely sundered areas, belong to genera not now
found in the intermediate tropical lowlands.

These brief remarks apply to plants alone; but some
few analogous facts could be given in regard to terrestrial animals. In marine productions, similar cases
likewise occur; as an example, I may quote a statement

by the highest authority, Prof. Dana, that "it is certainly a wonderful fact that New Zealand should have a closer resemblance in its crustacea to Great Britain, its antipode, than to any other part of the world." Sir J. Richardson, also, speaks of the reappearance on the shores of New Zealand, Tasmania, &c., of northern forms of fish. Dr. Hooker informs me that twenty-five species of Algæ are common to New Zealand and to Europe, but have not been found in the intermediate tropical seas.

From the foregoing facts, namely, the presence of temperate forms on the highlands across the whole of equatorial Africa, and along the Peninsula of India, to Ceylon and the Malay Archipelago, and in a less well-marked manner across the wide expanse of tropical South America, it appears almost certain that at some former period, no doubt during the most severe part of a Glacial period, the lowlands of these great continents were everywhere tenanted under the equator by a considerable number of temperate forms. At this period the equatorial climate at the level of the sea was probably about the same with that now experienced at the height of from five to six thousand feet under the same latitude, or perhaps even rather cooler. During this, the coldest period, the lowlands under the equator must have been clothed with a mingled tropical and temperate vegetation, like that described by Hooker as growing luxuriantly at the height of from four to five thousand feet on the lower slopes of the Himalaya, but with perhaps a still greater preponderance of temperate forms. So again in the mountainous island of Fernando Po, in the Gulf of Guinea, Mr. Mann found temperate European forms beginning to appear at the height of about five thousand feet. On the mountains of Panama, at the height of only two thousand feet, Dr. Seemann found the vegetation like that of Mexico, "with forms of the torrid zone harmoniously blended with those of the temperate."

Now let us see whether Mr. Croll's conclusion that when the northern hemisphere suffered from the extreme cold of the great Glacial period, the southern hemisphere was actually warmer, throws any clear light on the present apparently inexplicable distribution of various organisms in the temperate parts of both hemispheres, and on the mountains of the tropics. The Glacial period, as measured by years, must have been very long; and when we remember over what vast spaces some naturalised plants and animals have spread within a few centuries, this period will have been ample for any amount of migration. As the cold became more and more intense, we know that Arctic forms invaded the temperate regions; and, from the facts just given, there can hardly be a doubt that some of the more vigorous, dominant and widest-spreading temperate forms invaded the equatorial lowlands. The inhabitants of these hot lowlands would at the same time have migrated to the tropical and subtropical regions of the south, for the southern hemisphere was at this period warmer. On the decline of the Glacial period, as both hemispheres gradually recovered their former temperatures, the northern temperate forms living on the lowlands under the equator, would have been driven to their former homes or have been destroyed, being replaced by the equatorial forms returning from the south. Some, however, of the northern temperate forms would almost certainly have ascended any adjoining high land,

2 M

where, if sufficiently lofty, they would have long survived like the Arctic forms on the mountains of Europe.
They might have survived, even if the climate was not
perfectly fitted for them, for the change of temperature
must have been very slow, and plants undoubtedly
possess a certain capacity for acclimatisation, as shown
by their transmitting to their offspring different constitutional powers of resisting heat and cold.

In the regular course of events the southern hemisphere would in its turn be subjected to a severe Glacial
period, with the northern hemisphere rendered warmer;
and then the southern temperate forms would invade
the equatorial lowlands. The northern forms which
had before been left on the mountains would now
descend and mingle with the southern forms. These
latter, when the warmth returned, would return to their
former homes, leaving some few species on the mountains, and carrying southward with them some of the
northern temperate forms which had descended from
their mountain fastnesses. Thus, we should have some
few species identically the same in the northern and
southern temperate zones and on the mountains of the
intermediate tropical regions. But the species left
during a long time on these mountains, or in opposite
hemispheres, would have to compete with many new
forms and would be exposed to somewhat different
physical conditions; hence they would be eminently
liable to modification, and would generally now exist
as varieties or as representative species; and this is the
case. We must, also, bear in mind the occurrence in
both hemispheres of former Glacial periods; for these
will account, in accordance with the same principles, for
the many quite distinct species inhabiting the same

widely separated areas, and belonging to genera not now found in the intermediate torrid zones.

It is a remarkable fact strongly insisted on by Hooker in regard to America, and by Alph. de Candolle in regard to Australia, that many more identical or slightly modified species have migrated from the north to the south, than in a reversed direction, We see, however, a few southern forms on the mountains of Borneo and Abyssinia. I suspect that this preponderant migration from the north to the south is due to the greater extent of land in the north, and to the northern forms having existed in their own homes in greater numbers, and having consequently been advanced through natural selection and competition to a higher stage of perfection, or dominating power, than the southern forms. And thus, when the two sets became commingled in the equatorial regions, during the alternations of the Glacial periods, the northern forms were the more powerful and were able to hold their places on the mountains, and afterwards to migrate southward with the southern forms; but not so the southern in regard to the northern forms. In the same manner at the present day, we see that very many European productions cover the ground in La Plata, New Zealand, and to a lesser degree in Australia, and have beaten the natives; whereas extremely few southern forms have become naturalised in any part of the northern hemisphere, though hides, wool, and other objects likely to carry seeds have been largely imported into Europe during the last two or three centuries from La Plata and during the last forty or fifty years from Australia. The Neilgherrie mountains in India, however, offer a partial exception; for here, as I hear from Dr. Hooker,

Australian forms are rapidly sowing themselves and becoming naturalised. Before the last great Glacial period, no doubt the intertropical mountains were stocked with endemic Alpine forms; but these have almost everywhere yielded to the more dominant forms generated in the larger areas and more efficient workshops of the north. In many islands the native productions are nearly equalled, or even outnumbered, by those which have become naturalised; and this is the first stage towards their extinction. Mountains are islands on the land, and their inhabitants have yielded to those produced within the larger areas of the north, just in the same way as the inhabitants of real islands have everywhere yielded and are still yielding to continental forms naturalised through man's agency.

The same principles apply to the distribution of terrestrial animals and of marine productions, in the northern and southern temperate zones, and on the intertropical mountains. When, during the height of the Glacial period, the ocean-currents were widely different to what they now are, some of the inhabitants of the temperate seas might have reached the equator; of these a few would perhaps at once be able to migrate southward, by keeping to the cooler currents, whilst others might remain and survive in the colder depths until the southern hemisphere was in its turn subjected to a glacial climate and permitted their further progress; in nearly the same manner as, according to Forbes, isolated spaces inhabited by Arctic productions exist to the present day in the deeper parts of the northern temperate seas.

I am far from supposing that all the difficulties in regard to the distribution and affinities of the identical

and allied species, which now live so widely separated
in the north and south, and sometimes on the intermediate mountain-ranges, are removed on the views
above given. The exact lines of migration cannot be
indicated. We cannot say why certain species and not
others have migrated; why certain species have been
modified and have given rise to new forms, whilst others
have remained unaltered. We cannot hope to explain
such facts, until we can say why one species and not
another becomes naturalised by man's agency in a
foreign land; why one species ranges twice or thrice as
far, and is twice or thrice as common, as another species
within their own homes.

Various special difficulties also remain to be solved;
for instance, the occurrence, as shown by Dr. Hooker, of
the same plants at points so enormously remote as
Kerguelen Land, New Zealand, and Fuegia; but icebergs,
as suggested by Lyell, may have been concerned in their
dispersal. The existence at these and other distant
points of the southern hemisphere, of species, which,
though distinct, belong to genera exclusively confined
to the south, is a more remarkable case. Some of these
species are so distinct, that we cannot suppose that there
has been time since the commencement of the last
Glacial period for their migration and subsequent
modification to the necessary degree. The facts seem to
indicate that distinct species belonging to the same
genera have migrated in radiating lines from a common
centre; and I am inclined to look in the southern, as in
the northern hemisphere, to a former and warmer period,
before the commencement of the last Glacial period,
when the Antarctic lands, now covered with ice, supported
a highly peculiar and isolated flora. It may be suspected

that before this flora was exterminated during the last Glacial epoch, a few forms had been already widely dispersed to various points of the southern hemisphere by occasional means of transport, and by the aid as halting-places, of now sunken islands. Thus the southern shores of America, Australia, and New Zealand may have become slightly tinted by the same peculiar forms of life.

Sir C. Lyell in a striking passage has speculated, in language almost identical with mine, on the effects of great alterations of climate throughout the world on geographical distribution. And we have now seen that Mr. Croll's conclusion that successive Glacial periods in the one hemisphere coincide with warmer periods in the opposite hemisphere, together with the admission of the slow modification of species, explains a multitude of facts in the distribution of the same and of the allied forms of life in all parts of the globe. The living waters have flowed during one period from the north and during another from the south, and in both cases have reached the equator; but the stream of life has flowed with greater force from the north than in the opposite direction. and has consequently more freely inundated the south, As the tide leaves its drift in horizontal lines, rising higher on the shores where the tide rises highest, so have the living waters left their living drift on our mountain summits, in a line gently rising from the Arctic lowlands to a great altitude under the equator. The various beings thus left stranded may be compared with savage races of man, driven up and surviving in the mountain fastnesses of almost every land, which serves as a record, full of interest to us, of the former inhabitants of the surrounding lowlands.

CHAPTER XIII.

GEOGRAPHICAL DISTRIBUTION—*continued.*

Distribution of fresh-water productions—On the inhabitants of oceanic islands—Absence of Batrachians and of terrestrial Mammals—On the relation of the inhabitants of islands to those of the nearest mainland—On colonisation from the nearest source with subsequent modification—Summary of the last and present chapter.

Fresh-water Productions.

As lakes and river-systems are separated from each other by barriers of land, it might have been thought that fresh-water productions would not have ranged widely within the same country, and as the sea is apparently a still more formidable barrier, that they would never have extended to distant countries. But the case is exactly the reverse. Not only have many fresh-water species, belonging to different classes, an enormous range, but allied species prevail in a remarkable manner throughout the world. When first collecting in the fresh waters of Brazil, I well remember feeling much surprise at the similarity of the fresh-water insects, shells, &c., and at the dissimilarity of the surrounding terrestrial beings, compared with those of Britain.

But the wide ranging power of fresh-water productions can, I think, in most cases be explained by their having become fitted, in a manner highly useful to them, for

short and frequent migrations from pond to pond, or
from stream to stream, within their own countries;
and liability to wide dispersal would follow from this
capacity as an almost necessary consequence.   We can
here consider only a few cases; of these, some of the
most difficult to explain are presented by fish.   It was
formerly believed that the same fresh-water species
never existed on two continents distant from each other.
But Dr. Günther has lately shown that the Galaxias
attenuatus inhabits Tasmania, New Zealand, the Falkland Islands, and the mainland of South America.   This
is a wonderful case, and probably indicates dispersal
from an Antarctic centre during a former warm period.
This case, however, is rendered in some degree less
surprising by the species of this genus having the
power of crossing by some unknown means considerable
spaces of open ocean : thus there is one species common
to New Zealand and to the Auckland Islands, though
separated by a distance of about 230 miles.   On the same
continent fresh-water fish often range widely, and as if
capriciously; for in two adjoining river-systems some of
the species may be the same, and some wholly different.

It is probable that they are occasionally transported
by what may be called accidental means.   Thus fishes
still alive are not very rarely dropped at distant points
by whirlwinds; and it is known that the ova retain
their vitality for a considerable time after removal from
the water.   Their dispersal may, however, be mainly
attributed to changes in the level of the land within the
recent period, causing rivers to flow into each other.
Instances, also, could be given of this having occurred
during floods, without any change of level.   The wide
difference of the fish on the opposite sides of most

mountain-ranges, which are continuous, and which consequently must from an early period have completely
prevented the inosculation of the river-systems on the
two sides, leads to the same conclusion.   Some fresh-
water fish belong to very ancient forms, and in such cases
there will have been ample time for great geographical
changes, and consequently time and means for much
migration.   Moreover Dr. Günther has recently been
led by several considerations to infer that with fishes
the same forms have a long endurance.   Salt-water fish
can with care be slowly accustomed to live in fresh
water; and, according to Valenciennes, there is hardly a
single group of which all the members are confined to
fresh water, so that a marine species belonging to a
fresh-water group might travel far along the shores of
the sea, and could, it is probable, become adapted
without much difficulty to the fresh waters of a distant
land.

Some species of fresh-water shells have very wide
ranges, and allied species which, on our theory, are
descended from a common parent, and must have proceeded from a single source, prevail throughout the
world.   Their distribution at first perplexed me much,
as their ova are not likely to be transported by birds;
and the ova, as well as the adults, are immediately
killed by sea-water.   I could not even understand how
some naturalised species have spread rapidly throughout the same country.   But two facts, which I have
observed—and many others no doubt will be discovered
—throw some light on this subject.   When ducks
suddenly emerge from a pond covered with duck-weed,
I have twice seen these little plants adhering to their
backs; and it has happened to me, in removing a little

duck-weed from one aquarium to another, that I have unintentionally stocked the one with fresh-water shells from the other. But another agency is perhaps more effectual: I suspended the feet of a duck in an aquarium, where many ova of fresh-water shells were hatching; and I found that numbers of the extremely minute and just-hatched shells crawled on the feet, and clung to them so firmly that when taken out of the water they could not be jarred off, though at a somewhat more advanced age they would voluntarily drop off. These just-hatched molluscs, though aquatic in their nature, survived on the duck's feet, in damp air, from twelve to twenty hours; and in this length of time a duck or heron might fly at least six or seven hundred miles, and if blown across the sea to an oceanic island, or to any other distant point, would be sure to alight on a pool or rivulet. Sir Charles Lyell informs me that a Dytiscus has been caught with an Ancylus (a fresh-water shell like a limpet) firmly adhering to it; and a water-beetle of the same family, a Colymbetes, once flew on board the 'Beagle,' when forty-five miles distant from the nearest land: how much farther it might have been blown by a favouring gale no one can tell.

With respect to plants, it has long been known what enormous ranges many fresh-water, and even marsh species, have, both over continents and to the most remote oceanic islands. This is strikingly illustrated, according to Alph. de Candolle, in those large groups of terrestrial plants, which have very few aquatic members; for the latter seem immediately to acquire, as if in consequence, a wide range. I think favourable means of dispersal explain this fact. I have before mentioned that earth occasionally adheres in some quantity to the

feet and beaks of birds. Wading birds, which frequent
the muddy edges of ponds, if suddenly flushed, would be
the most likely to have muddy feet. Birds of this order
wander more than those of any other; and they are
occasionally found on the most remote and barren
islands of the open ocean; they would not be likely to
alight on the surface of the sea, so that any dirt on their
feet would not be washed off; and when gaining the
land, they would be sure to fly to their natural fresh-
water haunts. I do not believe that botanists are
aware how charged the mud of ponds is with seeds; I
have tried several little experiments, but will here give
only the most striking case: I took in February three
table-spoonfuls of mud from three different points,
beneath water, on the edge of a little pond: this mud
when dried weighed only 6¾ ounces; I kept it covered
up in my study for six months, pulling up and counting
each plant as it grew; the plants were of many kinds,
and were altogether 537 in number; and yet the viscid
mud was all contained in a breakfast cup! Considering
these facts, I think it would be an inexplicable circumstance if water-birds did not transport the seeds
of fresh-water plants to unstocked ponds and streams,
situated at very distant points. The same agency may
have come into play with the eggs of some of the
smaller fresh-water animals.

Other and unknown agencies probably have also
played a part. I have stated that fresh-water fish eat
some kinds of seeds, though they reject many other
kinds after having swallowed them; even small fish
swallow seeds of moderate size, as of the yellow water-
lily and Potamogeton. Herons and other birds, century
after century, have gone on daily devouring fish; they

then take flight and go to other waters, or are blown across the sea; and we have seen that seeds retain their power of germination, when rejected many hours afterwards in pellets or in the excrement. When I saw the great size of the seeds of that fine water-lily, the Nelumbium, and remembered Alph. de Candolle's remarks on the distribution of this plant, I thought that the means of its dispersal must remain inexplicable; but Audubon states that he found the seeds of the great southern water-lily (probably, according to Dr. Hooker, the Nelumbium luteum) in a heron's stomach. Now this bird must often have flown with its stomach thus well stocked to distant ponds, and then getting a hearty meal of fish, analogy makes me believe that it would have rejected the seeds in a pellet in a fit state for germination.

In considering these several means of distribution, it should be remembered that when a pond or stream is first formed, for instance, on a rising islet, it will be unoccupied; and a single seed or egg will have a good chance of succeeding. Although there will always be a struggle for life between the inhabitants of the same pond, however few in kind, yet as the number even in a well-stocked pond is small in comparison with the number of species inhabiting an equal area of land, the competition between them will probably be less severe than between terrestrial species; consequently an intruder from the waters of a foreign country would have a better chance of seizing on a new place, than in the case of terrestrial colonists. We should also remember that many fresh-water productions are low in the scale of nature, and we have reason to believe that such beings become modified . more slowly than the high;

and this will give time for the migration of aquatic species. We should not forget the probability of many fresh-water forms having formerly ranged continuously over immense areas, and then having become extinct at intermediate points. But the wide distribution of fresh-water plants and of the lower animals, whether retaining the same identical form or in some degree modified, apparently depends in main part on the wide dispersal of their seeds and eggs by animals, more especially by fresh-water birds, which have great powers of flight, and naturally travel from one piece of water to another.

On the Inhabitants of Oceanic Islands.

We now come to the last of the three classes of facts, which I have selected as presenting the greatest amount of difficulty with respect to distribution, on the view that not only all the individuals of the same species have migrated from some one area, but that allied species, although now inhabiting the most distant points, have proceeded from a single area,—the birth-place of their early progenitors. I have already given my reasons for disbelieving in continental extensions within the period of existing species, on so enormous a scale that all the many islands of the several oceans were thus stocked with their present terrestrial inhabitants. This view removes many difficulties, but it does not accord with all the facts in regard to the productions of islands. In the following remarks I shall not confine myself to the mere question of dispersal, but shall consider some other cases bearing on the truth of the two theories of independent creation and of descent with modification.

The species of all kinds which inhabit oceanic islands are few in number compared with those on equal continental areas: Alph. de Candolle admits this for plants, and Wollaston for insects. New Zealand, for instance, with its lofty mountains and diversified stations, extending over 780 miles of latitude, together with the outlying islands of Auckland, Campbell and Chatham, contain altogether only 960 kinds of flowering plants; if we compare this moderate number with the species which swarm over equal areas in South-Western Australia or at the Cape of Good Hope, we must admit that some cause, independently of different physical conditions, has given rise to so great a difference in number. Even the uniform county of Cambridge has 847 plants, and the little island of Anglesea 764, but a few ferns and a few introduced plants are included in these numbers, and the comparison in some other respects is not quite fair. We have evidence that the barren island of Ascension aboriginally possessed less than half-a-dozen flowering plants; yet many species have now become naturalised on it, as they have in New Zealand and on every other oceanic island which can be named. In St. Helena there is reason to believe that the naturalised plants and animals have nearly or quite exterminated many native productions. He who admits the doctrine of the creation of each separate species, will have to admit that a sufficient number of the best adapted plants and animals were not created for oceanic islands; for man has unintentionally stocked them far more fully and perfectly than did nature.

Although in oceanic islands the species are few in number, the proportion of endemic kinds (*i. e.* those

found nowhere else in the world) is often extremely
large. If we compare, for instance, the number of
endemic land-shells in Madeira, or of endemic birds in
the Galapagos Archipelago, with the number found on
any continent, and then compare the area of the island
with that of the continent, we shall see that this is
true. This fact might have been theoretically expected,
for, as already explained, species occasionally arriving
after long intervals of time in the new and isolated
district, and having to compete with new associates,
would be eminently liable to modification, and would
often produce groups of modified descendants. But it
by no means follows that, because in an island nearly
all the species of one class are peculiar, those of
another class, or of another section of the same class,
are peculiar; and this difference seems to depend
partly on the species which are not modified having
immigrated in a body, so that their mutual relations
have not been much disturbed; and partly on the
frequent arrival of unmodified immigrants from the
mother-country, with which the insular forms have
intercrossed. It should be borne in mind that the
offspring of such crosses would certainly gain in
vigour; so that even an occasional cross would produce
more effect than might have been anticipated. I will
give a few illustrations of the foregoing remarks: in
the Galapagos Islands there are 26 land-birds; of these
21 (or perhaps 23) are peculiar, whereas of the 11
marine birds only 2 are peculiar; and it is obvious that
marine birds could arrive at these islands much more
easily and frequently than land-birds. Bermuda, on
the other hand, which lies at about the same distance
from North America as the Galapagos Islands do from

South America, and which has a very peculiar soil, does not possess a single endemic land-bird; and we know from Mr. J. M. Jones's admirable account of Bermuda, that very many North American birds occasionally or even frequently visit this island. Almost every year, as I am informed by Mr. E. V. Harcourt, many European and African birds are blown to Madeira; this island is inhabited by 99 kinds, of which one alone is peculiar, though very closely related to a European form; and three or four other species are confined to this island and to the Canaries. So that the Islands of Bermuda and Madeira have been stocked from the neighbouring continents with birds, which for long ages have there struggled together, and have become mutually co-adapted. Hence when settled in their new homes, each kind will have been kept by the others to its proper place and habits, and will consequently have been but little liable to modification. Any tendency to modification will also have been checked by intercrossing with the unmodified inmigrants, often arriving from the mother-country. Madeira again is inhabited by a wonderful number of peculiar land-shells, whereas not one species of sea-shell is peculiar to its shores : now, though we do not know how sea-shells are dispersed, yet we can see that their eggs or larvæ, perhaps attached to seaweed or floating timber, or to the feet of wading-birds, might be transported across three or four hundred miles of open sea far more easily than land-shells. The different orders of insects inhabiting Madeira present nearly parallel cases.

Oceanic islands are sometimes deficient in animals of certain whole classes, and their places are occupied by other classes; thus in the Galapagos Islands reptiles, and in New Zealand gigantic wingless birds, take, or recently took, the place of mammals. Although New Zealand is here spoken of as an oceanic island, it is in some degree doubtful whether it should be so ranked; it is of large size, and is not separated from Australia by a profoundly deep sea; from its geological character and the direction of its mountain-ranges, the Rev. W. B. Clarke has lately maintained that this island, as well as New Caledonia, should be considered as appurtenances of Australia. Turning to plants, Dr. Hooker has shown that in the Galapagos Islands the proportional numbers of the different orders are very different from what they are elsewhere. All such differences in number, and the absence of certain whole groups of animals and plants, are generally accounted for by supposed differences in the physical conditions of the islands; but this explanation is not a little doubtful. Facility of immigration seems to have been fully as important as the nature of the conditions.

Many remarkable little facts could be given with respect to the inhabitants of oceanic islands. For instance, in certain islands not tenanted by a single mammal, some of the endemic plants have beautifully hooked seeds; yet few relations are more manifest than that hooks serve for the transportal of seeds in the wool or fur of quadrupeds. But a hooked seed might be carried to an island by other means; and the plant then becoming modified would form an endemic species, still retaining its hooks, which would form a useless appendage like the shrivelled wings under the soldered wing-covers of many insular beetles. Again, islands often possess trees or bushes belonging to orders which elsewhere include only herbaceous species; now trees, as Alph. de Candolle has shown, generally have, whatever the cause may be, confined ranges   Hence trees would be little likely to reach distant oceanic islands; and an herbaceous plant, which had no chance of successfully competing with the many fully developed trees growing on a continent, might, when established on an island, gain an advantage over other herbaceous plants by growing taller and taller and overtopping them.   In this case, natural selection would tend to add to the stature of the plant, to whatever order it belonged, and thus first convert it into a bush and then into a tree.

Absence of Batrachians and Terrestrial Mammals on Oceanic Islands.

With respect to the absence of whole orders of animals on oceanic islands, Bory St. Vincent long ago remarked that Batrachians (frogs, toads, newts) are never found on any of the many islands with which the great oceans are studded.   I have taken pains to verify this assertion, and have found it true, with the exception of New Zealand, New Caledonia, the Andaman Islands, and perhaps the Salomon Islands and the Seychelles.   But I have already remarked that it is doubtful whether New Zealand and New Caledonia ought to be classed as oceanic islands; and this is still more doubtful with respect to the Andaman and Salomon groups and the Seychelles.   This general absence of frogs, toads, and newts on so many true oceanic islands cannot be accounted for by their physical conditions: indeed it seems that islands are peculiarly fitted for these animals; for frogs have been introduced into Madeira, the Azores, and Mauritius, and have multiplied so as to become a nuisance. But as these animals and their spawn are immediately killed (with the exception, as far as known, of one Indian species) by sea-water, there would be great difficulty in their transportal across the sea, and therefore we can see why they do not exist on strictly oceanic islands. But why, on the theory of creation, they should not have been created there, it would be very difficult to explain.

Mammals offer another and similar case. I have carefully searched the oldest voyages, and have not found a single instance, free from doubt, of a terrestrial mammal (excluding domesticated animals kept by the natives) inhabiting an island situated above 300 miles from a continent or great continental island; and many islands situated at a much less distance are equally barren. The Falkland Islands, which are inhabited by a wolf-like fox, come nearest to an exception; but this group cannot be considered as oceanic, as it lies on a bank in connection with the mainland at the distance of about 280 miles; moreover, icebergs formerly brought boulders to its western shores, and they may have formerly transported foxes, as now frequently happens in the arctic regions. Yet it cannot be said that small islands will not support at least small mammals, for they occur in many parts of the world on very small islands, when lying close to a continent; and hardly an island can be named on which our smaller quadrupeds have not become naturalised and greatly multiplied. It cannot be said, on the ordinary view of creation, that there has not been time for the creation of mammals; many volcanic islands are sufficiently

ancient, as shown by the stupendous degradation which
they have suffered, and by their tertiary strata : there
has also been time for the production of endemic species
belonging to other classes; and on continents it is
known that new species of mammals appear and
disappear at a quicker rate than other and lower
animals. Although terrestrial mammals do not occur
on oceanic islands, aerial mammals do occur on almost
every island. New Zealand possesses two bats found
nowhere else in the world: Norfolk Island, the Viti
Archipelago, the Bonin Islands, the Caroline and
Marianne Archipelagoes, and Mauritius, all possess
their peculiar bats. Why, it may be asked, has the
supposed creative force produced bats and no other
mammals on remote islands ? On my view this
question can easily be answered; for no terrestrial
mammal can be transported across a wide space of sea,
but bats can fly across. Bats have been seen wandering
by day far over the Atlantic Ocean ; and two North
American species either regularly or occasionally visit
Bermuda, at the distance of 600 miles from the mainland. I hear from Mr. Tomes, who has specially
studied this family, that many species have enormous
ranges, and are found on continents and on far distant
islands. Hence we have only to suppose that such
wandering species have been modified in their new
homes in relation to their new position, and we can
understand the presence of endemic bats on oceanic
islands, with the absence of all other terrestrial mammals.

Another interesting relation exists, namely between
the depth of the sea separating islands from each other
or from the nearest continent, and the degree of affinity

of their mammalian inhabitants.   Mr. Windsor Earl
has made some striking observations on this head, since
greatly extended by Mr. Wallace's admirable researches,
in regard to the great Malay Archipelago, which is
traversed near Celebes by a space of deep ocean, and
this separates two widely distinct mammalian faunas.
On either side the islands stand on a moderately shallow
submarine bank, and these islands are inhabited by the
same or by closely allied quadrupeds.   I have not as
yet had time to follow up this subject in all quarters of
the world ; but as far as I have gone, the relation holds
good.   For instance, Britain is separated by a shallow
channel from Europe, and the mammals are the same
on both sides; and so it is with all the islands near the
shores of Australia.   The West Indian Islands, on the
other hand, stand on a deeply submerged bank, nearly
1000 fathoms in depth, and here we find American
forms, but the species and even the genera are quite
distinct.   As the amount of modification which animals
of all kinds undergo partly depends on the lapse of
time, and as the islands which are separated from each
other or from the mainland by shallow channels, are
more likely to have been continuously united within a
recent period than the islands separated by deeper
channels, we can understand how it is that a relation
exists between the depth of the sea separating two
mammalian faunas, and the degree of their affinity,—
a relation which is quite inexplicable on the theory of
independent acts of creation.

The foregoing statements in regard to the inhabitants
of oceanic islands,—namely, the fewness of the species,
with a large proportion consisting of endemic forms—
the members of certain groups, but not those of other

groups in the same class, having been modified—the absence of certain whole orders, as of batrachians and of terrestrial mammals, notwithstanding the presence of aerial bats,—the singular proportions of certain orders of plants,—herbaceous forms having been developed into trees, &c.,—seem to me to accord better with the belief in the efficiency of occasional means of transport, carried on during a long course of time, than with the belief in the former connection of all oceanic islands with the nearest continent; for on this latter view it is probable that the various classes would have immigrated more uniformly, and from the species having entered in a body their mutual relations would not have been much disturbed, and consequently they would either have not been modified, or all the species in a more equable manner.

I do not deny that there are many and serious difficulties in understanding how many of the inhabitants of the more remote islands, whether still retaining the same specific form or subsequently modified, have reached their present homes. But the probability of other islands having once existed as halting-places, of which not a wreck now remains, must not be overlooked. I will specify one difficult case. Almost all oceanic islands, even the most isolated and smallest, are inhabited by land-shells, generally by endemic species, but sometimes by species found elsewhere,—striking instances of which have been given by Dr. A. A. Gould in relation to the Pacific. Now it is notorious that land-shells are easily killed by sea-water; their eggs, at least such as I have tried, sink in it and are killed. Yet there must be some unknown, but occasionally efficient means for their transportal.

Would the just-hatched young sometimes adhere to the feet of birds roosting on the ground, and thus get transported ? It occurred to me that land-shells, when hybernating and having a membranous diaphragm over the mouth of the shell, might be floated in chinks of drifted timber across moderately wide arms of the sea. And I find that several species in this state withstand uninjured an immersion in sea-water during seven days: one shell, the Helix pomatia, after having been thus treated and again hybernating was put into sea-water for twenty days, and perfectly recovered. During this length of time the shell might have been carried by a marine current of average swiftness, to a distance of 660 geographical miles. As this Helix has a thick calcareous operculum, I removed it, and when it had formed a new membranous one, I again immersed it for fourteen days in sea-water, and again it recovered and crawled away. Baron Aucapitaine has since tried similar experiments : he placed 100 land-shells, belonging to ten species, in a box pierced with holes, and immersed it for a fortnight in the sea. Out of the hundred shells, twenty-seven recovered. The presence of an operculum seems to have been of importance, as out of twelve specimens of Cyclostoma elegans, which is thus furnished, eleven revived. It is remarkable, seeing how well the Helix pomatia resisted with me the salt-water, that not one of fifty-four specimens belonging to four other species of Helix tried by Aucapitaine, recovered. It is, however, not at all probable that land-shells have often been thus transported ; the feet of birds offer a more probable method.

*On the Relations of the Inhabitants of Islands to those of
the nearest Mainland.*

The most striking and important fact for us is the
affinity of the species which inhabit islands to those of
the nearest mainland, without being actually the same.
Numerous instances could be given. The Galapagos
Archipelago, situated under the equator, lies at the
distance of between 500 and 600 miles from the shores
of South America. Here almost every product of the
land and of the water bears the unmistakable stamp of
the American continent. There are twenty-six land-
birds; of these, twenty-one, or perhaps twenty-three
are ranked as distinct species, and would commonly be
assumed to have been here created; yet the close affinity
of most of these birds to American species is manifest
in every character, in their habits, gestures, and tones
of voice. So it is with the other animals, and with a
large proportion of the plants, as shown by Dr. Hooker
in his admirable Flora of this archipelago. The naturalist, looking at the inhabitants of these volcanic
islands in the Pacific, distant several hundred miles
from the continent, feels that he is standing on American
land. Why should this be so? why should the species
which are supposed to have been created in the Galapagos Archipelago, and nowhere else, bear so plainly
the stamp of affinity to those created in America?
There is nothing in the conditions of life, in the
geological nature of the islands, in their height or
climate, or in the proportions in which the several
classes are associated together, which closely resembles
the conditions of the South American coast: in fact,
there is a considerable dissimilarity in all these respects.

On the other hand, there is a considerable degree of resemblance in the volcanic nature of the soil, in the climate, height, and size of the islands, between the Galapagos and Cape Verde Archipelagoes : but what an entire and absolute difference in their inhabitants ! The inhabitants of the Cape Verde Islands are related to those of Africa, like those of the Galapagos to America. Facts such as these, admit of no sort of explanation on the ordinary view of independent creation ; whereas on the view here maintained, it is obvious that the Galapagos Islands would be likely to receive colonists from America, whether by occasional means of transport or (though I do not believe in this doctrine) by formerly continuous land, and the Cape Verde Islands from Africa ; such colonists would be liable to modification,—the principle of inheritance still betraying their original birthplace.

Many analogous facts could be given : indeed it is an almost universal rule that the endemic productions of islands are related to those of the nearest continent, or of the nearest large island. The exceptions are few, and most of them can be explained. Thus although Kerguelen Land stands nearer to Africa than to America, the plants are related, and that very closely, as we know from Dr. Hooker's account, to those of America : but on the view that this island has been mainly stocked by seeds brought with earth and stones on icebergs, drifted by the prevailing currents, this anomaly disappears. New Zealand in its endemic planes is much more closely related to Australia, the nearest mainland, than to any other region : and this is what might have been expected ; but it is also plainly related to South America, which, although the next nearest continent, is so enormously remote, that the fact becomes an anomaly. But this difficulty partially disappears on the view that New Zealand, South America, and the other southern lands have been stocked in part from a nearly intermediate though distant point, namely from the antarctic islands, when they were clothed with vegetation, during a warmer tertiary period, before the commencement of the last Glacial period. The affinity, which though feeble, I am assured by Dr. Hooker is real, between the flora of the south-western corner of Australia and of the Cape of Good Hope, is a far more remarkable case; but this affinity is confined to the plants, and will, no doubt, some day be explained.

The same law which has determined the relationship between the inhabitants of islands and the nearest mainland, is sometimes displayed on a small scale, but in a most interesting manner, within the limits of the same archipelago. Thus each separate island of the Galapagos Archipelago is tenanted, and the fact is a marvellous one, by many distinct species; but these species are related to each other in a very much closer manner than to the inhabitants of the American continent, or of any other quarter of the world. This is what might have been expected, for islands situated so near to each other would almost necessarily receive immigrants from the same original source, and from each other. But how is it that many of the immigrants have been differently modified, though only in a small degree, in islands situated within sight of each other, having the same geological nature, the same height, climate, &c.? This long appeared to me a great difficulty: but it arises in chief part from the deeply-seated error of considering the physical conditions of a country as the most important; whereas it cannot be disputed that the nature of the other species with which each has to compete, is at least as important, and generally a far more important element of success. Now if we look to the species which inhabit the Galapagos Archipelago, and are likewise found in other parts of the world, we find that they differ considerably in the several islands. This difference might indeed have been expected if the islands have been stocked by occasional means of transport—a seed, for instance, of one plant having been brought to one island, and that of another plant to another island, though all proceeding from the same general source. Hence, when in former times an immigrant first settled on one of the islands, or when it subsequently spread from one to another, it would undoubtedly be exposed to different conditions in the different islands, for it would have to compete with a different set of organisms; a plant, for instance, would find the ground best fitted for it occupied by somewhat different species in the different islands, and would be exposed to the attacks of somewhat different enemies. If then it varied, natural selection would probably favour different varieties in the different islands. Some species, however, might spread and yet retain the same character throughout the group, just as we see some species spreading widely throughout a continent and remaining the same.

The really surprising fact in this case of the Galapagos Archipelago, and in a lesser degree in some analogous cases, is that each new species after being formed in any one island, did not spread quickly to the other islands. But the islands, though in sight of each other, are separated by deep arms of the sea, in most cases wider

than the British Channel, and there is no reason to
suppose that they have at any former period been
continuously united.   The currents of the sea are rapid
and sweep between the islands, and gales of wind are
extraordinarily rare; so that the islands are far more
effectually separated from each other than they appear
on a map.   Nevertheless some of the species, both of
those found in other parts of the world and of those
confined to the archipelago, are common to the several
islands; and we may infer from their present manner
of distribution, that they have spread from one island
to the others.   But we often take, I think, an erroneous
view of the probability of closely-allied species invading
each other's territory, when put into free intercommuni-
cation.   Undoubtedly, if one species has any advantage
over another, it will in a very brief time wholly or in
part supplant it; but if both are equally well fitted for
their own places, both will probably hold their separate
places for almost any length of time.   Being familiar
with the fact that many species, naturalised through
man's agency, have spread with astonishing rapidity
over wide areas, we are apt to infer that most species
would thus spread; but we should remember that the
species which become naturalised in new countries are
not generally closely allied to the aboriginal inhabitants,
but are very distinct forms, belonging in a large proportion of cases, as shown by Alph. de Candolle, to
distinct genera.   In the Galapagos Archipelago, many
even of the birds, though so well adapted for flying
from island to island, differ on the different islands;
thus there are three closely-allied species of mocking-
thrush, each confined to its own island.   Now let us
suppose the mocking-thrush of Chatham Island to be

blown to Charles Island, which has its own mocking-thrush; why should it succeed in establishing itself there? We may safely infer that Charles Island is well stocked with its own species, for annually more eggs are laid and young birds hatched, than can possibly be reared; and we may infer that the mocking-thrush peculiar to Charles Island is at least as well fitted for its home as is the species peculiar to Chatham Island. Sir C. Lyell and Mr. Wollaston have communicated to me a remarkable fact bearing on this subject; namely, that Madeira and the adjoining islet of Porto Santo possess many distinct but representative species of land-shells, some of which live in crevices of stone; and although large quantities of stone are annually transported from Porto Santo to Madeira, yet this latter island has not become colonised by the Porto Santo species; nevertheless both islands have been colonised by European land-shells, which no doubt had some advantage over the indigenous species. From these considerations I think we need not greatly marvel at the endemic species which inhabit the several islands of the Galapagos Archipelago, not having all spread from island to island. On the same continent, also, pre-occupation has probably played an important part in checking the commingling of the species which inhabit different districts with nearly the same physical conditions. Thus, the south-east and south-west corners of Australia have nearly the same physical conditions, and are united by continuous land, yet they are inhabited by a vast number of distinct mammals, birds, and plants; so it is, according to Mr. Bates, with the butterflies and other animals inhabiting the great, open, and continuous valley of the Amazons.

The same principle which governs the general character of the inhabitants of oceanic islands, namely, the relation to the source whence colonists could have been mostly easily derived, together with their subsequent modification, is of the widest application throughout nature. We see this on every mountain-summit, in every lake and marsh. For Alpine species, excepting in as far as the same species have become widely spread during the Glacial epoch, are related to those of the surrounding lowlands; thus we have in South America, Alpine humming-birds, Alpine rodents, Alpine plants, &c., all strictly belonging to American forms; and it is obvious that a mountain, as it became slowly upheaved, would be colonised from the surrounding lowlands. So it is with the inhabitants of lakes and marshes, excepting in so far as great facility of transport has allowed the same forms to prevail throughout large portions of the world. We see this same principle in the character of most of the blind animals inhabiting the caves of America and of Europe. Other analogous facts could be given. It will, I believe, be found universally true, that wherever in two regions, let them be ever so distant, many closely allied or representative species occur, there will likewise be found some identical species; and wherever many closely-allied species occur, there will be found many forms which some naturalists rank as distinct species, and others as mere varieties; these doubtful forms showing us the steps in the progress of modification.

The relation between the power and extent of migration in certain species, either at the present or at some former period, and the existence at remote points of the world of closely-allied species, is shown

in another and more general way. Mr. Gould remarked to me long ago, that in those genera of birds
which range over the world, many of the species have
very wide ranges. I can hardly doubt that this rule
is generally true, though difficult of proof. Amongst
mammals, we see it strikingly displayed in Bats, and
in a lesser degree in the Felidæ and Canidæ. We see
the same rule in the distribution of butterflies and
beetles. So it is with most of the inhabitants of fresh
water, for many of the genera in the most distinct
classes range over the world, and many of the species
have enormous ranges. It is not meant that all, but
that some of the species have very wide ranges in the
genera which range very widely. Nor is it meant that
the species in such genera have on an average a very
wide range; for this will largely depend on how far
the process of modification has gone; for instance, two
varieties of the same species inhabit America and
Europe, and thus the species has an immense range;
but, if variation were to be carried a little further,
the two varieties would be ranked as distinct species,
and their range would be greatly reduced. Still less is
it meant, that species which have the capacity of
crossing barriers and ranging widely, as in the case
of certain powerfully-winged birds, will necessarily
range widely; for we should never forget that to range
widely implies not only the power of crossing barriers,
but the more important power of being victorious in
distant lands in the struggle for life with foreign
associates. But according to the view that all the
species of a genus, though distributed to the most
remote points of the world, are descended from a single
progenitor, we ought to find, and I believe as a general

rule we do find, that some at least of the species range very widely.

We should bear in mind that many genera in all classes are of ancient origin, and the species in this case will have had ample time for dispersal and subsequent modification.   There is also reason to believe from geological evidence, that within each great class the lower organisms change at a slower rate than the higher'; consequently they will have had a better chance of ranging widely and of still retaining the same specific character.   This fact, together with that of the seeds and eggs of most lowly organised forms being very minute and better fitted for distant transportal, probably accounts for a law which has long been observed, and which has lately been discussed by Alph. de Candolle in regard to plants, namely, that the lower any group of organisms stands the more widely it ranges.

The relations just discussed,—namely, lower organisms ranging more widely than the higher,—some of the species of widely-ranging genera themselves ranging widely,—such facts, as alpine, lacustrine, and marsh productions being generally related to those which live on the surrounding low lands and dry lands,—the striking relationship between the inhabitants of islands and those of the nearest mainland—the still closer relationship of the distinct inhabitants of the islands in the same archipelago—are inexplicable on the ordinary view of the independent creation of each species, but are explicable if we admit colonisation from the nearest or readiest source, together with the subsequent adaptation of the colonists to their new homes.

Summary of the last and present Chapters.

In these chapters I have endeavoured to show, that if we make due allowance for our ignorance of the full effects of changes of climate and of the level of the land, which have certainly occurred within the recent period, and of other changes which have probably occurred,—if we remember how ignorant we are with respect to the many curious means of occasional transport,—if we bear in mind, and this is a very important consideration, how often a species may have ranged continuously over a wide area, and then have become extinct in the intermediate tracts,—the difficulty is not insuperable in believing that all the individuals of the same species, wherever found, are descended from common parents. And we are led to this conclusion, which has been arrived at by many naturalists under the designation of single centres of creation, by various general considerations, more especially from the importance of barriers of all kinds, and from the analogical distribution of sub-genera, genera, and families.

With respect to distinct species belonging to the same genus, which on our theory have spread from one parent-source; if we make the same allowances as before for our ignorance, and remember that some forms of life have changed very slowly, enormous periods of time having been thus granted for their migration, the difficulties are far from insuperable; though in this case, as in that of the individuals of the same species, they are often great.

As exemplifying the effects of climatal changes on distribution, I have attempted to show how important a part the last Glacial period has played, which affected

2 o

even the equatorial regions, and which, during the alternations of the cold in the north and south, allowed the productions of opposite hemispheres to mingle, and left some of them stranded on the mountain-summits in all parts of the world. As showing how diversified are the means of occasional transport, I have discussed at some little length the means of dispersal of fresh-water productions.

If the difficulties be not insuperable in admitting that in the long course of time all the individuals of the same species, and likewise of the several species belonging to the same genus, have proceeded from some one source; then all the grand leading facts of geographical distribution are explicable on the theory of migration, together with subsequent modification and the multiplication of new forms. We can thus understand the high importance of barriers, whether of land or water, in not only separating, but in apparently forming the several zoological and botanical provinces. We can thus understand the concentration of related species within the same areas; and how it is that under different latitudes, for instance in South America, the inhabitants of the plains and mountains, of the forests, marshes, and deserts, are linked together in so mysterious a manner, and are likewise linked to the extinct beings which formerly inhabited the same continent. Bearing in mind that the mutual relation of organism to organism is of the highest importance, we can see why two areas having nearly the same physical conditions should often be inhabited by very different forms of life; for according to the length of time which has elapsed since the colonists entered one of the regions, or both; according to the nature of the communication which allowed certain forms and not others to enter, either in greater or lesser numbers; according or not, as those which entered happened to come into more or less direct competition with each other and with the aborigines; and according as the immigrants were capable of varying more or less rapidly, there would ensue in the two or more regions, independently of their physical conditions, infinitely diversified conditions of life,—there would be an almost endless amount of organic action and reaction,—and we should find some groups of beings greatly, and some only slightly modified,—some developed in great force, some existing in scanty numbers—and this we do find in the several great geographical provinces of the world.

On these same principles we can understand, as I have endeavoured to show, why oceanic islands should have few inhabitants, but that of these, a large proportion should be endemic or peculiar; and why, in relation to the means of migration, one group of beings should have all its species peculiar, and another group, even within the same class, should have all its species the same with those in an adjoining quarter of the world. We can see why whole groups of organisms, as batrachians and terrestrial mammals, should be absent from oceanic islands, whilst the most isolated islands should possess their own peculiar species of aerial mammals or bats. We can see why, in islands, there should be some relation between the presence of mammals, in a more or less modified condition, and the depth of the sea between such islands and the mainland. We can clearly see why all the inhabitants of an archipelago, though specifically distinct on the several islets, should

be closely related to each other; and should likewise
be related, but less closely, to those of the nearest
continent, or other source whence immigrants might
have been derived.   We can see why, if there exist
very closely allied or representative species in two
areas, however distant from each other, some identical
species will almost always there be found.

As the late Edward Forbes often insisted, there is a
striking parallelism in the laws of life throughout time
and space; the laws governing the succession of forms
in past times being nearly the same with those governing at the present time the differences in different
areas.   We see this in many facts.   The endurance of
each species and group of species is continuous in time;
for the apparent exceptions to the rule are so few, that
they may fairly be attributed to our not having as yet
discovered in an intermediate deposit certain forms
which are absent in it, but which occur both above and
below: so in space, it certainly is the general rule that
the area inhabited by a single species, or by a group of
species, is continuous, and the exceptions, which are not
rare, may, as I have attempted to show, be accounted for
by former migrations under different circumstances, or
through occasional means of transport, or by the species
having become extinct in the intermediate tracts.
Both in time and space species and groups of species
have their points of maximum development.   Groups
of species, living during the same period of time, or
living within the same area, are often characterised by
trifling features in common, as of sculpture or colour.
In looking to the long succession of past ages, as in
looking to distant provinces throughout the world, we
find that species in certain classes differ little from each

other, whilst those in another class, or only in a different
section of the same order, differ greatly from each other.
In both time and space the lowly organised members
of each class generally change less than the highly
organised; but there are in both cases marked exceptions to the rule.   According to our theory, these
several relations throughout time and space are intelligible; for whether we look to the allied forms of life
which have changed during successive ages, or to
those which have changed after having migrated into
distant quarters, in both cases they are connected by
the same bond of ordinary generation; in both cases
the laws of variation have been the same, and modifications have been accumulated by the same means of
natural selection.

CHAPTER XIV.

MUTUAL AFFINITIES OF ORGANIC BEINGS: MORPHOLOGY: EMBRYOLOGY: RUDIMENTARY ORGANS.

CLASSIFICATION, groups subordinate to groups—Natural system— Rules and difficulties in classification, explained on the theory of descent with modification—Classification of varieties—Descent always used in classification—Analogical or adaptive characters —Affinities, general, complex, and radiating—Extinction separates and defines groups—MORPHOLOGY, between members of the same class, between parts of the same individual— EMBRYOLOGY, laws of, explained by variations not supervening at an early age, and being inherited at a corresponding age— RUDIMENTARY ORGANS; their origin explained—Summary.

Classification.

FROM the most remote period in the history of the world organic beings have been found to resemble each other in descending degrees, so that they can be classed in groups under groups. This classification is not arbitrary like the grouping of the stars in constellations. The existence of groups would have been of simple significance, if one group had been exclusively fitted to inhabit the land, and another the water; one to feed on flesh, another on vegetable matter, and so on; but the case is widely different, for it is notorious how commonly members of even the same sub-group have different habits. In the second and fourth chapters, on Variation and on Natural Selection, I have attempted

to show that within each country it is the widely
ranging, the much diffused and common, that is the
dominant species, belonging to the larger genera in each
class, which vary most. The varieties, or incipient
species, thus produced, ultimately become converted
into new and distinct species; and these, on the principle of inheritance, tend to produce other new and
dominant species. Consequently the groups which are
now large, and which generally include many dominant
species, tend to go on increasing in size. I further
attempted to show that from the varying descendants
of each species trying to occupy as many and as different
places as possible in the economy of nature, they
constantly tend to diverge in character. This latter
conclusion is supported by observing the great diversity
of forms which, in any small area, come into the closest
competition, and by certain facts in naturalisation.

I attempted also to show that there is a steady
tendency in the forms which are increasing in number
and diverging in character, to supplant and exterminate
the preceding, less divergent and less improved forms.
I request the reader to turn to the diagram illustrating
the action, as formerly explained, of these several principles; and he will see that the inevitable result is,
that the modified descendants proceeding from one
progenitor become broken up into groups subordinate
to groups. In the diagram each letter on the uppermost
line may represent a genus including several species;
and the whole of the genera along this upper line form
together one class, for all are descended from one ancient
parent, and, consequently, have inherited something in
common. But the three genera on the left hand have,
on this same principle, much in common, and form a

sub-family, distinct from that containing the next two
genera on the right hand, which diverged from a common
parent at the fifth stage of descent. These five genera
have also much in common, though less than when
grouped in sub-families; and they form a family distinct
from that containing the three genera still farther to
the right hand, which diverged at an earlier period.
And all these genera, descended from (A), form an
order distinct from the genera descended from (I). So
that we here have many species descended from a single
progenitor grouped into genera; and the genera into
sub-families, families, and orders, all under one great
class. The grand fact of the natural subordination of
organic beings in groups under groups, which, from its
familiarity, does not always sufficiently strike us, is in
my judgment thus explained. No doubt organic beings,
like all other objects, can be classed in many ways,
either artificially by single characters, or more naturally
by a number of characters. We know, for instance, that
minerals and the elemental substances can be thus
arranged. In this case there is of course no relation to
genealogical succession, and no cause can at present be
assigned for their falling into groups. But with organic
beings the case is different, and the view above given
accords with their natural arrangement in group under
group; and no other explanation has ever been
attempted.

Naturalists, as we have seen, try to arrange the
species, genera, and families in each class, on what is
called the Natural System. But what is meant by this
system ? Some authors look at it merely as a scheme
for arranging together those living objects which are
most alike, and for separating those which are most

unlike; or as an artificial method of enunciating, as briefly as possible, general propositions,—that is, by one sentence to give the characters common, for instance, to all mammals, by another those common to all carnivora, by another those common to the dog-genus, and then, by adding a single sentence, a full description is given of each kind of dog. The ingenuity and utility of this system are indisputable. But many naturalists think that something more is meant by the Natural System; they believe that it reveals the plan of the Creator; but unless it be specified whether order in time or space, or both, or what else is meant by the plan of the Creator, it seems to me that nothing is thus added to our knowledge. Expressions such as that famous one by Linnæus, which we often meet with in a more or less concealed form, namely, that the characters do not make the genus, but that the genus gives the characters, seem to imply that some deeper bond is included in our classifications than mere resemblance. I believe that this is the case, and that community of descent—the one known cause of close similarity in organic beings— is the bond, which though observed by various degrees of modification, is partially revealed to us by our classifications.

Let us now consider the rules followed in classification, and the difficulties which are encountered on the view that classification either gives some unknown plan of creation, or is simply a scheme for enunciating general propositions and of placing together the forms most like each other. It might have been thought (and was in ancient times thought) that those parts of the structure which determined the habits of life, and the general place of each being in the economy of nature, would be of very high importance in classification. Nothing can be more false. No one regards the external similarity of a mouse to a shrew, of a dugong to a whale, of a whale to a fish, as of any importance. These resemblances, though so intimately connected with the whole life of the being, are ranked as merely "adaptive or analogical characters;" but to the consideration of these resemblances we shall recur. It may even be given as a general rule, that the less any part of the organisation is concerned with special habits, the more important it becomes for classification. As an instance: Owen, in speaking of the dugong, says, "The generative organs, being those which are most remotely related to the habits and food of an animal, I have always regarded as affording very clear indications of its true affinities. We are least likely in the modifications of these organs to mistake a merely adaptive for an essential character." With plants how remarkable it is that the organs of vegetation, on which their nutrition and life depend, are of little signification; whereas the organs of reproduction, with their product the seed and embryo, are of paramount importance! So again in formerly discussing certain morphological characters which are not functionally important, we have seen that they are often of the highest service in classification. This depends on their constancy throughout many allied groups; and their constancy chiefly depends on any slight deviations not having been preserved and accumulated by natural selection, which acts only on serviceable characters.

That the mere physiological importance of an organ does not determine its classificatory value, is almost proved by the fact, that in allied groups, in which the same organ, as we have every reason to suppose, has nearly the same physiological value, its classificatory value is widely different. No naturalist can have worked long at any group without being struck with this fact; and it has been fully acknowledged in the writings of almost every author. It will suffice to quote the highest authority, Robert Brown, who, in speaking of certain organs in the Proteaceæ, says their generic importance, "like that of all their parts, not only in this, but, as I apprehend, in every natural family, is very unequal, and in some cases seems to be entirely lost." Again, in another work he says, the genera of the Connaraceæ "differ in having one or more ovaria, in the existence or absence of albumen, in the imbricate or valvular æstivation. Any one of these characters singly is frequently of more than generic importance, though here even when all taken together they appear insufficient to separate Cnestis from Connarus." To give an example amongst insects: in one great division of the Hymenoptera, the antennæ, as Westwood has remarked, are most constant in structure; in another division they differ much, and the differences are of quite subordinate value in classification; yet no one will say that the antennæ in these two divisions of the same order are of unequal physiological importance. Any number of instances could be given of the varying importance for classification of the same important organ within the same group of beings.

Again, no one will say that rudimentary or atrophied organs are of high physiological or vital importance; yet, undoubtedly, organs in this condition are often of much value in classification. No one will dispute that the rudimentary teeth in the upper jaws of young ruminants, and certain rudimentary bones of the leg, are highly serviceable in exhibiting the close affinity between ruminants and pachyderms. Robert Brown has strongly insisted on the fact that the position of the rudimentary florets is of the highest importance in the classification of the grasses.

Numerous instances could be given of characters derived from parts which must be considered of very trifling physiological importance, but which are universally admitted as highly serviceable in the definition of whole groups. For instance, whether or not there is an open passage from the nostrils to the mouth, the only character, according to Owen, which absolutely distinguishes fishes and reptiles—the inflection of the angle of the lower jaw in Marsupials—the manner in which the wings of insects are folded—mere colour in certain Algæ—mere pubescence on parts of the flower in grasses—the nature of the dermal covering, as hair or feathers, in the Vertebrata. If the Ornithorhynchus had been covered with feathers instead of hair, this external and trifling character would have been considered by naturalists as an important aid in determining the degree of affinity of this strange creature to birds.

The importance, for classification, of trifling characters, mainly depends on their being correlated with many other characters of more or less importance. The value indeed of an aggregate of characters is very evident in natural history. Hence, as has often been remarked, a species may depart from its allies in several characters, both of high physiological importance, and of almost universal prevalence, and yet leave us in no doubt where it should be ranked. Hence, also, it has been found that a classification founded on any single character, however important that may be, has always failed; for no part of the organisation is invariably constant. The importance of an aggregate of characters, even when none are important, alone explains the aphorism enunciated by Linnæus, namely, that the characters do not give the genus, but the genus gives the characters; for this seems founded on the appreciation of many trifling points of resemblance, too slight to be defined. Certain plants, belonging to the Malpighiaceæ, bear perfect and degraded flowers; in the latter, as A. de Jussieu has remarked, "the greater number of the characters proper to the species, to the genus, to the family, to the class, disappear, and thus laugh at our classification." When Aspicarpa produced in France, during several years, only these degraded flowers, departing so wonderfully in a number of the most important points of structure from the proper type of the order, yet M. Richard sagaciously saw, as Jussieu observes, that this genus should still be retained amongst the Malpighiaceæ. This case well illustrates the spirit of our classifications.

Practically, when naturalists are at work, they do not trouble themselves about the physiological value of the characters which they use in defining a group or in allocating any particular species. If they find a character nearly uniform, and common to a great number of forms, and not common to others, they use it as one of high value; if common to some lesser number, they use it as of subordinate value. This principle has been broadly confessed by some naturalists to be the true one; and by none more clearly than by that excellent botanist, Aug. St. Hilaire. If several trifling characters are always found in combination, though no apparent bond of connection can be discovered between them, especial value is set on them. As in most groups of animals, important organs, such as those for propelling the blood, or for aerating it, or those for propagating the race, are found nearly uniform, they are considered as highly serviceable in classification; but in some groups all these, the most important vital organs, are found to offer characters of quite subordinate value. Thus, as Fritz Müller has lately remarked, in the same group of crustaceans, Cypridina is furnished with a heart, whilst in two closely allied genera, namely Cypris and Cytherea, there is no such organ; one species of Cypridina has well-developed branchiæ, whilst another species is destitute of them.

We can see why characters derived from the embryo should be of equal importance with those derived from the adult, for a natural classification of course includes all ages. But it is by no means obvious, on the ordinary view, why the structure of the embryo should be more important for this purpose than that of the adult, which alone plays its full part in the economy of nature. Yet it has been strongly urged by those great naturalists, Milne Edwards and Agassiz, that embryological characters are the most important of all; and this doctrine has very generally been admitted as true. Nevertheless, their importance has sometimes been exaggerated, owing to the adaptive characters of larvæ not having been excluded; in order to show this, Fritz Müller arranged by the aid of such characters alone the great class of crustaceans, and the arrangement did not prove a natural one. But there can be no doubt that embryonic, excluding larval characters, are of the highest value for classification, not only with animals but with plants. Thus the main divisions of flowering plants are founded on differences in the embryo,—on the number and position of the cotyledons, and on the mode of development of the plumule and radicle. We shall immediately see why these characters possess so high a value in classification, namely, from the natural system being genealogical in its arrangement.

Our classifications are often plainly influenced by chains of affinities. Nothing can be easier than to define a number of characters common to all birds; but with crustaceans, any such definition has hitherto been found impossible. There are crustaceans at the opposite ends of the series, which have hardly a character in common; yet the species at both ends, from being plainly allied to others, and these to others, and so onwards, can be recognised as unequivocally belonging to this, and to no other class of the Articulata.

Geographical distribution has often been used, though perhaps not quite logically, in classification, more especially in very large groups of closely allied forms. Temminck insists on the utility or even necessity of this practice in certain groups of birds; and it has been followed by several entomologists and botanists.

Finally, with respect to the comparative value of the various groups of species, such as orders, sub-orders, families, sub-families, and genera, they seem to be, at least at present, almost arbitrary. Several of the best botanists, such as Mr. Bentham and others, have strongly insisted on their arbitrary value. Instances could be given amongst plants and insects, of a group first ranked by practised naturalists as only a genus, and then raised to the rank of a sub-family or family; and

this has been done, not because further research has
detected important structural differences, at first overlooked, but because numerous allied species with slightly
different grades of difference, have been subsequently
discovered.

All the foregoing rules and aids and difficulties in
classification may be explained, if I do not greatly
deceive myself, on the view that the Natural System is
founded on descent with modification;—that the characters which naturalists consider as showing true
affinity between any two or more species, are those
which have been inherited from a common parent, all
true classification being genealogical;—that community
of descent is the hidden bond which naturalists have
been unconsciously seeking, and not some unknown
plan of creation, or the enunciation of general propositions, and the mere putting together and separating
objects more or less alike.

But I must explain my meaning more fully. I
believe that the *arrangement* of the groups within each
class, in due subordination and relation to each other,
must be strictly genealogical in order to be natural;
but that the *amount* of difference in the several branches
or groups, though allied in the same degree in blood to
their common progenitor, may differ greatly, being due
to the different degrees of modification which they have
undergone; and this is expressed by the forms being
ranked under different genera, families, sections, or
orders.   The reader will best understand what is meant,
if he will take the trouble to refer to the diagram in the
fourth chapter.   We will suppose the letters A to L to
represent allied genera existing during the Silurian
epoch, and descended from some still earlier form.   In

three of these genera (A, F, and I), a species has transmitted modified descendants to the present day, represented by the fifteen genera (a^{14} to z^{14}) on the uppermost horizontal line. Now all these modified descendants from a single species, are related in blood or descent in the same degree; they may metaphorically be called cousins to the same millionth degree; yet they differ widely and in different degrees from each other. The forms descended from A, now broken up into two or three families, constitute a distinct order from those descended from I, also broken up into two families. Nor can the existing species, descended from A, be ranked in the same genus with the parent A; or those from I, with the parent I. But the existing genus F^{14} may be supposed to have been but slightly modified; and it will then rank with the parent-genus F; just as some few still living organisms belong to Silurian genera. So that the comparative value of the differences between these organic beings, which are all related to each other in the same degree in blood, has come to be widely different. Nevertheless their genealogical *arrangement* remains strictly true, not only at the present time, but at each successive period of descent. All the modified descendants from A will have inherited something in common from their common parent, as will all the descendants from I; so will it be with each subordinate branch of descendants, at each successive stage. If, however, we suppose any descendant of A, or of I, to have become so much modified as to have lost all traces of its parentage, in this case, its place in the natural system will be lost, as seems to have occurred with some few existing organisms. All the descendants of the genus F, along its whole line of descent, are supposed to have been but little modified, and they form a single genus. But this genus, though much isolated, will still occupy its proper intermediate position. The representation of the groups, as here given in the diagram on a flat surface, is much too simple. The branches ought to have diverged in all directions. If the names of the groups had been simply written down in a linear series, the representation would have been still less natural; and it is notoriously not possible to represent in a series, on a flat surface, the affinities which we discover in nature amongst the beings of the same group. Thus, the natural system is genealogical in its arrangement, like a pedigree : but the amount of modification which the different groups have undergone has to be expressed by ranking them under different so-called genera, sub-families, families, sections, orders, and classes.

It may be worth while to illustrate this view of classification, by taking the case of languages. If we possessed a perfect pedigree of mankind, a genealogical arrangement of the races of man would afford the best classification of the various languages now spoken throughout the world; and if all extinct languages, and all intermediate and slowly changing dialects, were to be included, such an arrangement would be the only possible one. Yet it might be that some ancient languages had altered very little and had given rise to few new languages, whilst others had altered much owing to the spreading, isolation, and state of civilisation of the several co-descended races, and had thus given rise to many new dialects and languages. The various degrees of difference between the languages of the same stock, would have to be expressed by groups subordinate

to groups; but the proper or even the only possible
arrangement would still be genealogical; and this would
be strictly natural, as it would connect together all
languages, extinct and recent, by the closest affinities,
and would give the filiation and origin of each tongue.

In confirmation of this view, let us glance at the
classification of varieties, which are known or believed
to be descended from a single species. These are
grouped under the species, with the sub-varieties under
the varieties; and in some cases, as with the domestic
pigeon, with several other grades of difference. Nearly
the same rules are followed as in classifying species.
Authors have insisted on the necessity of arranging
varieties on a natural instead of an artificial system; we
are cautioned, for instance, not to class two varieties of
the pine-apple together, merely because their fruit,
though the most important part, happens to be nearly
identical; no one puts the Swedish and common turnip
together, though the esculent and thickened stems are
so similar. Whatever part is found to be most constant, is used in classing varieties: thus the great
agriculturist Marshall says the horns are very useful for
this purpose with cattle, because they are less variable
than the shape or colour of the body, &c.; whereas with
sheep the horns are much less serviceable, because less
constant. In classing varieties, I apprehend that if we
had a real pedigree, a genealogical classification would
be universally preferred; and it has been attempted in
some cases. For we might feel sure, whether there had
been more or less modification, that the principle of
inheritance would keep the forms together which were
allied in the greatest number of points. In tumbler
pigeons, though some of the sub-varieties differ in the

important character of the length of the beak, yet all are kept together from having the common habit of tumbling; but the short-faced breed has nearly or quite lost this habit: nevertheless, without any thought on the subject, these tumblers are kept in the same group, because allied in blood and alike in some other respects.

With species in a state of nature, every naturalist has in fact brought descent into his classification; for he includes in his lowest grade, that of species, the two sexes; and how enormously these sometimes differ in the most important characters, is known to every naturalist: scarcely a single fact can be predicated in common of the adult males and hermaphrodites of certain cirripedes, and yet no one dreams of separating them. As soon as the three Orchidean forms, Monachanthus, Myanthus, and Catasetum, which had previously been ranked as three distinct genera, were known to be sometimes produced on the same plant, they were immediately considered as varieties; and now I have been able to show that they are the male, female, and hermaphrodite forms of the same species. The naturalist includes as one species the various larval stages of the same individual, however much they may differ from each other and from the adult, as well as the so-called alternate generations of Steenstrup, which can only in a technical sense be considered as the same individual. He includes monsters and varieties, not from their partial resemblance to the parent-form, but because they are descended from it.

As descent has universally been used in classing together the individuals of the same species, though the males and females and larvæ are sometimes extremely different; and as it has been used in classing varieties

which have undergone a certain, and sometimes a
considerable amount of modification, may not this same
element of descent have been unconsciously used in
grouping species under genera, and genera under higher
groups, all under the so-called natural system ? 1
believe it has been unconsciously used; and thus only
can I understand the several rules and guides which
have been followed by our best systematists. As we
have no written pedigrees, we are forced to trace
community of descent by resemblances of any kind.
Therefore we choose those characters which are the least
likely to have been modified, in relation to the conditions of life to which each species has been recently
exposed. Rudimentary structures on this view are as
good as, or even sometimes better than, other parts of
the organisation. We care not how trifling a character
may be—let it be the mere inflection of the angle of the
jaw, the manner in which an insect's wing is folded,
whether the skin be covered by hair or feathers—if it
prevail throughout many and different species, especially
those having very different habits of life, it assumes high
value; for we can account for its presence in so many
forms with such different habits, only by inheritance
from a common parent. We may err in this respect in
regard to single points of structure, but when several
characters, let them be ever so trifling, concur throughout a large group of beings having different habits, we
may feel almost sure, on the theory of descent, that
these characters have been inherited from a common
ancestor; and we know that such aggregated characters
have especial value in classification.

We can understand why a species or a group of
species may depart from its allies, in several of its most

important characteristics, and yet be safely classed with
them.   This may be safely done, and is often done, as
long as a sufficient number of characters, let them be ever
so unimportant, betrays the hidden bond of community
of descent.   Let two forms have not a single character
in common, yet, if these extreme forms are connected
together by a chain of intermediate groups, we may at
once infer their community of descent, and we put them
all into the same class.   As we find organs of high
physiological importance—those which serve to preserve
life under the most diverse conditions of existence—are
generally the most constant, we attach especial value to
them; but if these same organs, in another group or
section of a group, are found to differ much, we at once
value them less in our classification.   We shall presently
see why embryological characters are of such high classificatory importance.   Geographical distribution may
sometimes be brought usefully into play in classing
large genera, because all the species of the same genus,
inhabiting any distinct and isolated region, are in all
probability descended from the same parents.

Analogical Resemblances.—We can understand, on
the above views, the very important distinction between
real affinities and analogical or adaptive resemblances.
Lamarck first called attention to this subject, and he
has been ably followed by Macleay and others.   The
resemblance in the shape of the body and in the fin-like
anterior limbs between dugongs and whales, and between
these two orders of mammals and fishes, are analogical.
So is the resemblance between a mouse and a shrew-
mouse (Sorex), which belong to different orders; and
the still closer resemblance, insisted on by Mr. Mivart,
between the mouse and a small marsupial animal

(Antechinus) of Australia. These latter resemblances may be accounted for, as it seems to me, by adaptation for similarly active movements through thickets and herbage, together with concealment from enemies.

Amongst insects there are innumerable similar instances; thus Linnæus, misled by external appearances, actually classed an homopterous insect as a moth. We see something of the same kind even with our domestic varieties, as in the strikingly similar shape of the body in the improved breeds of the Chinese and common pig, which are descended from distinct species; and in the similarly thickened stems of the common and specifically distinct Swedish turnip. The resemblance between the greyhound and the racehorse is hardly more fanciful than the analogies which have been drawn by some authors between widely different animals.

On the view of characters being of real importance for classification, only in so far as they reveal descent, we can clearly understand why analogical or adaptive characters, although of the utmost importance to the welfare of the being, are almost valueless to the systematist. For animals, belonging to two most distinct lines of descent, may have become adapted to similar conditions, and thus have assumed a close external resemblance; but such resemblances will not reveal—will rather tend to conceal their blood-relationship. We can thus also understand the apparent paradox, that the very same characters are analogical when one group is compared with another, but give true affinities when the members of the same group are compared together: thus, the shape of the body and fin-like limbs are only analogical when whales are compared with fishes, being adaptations in both classes for swimming through the water; but between the several members of the whale family, the shape of the body and the fin-like limbs offer characters exhibiting true affinity; for as these parts are so nearly similar throughout the whole family, we cannot doubt that they have been inherited from a common ancestor. So it is with fishes.

Numerous cases could be given of striking resemblances in quite distinct beings between single parts or organs, which have been adapted for the same functions. A good instance is afforded by the close resemblance of the jaws of the dog and Tasmanian wolf or Thylacinus,—animals which are widely sundered in the natural system. But this resemblance is confined to general appearance, as in the prominence of the canines, and in the cutting shape of the molar teeth. For the teeth really differ much: thus the dog has on each side of the upper jaw four pre-molars and only two molars; whilst the Thylacinus has three pre-molars and four molars. The molars also differ much in the two animals in relative size and structure. The adult dentition is preceded by a widely different milk dentition. Any one may of course deny that the teeth in either case have been adapted for tearing flesh, through the natural selection of successive variations; but if this be admitted in the one case, it is unintelligible to me that it should be denied in the other. I am glad to find that so high an authority as Professor Flower has come to this same conclusion.

The extraordinary cases given in a former chapter, of widely different fishes possessing electric organs,—of widely different insects possessing luminous organs,—and of orchids and asclepiads having pollen-masses

with viscid discs, come under this same head of analogical resemblances.   But these cases are so wonderful
that they were introduced as difficulties or objections
to our theory.   In all such· cases some fundamental
difference in the growth or development of the parts,
and generally in their matured structure, can be
detected.   The end gained is the same, but the means,
though appearing superficially to be the same, are
essentially different.   The principle formerly alluded to
under the term of *analogical variation* has probably in
these cases often come into play; that is, the members
of the same class, although only distantly allied, have
inherited so much in common in their constitution,
that they are apt to vary under similar exciting causes
in a similar manner; and this would obviously aid in
the acquirement through natural selection of parts or
organs, strikingly like each other, independently of
their direct inheritance from a common progenitor.

As species belonging to distinct classes have often been
adapted by successive slight modifications to live under
nearly similar circumstances,—to inhabit, for instance,
the three elements of land, air, and water,—we can
perhaps understand how it is that a numerical parallelism has sometimes been observed between the sub-
groups of distinct classes.   A naturalist, struck with a
parallelism of this nature, by arbitrarily raising or
sinking the value of the groups in several classes (and
all our experience shows that their valuation is as yet
arbitrary), could easily extend the parallelism over a
wide range; and thus the septenary, quinary, quater-
nary and ternary classifications have probably arisen.

There is another and curious class of cases in which
close external resemblance does not depend on adapta-

tion to similar habits of life, but has been gained for
the sake of protection. I allude to the wonderful
manner in which certain butterflies imitate, as first
described by Mr. Bates, other and quite distinct species·
This excellent observer has shown that in some districts
of S. America, where, for instance, an Ithomia abounds
in gaudy swarms, another butterfly, namely, a Leptalis,
is often found mingled in the same flock; and the
latter so closely resembles the Ithomia in every shade
and stripe of colour and even in the shape of its wings,
that Mr. Bates, with his eyes sharpened by collecting
during eleven years, was, though always on his guard,
continually deceived. When the mockers and the
mocked are caught and compared, they are found to be
very different in essential structure, and to belong not
only to distinct genera, but often to distinct families.
Had this mimicry occurred in only one or two instances, it might have been passed over as a strange
coincidence. But, if we proceed from a district where
one Leptalis imitates an Ithomia, another mocking and
mocked species belonging to the same two genera,
equally close in their resemblance, may be found.
Altogether no less than ten genera are enumerated,
which include species that imitate other butterflies.
The mockers and mocked always inhabit the same
region; we never find an imitator living remote from
the form which it imitates. The mockers are almost
invariably rare insects; the mocked in almost every
case abound in swarms. In the same district in which
a species of Leptalis closely imitates an Ithomia, there
are sometimes other Lepidoptera mimicking the same
Ithomia: so that in the same place, species of three
genera of butterflies and even a moth are found all

closely resembling a butterfly belonging to a fourth
genus. It deserves especial notice that many of the
mimicking forms of the Leptalis, as well as of the
mimicked forms, can be shown by a graduated series to
be merely varieties of the same species; whilst others
are undoubtedly distinct species. But why, it may be
asked, are certain forms treated as the mimicked and
others as the mimickers? Mr. Bates satisfactorily
answers this question, by showing that the form which
is imitated keeps the usual dress of the group to which
it belongs, whilst the counterfeiters have changed their
dress and do not resemble their nearest allies.

We are next led to inquire what reason can be
assigned for certain butterflies and moths so often
assuming the dress of another and quite distinct form;
why, to the perplexity of naturalists, has nature con-
descended to the tricks of the stage? Mr. Bates has,
no doubt, hit on the true explanation. The mocked
forms, which always abound in numbers, must habitually escape destruction to a large extent, otherwise they
could not exist in such swarms; and a large amount of
evidence has now been collected, showing that they are
distasteful to birds and other insect-devouring animals.
The mocking forms, on the other hand, that inhabit the
same district, are comparatively rare, and belong to
rare groups; hence they must suffer habitually from
some danger, for otherwise, from the number of eggs
laid by all butterflies, they would in three or four
generations swarm over the whole country. Now if a
member of one of these persecuted and rare groups were
to assume a dress so like that of a well-protected species
that it continually deceived the practised eyes of an
entomologist, it would often deceive predaceous birds

and insects, and thus often escape destruction. Mr.
Bates may almost be said to have actually witnessed
the process by which the mimickers have come so
closely to resemble the mimicked; for he found that
some of the forms of Leptalis which mimic so many
other butterflies, varied in an extreme degree. In one
district several varieties occurred, and of these one
alone resembled to a certain extent, the common
Ithomia of the same district. In another district there
were two or three varieties, one of which was much
commoner than the others, and this closely mocked
another form of Ithomia. From facts of this nature,
Mr. Bates concludes that the Leptalis first varies; and
when a variety happens to resemble in some degree
any common butterfly inhabiting the same district,
this variety, from its resemblance to a flourishing and
little-persecuted kind, has a better chance of escaping
destruction from predaceous birds and insects, and is
consequently oftener preserved;—"the less perfect
degrees of resemblance being generation after generation
eliminated, and only the others left to propagate their
kind." So that here we have an excellent illustration
of natural selection.

Messrs. Wallace and Trimen have likewise described
several equally striking cases of imitation in the
Lepidoptera of the Malay Archipelago and Africa, and
with some other insects. Mr. Wallace has also detected one such case with birds, but we have none
with the larger quadrupeds. The much greater frequency of imitation with insects than with other
animals, is probably the consequence of their small
size; insects cannot defend themselves, excepting
indeed the kinds furnished with a sting, and I have

never heard of an instance of such kinds mocking other insects, though they are mocked; insects cannot easily escape by flight from the larger animals which prey on them; therefore, speaking metaphorically, they are reduced, like most weak creatures, to trickery and dissimulation.

It should be observed that the process of imitation probably never commenced between forms widely dissimilar in colour. But starting with species already somewhat like each other, the closest resemblance, if beneficial, could readily be gained by the above means; and if the imitated form was subsequently and gradually modified through any agency, the imitating form would be led along the same track, and thus be altered to almost any extent, so that it might ultimately assume an appearance or colouring wholly unlike that of the other members of the family to which it belonged. There is, however, some difficulty on this head, for it is necessary to suppose in some cases that ancient members belonging to several distinct groups, before they had diverged to their present extent, accidentally resembled a member of another and protected group in a sufficient degree to afford some slight protection; this having given the basis for the subsequent acquisition of the most perfect resemblance.

On the Nature of the Affinities connecting Organic Beings.—As the modified descendants of dominant species, belonging to the larger genera, tend to inherit the advantages which made the groups to which they belong large and their parents dominant, they are almost sure to spread widely, and to seize on more and more places in the economy of nature. The larger and more dominant groups within each class thus tend to go on increasing in size; and they consequently supplant many smaller and feebler groups. Thus we can account for the fact that all organisms, recent and extinct, are included under a few great orders, and under still fewer classes. As showing how few the higher groups are in number, and how widely they are spread throughout the world, the fact is striking that the discovery of Australia has not added an insect belonging to a new class; and that in the vegetable kingdom, as I learn from Dr. Hooker, it has added only two or three families of small size.

In the chapter on Geological Succession I attempted to show, on the principle of each group having generally diverged much in character during the long-continued process of modification, how it is that the more ancient forms of life often present characters in some degree intermediate between existing groups. As some few of the old and intermediate forms have transmitted to the present day descendants but little modified, these constitute our so-called osculant or aberrant species. The more aberrant any form is, the greater must be the number of connecting forms which have been exterminated and utterly lost. And we have some evidence of aberrant groups having suffered severely from extinction, for they are almost always represented by extremely few species; and such species as do occur are generally very distinct from each other, which again implies extinction. The genera Ornithorhynchus and Lepidosiren, for example, would not have been less aberrant had each been represented by a dozen species, instead of as at present by a single one, or by two or three. We can, I think, account for this fact only by looking at aberrant groups as forms which have been conquered by more successful competitors, with a few members still preserved under unusually favourable conditions.

Mr. Waterhouse has remarked that, when a member belonging to one group of animals exhibits an affinity to a quite distinct group, this affinity in most cases is general and not special; thus, according to Mr. Waterhouse, of all Rodents, the bizcacha is most nearly related to Marsupials; but in the points in which it approaches this order, its relations are general, that is, not to any one marsupial species more than to another. As these points of affinity are believed to be real and not merely adaptive, they must be due in accordance with our view to inheritance from a common progenitor. Therefore we must suppose either that all Rodents, including the bizcacha, branched off from some ancient Marsupial, which will naturally have been more or less intermediate in character with respect to all existing Marsupials; or that both Rodents and Marsupials branched off from a common progenitor, and that both groups have since undergone much modification in divergent directions. On either view we must suppose that the bizcacha has retained, by inheritance, more of the characters of its ancient progenitor than have other Rodents; and therefore it will not be specially related to any one existing Marsupial, but indirectly to all or nearly all Marsupials, from having partially retained the character of their common progenitor, or of some early member of the group. On the other hand, of all Marsupials, as Mr. Waterhouse has remarked, the Phascolomys resembles most nearly, not any one species, but the general order of Rodents. In this case, however, it may be strongly suspected that the resemblance is only analogical, owing to the Phascolomys having become adapted to habits like those of a Rodent. The elder De Candolle has made nearly similar observations on the general nature of the affinities of distinct families of plants.

On the principle of the multiplication and gradual divergence in character of the species descended from a common progenitor, together with their retention by inheritance of some characters in common, we can understand the excessively complex and radiating affinities by which all the members of the same family or higher group are connected together. For the common progenitor of a whole family, now broken up by extinction into distinct groups and sub-groups, will have transmitted some of its characters, modified in various ways and degrees, to all the species; and they will consequently be related to each other by circuitous lines of affinity of various lengths (as may be seen in the diagram so often referred to), mounting up through many predecessors. As it is difficult to show the blood-relationship between the numerous kindred of any ancient and noble family even by the aid of a genea-logical tree, and almost impossible to do so without this aid, we can understand the extraordinary difficulty which naturalists have experienced in describing, with-out the aid of a diagram, the various affinities which they perceive between the many living and extinct members of the same great natural class.

Extinction, as we have seen in the fourth chapter, has played an important part in defining and widening the intervals between the several groups in each class. We may thus account for the distinctness of whole classes from each other—for instance, of birds from all other vertebrate animals—by the belief that many ancient

forms of life have been utterly lost, through which the
early progenitors of birds were formerly connected with
the early progenitors of the other and at that time less
differentiated vertebrate classes.   There has been much
less extinction of the forms of life which once connected
fishes with batrachians.   There has been still less
within some whole classes, for instance the Crustacea,
for here the most wonderfully diverse forms are still
linked together by a long and only partially broken
chain of affinities.   Extinction has only defined the
groups : it has by no means made them ; for if every
form which has ever lived on this earth were suddenly
to reappear, though it would be quite impossible to give
definitions by which each group could be distinguished,
still a natural classification, or at least a natural
arrangement, would be possible.   We shall see this by
turning to the diagram ; the letters, A to L, may represent
eleven Silurian genera, some of which have produced
large groups of modified descendants, with every link in
each branch and sub-branch still alive ; and the links
not greater than those between existing varieties.   In
this case it would be quite impossible to give definitions
by which the several members of the several groups
could be distinguished from their more immediate
parents and descendants.   Yet the arrangement in the
diagram would still hold good and would be natural ;
for, on the principle of inheritance, all the forms
descended, for instance, from A, would have something
in common.   In a tree we can distinguish this or that
branch, though at the actual fork the two unite and
blend together.   We could not, as I have said, define
the several groups ; but we could pick out types, or
forms, representing most of the characters of each group,

2 Q

whether large or small, and thus give a general idea
of the value of the differences between them. This is
what we should be driven to, if we were ever to succeed
in collecting all the forms in any one class which have
lived throughout all time and space. Assuredly we
shall never succeed in making so perfect a collection :
nevertheless, in certain classes, we are tending towards
this end ; and Milne Edwards has lately insisted, in an
able paper, on the high importance of looking to types,
whether or not we can separate and define the groups
to which such types belong.

Finally, we have seen that natural selection, which
follows from the struggle for existence, and which almost
inevitably leads to extinction and divergence of character in the descendants from any one parent-species,
explains that great and universal feature in the affinities
of all organic beings, namely, their subordination in
group under group. We use the element of descent in
classing the individuals of both sexes and of all ages
under one species, although they may have but few
characters in common; we use descent in classing
acknowledged varieties, however different they may be
from their parents; and I believe that this element of
descent is the hidden bond of connexion which naturalists have sought under the term of the Natural System.
On this idea of the natural system being, in so far as it
has been perfected, genealogical in its arrangement,
with the grades of difference expressed by the terms
genera, families, orders, &c., we can understand the
rules which we are compelled to follow in our classification. We can understand why we value certain
resemblances far more than others; why we use
rudimentary and useless organs, or others of trifling

physiological importance; why, in finding the relations between one group and another, we summarily reject analogical or adaptive characters, and yet use these same characters within the limits of the same group. We can clearly see how it is that all living and extinct forms can be grouped together within a few great classes; and how the several members of each class are connected together by the most complex and radiating lines of affinities. We shall never, probably, disentangle the inextricable web of the affinities between the members of any one class; but when we have a distinct object in view, and do not look to some unknown plan of creation, we may hope to make sure but slow progress.

Professor Häckel in his ' Generelle Morphologie' and in other works, has recently brought his great knowledge and abilities to bear on what he calls phylogeny, or the lines of descent of all organic beings. In drawing up the several series he trusts chiefly to embryological characters, but receives aid from homologous and rudimentary organs, as well as from the successive periods at which the various forms of life are believed to have first appeared in our geological formations. He has thus boldly made a great beginning, and shows us how classification will in the future be treated.

Morphology.

We have seen that the members of the same class, independently of their habits of life, resemble each other in the general plan of their organisation. This resemblance is often expressed by the term " unity of type;" or by saying that the several parts and organs in the different species of the class are homologous. The

whole subject is included under the general term of
Morphology. This is one of the most interesting departments of natural history, and may almost be said to
be its very soul. What can be more curious than that
the hand of a man, formed for grasping, that of a mole
for digging, the leg of the horse, the paddle of the porpoise, and the wing of the bat, should all be constructed
on the same pattern, and should include similar bones,
in the same relative positions? How curious it is, to
give a subordinate though striking instance, that the
hind-feet of the kangaroo, which are so well fitted for
bounding over the open plains,—those of the climbing,
leaf-eating koala, equally well fitted for grasping the
branches of trees,—those of the ground-dwelling, insect
or root eating, bandicoots,—and those of some other
Australian marsupials,—should all be constructed on
the same extraordinary type, namely with the bones of
the second and third digits extremely slender and
enveloped within the same skin, so that they appear like
a single toe furnished with two claws. Notwithstanding this similarity of pattern, it is obvious that the
hind feet of these several animals are used for as widely
different purposes as it is possible to conceive. The
case is rendered all the more striking by the American
opossums, which follow nearly the same habits of life as
some of their Australian relatives, having feet constructed on the ordinary plan. Professor Flower, from
whom these statements are taken, remarks in conclusion:
"We may call this conformity to type, without getting
much nearer to an explanation of the phenomenon;"
and he then adds "but is it not powerfully suggestive
of true relationship, of inheritance from a common
ancestor?"

Geoffroy St. Hilaire has strongly insisted on the high importance of relative position or connexion in homologous parts; they may differ to almost any extent in form and size, and yet remain connected together in the same invariable order. We never find, for instance, the bones of the arm and fore-arm, or of the thigh and leg, transposed. Hence the same names can be given to the homologous bones in widely different animals. We see the same great law in the construction of the mouths of insects: what can be more different than the immensely long spiral proboscis of a sphinx-moth, the curious folded one of a bee or bug, and the great jaws of a beetle?—yet all these organs, serving for such widely different purposes, are formed by infinitely numerous modifications of an upper lip, mandibles, and two pairs of maxillæ. The same law governs the construction of the mouths and limbs of crustaceans. So it is with the flowers of plants.

Nothing can be more hopeless than to attempt to explain this similarity of pattern in members of the same class, by utility or by the doctrine of final causes. The hopelessness of the attempt has been expressly admitted by Owen in his most interesting work on the 'Nature of Limbs.' On the ordinary view of the independent creation of each being, we can only say that so it is;—that it has pleased the Creator to construct all the animals and plants in each great class on a uniform plan; but this is not a scientific explanation.

The explanation is to a large extent simple on the theory of the selection of successive slight modifications,—each modification being profitable in some way to the modified form, but often affecting by correlation other parts of the organisation. In changes of this

nature, there will be little or no tendency to alter the
original pattern, or to transpose the parts. The bones
of a limb might be shortened and flattened to any
extent, becoming at the same time enveloped in thick
membrane, so as to serve as a fin; or a webbed hand
might have all its bones, or certain bones, lengthened
to any extent, with the membrane connecting them
increased, so as to serve as a wing; yet all these
modifications would not tend to alter the framework of
the bones or the relative connexion of the parts. If
we suppose that an early progenitor—the archetype as
it may be called—of all mammals, birds, and reptiles,
had its limbs constructed on the existing general
pattern, for whatever purpose they served, we can at
once perceive the plain signification of the homologous
construction of the limbs throughout the class. So
with the mouths of insects, we have only to suppose
that their common progenitor had an upper lip,
mandibles, and two pairs of maxillæ, these parts
being perhaps very simple in form; and then natural
selection will account for the infinite diversity in the
structure and functions of the mouths of insects.
Nevertheless, it is conceivable that the general pattern
of an organ might become so much obscured as to be
finally lost, by the reduction and ultimately by the
complete abortion of certain parts, by the fusion of
other parts, and by the doubling or multiplication of
others,—variations which we know to be within the
limits of possibility. In the paddles of the gigantic
extinct sea-lizards, and in the mouths of certain
suctorial crustaceans, the general pattern seems thus
to have become partially obscured.

There is another and equally curious branch of our

subject; namely, serial homologies, or the comparison
of the different parts or organs in the same individual,
and not of the same parts or organs in different
members of the same class.   Most physiologists believe
that the bones of the skull are homologous—that is,
correspond in number and in relative connexion—with
the elemental parts of a certain number of vertebræ.
The anterior and posterior limbs in all the higher
vertebrate classes are plainly homologous.   So it is
with the wonderfully complex jaws and legs of crustaceans.   It is familiar to almost every one, that in
a flower the relative position of the sepals, petals,
stamens, and pistils, as well as their intimate structure,
are intelligible on the view that they consist of metamorphosed leaves, arranged in a spire.   In monstrous
plants, we often get direct evidence of the possibility of
one organ being transformed into another; and we can
actually see, during the early or embryonic stages of
development in flowers, as well as in crustaceans and
many other animals, that organs, which when mature
become extremely different are at first exactly alike.

How inexplicable are the cases of serial homologies
on the ordinary view of creation!   Why should the
brain be enclosed in a box composed of such numerous
and such extraordinarily shaped pieces of bone, apparently representing vertebræ?   As Owen has remarked,
the benefit derived from the yielding of the separate
pieces in the act of parturition by mammals, will by
no means explain the same construction in the skulls
of birds and reptiles.   Why should similar bones have
been created to form the wing and the leg of a bat,
used as they are for such totally different purposes,
namely flying and walking?   Why should one crus-

tacean, which has an extremely complex mouth formed of many parts, consequently always have fewer legs; or conversely, those with many legs have simpler mouths? Why should the sepals, petals, stamens, and pistils, in each flower, though fitted for such distinct purposes, be all constructed on the same pattern?

On the theory of natural selection, we can, to a certain extent, answer these questions. We need not here consider how the bodies of some animals first became divided into a series of segments, or how they became divided into right and left sides, with corresponding organs, for such questions are almost beyond investigation. It is, however, probable that some serial structures are the result of cells multiplying by division, entailing the multiplication of the parts developed from such cells. It must suffice for our purpose to bear in mind that an indefinite repetition of the same part or organ is the common characteristic, as Owen has remarked, of all low or little specialised forms; therefore the unknown progenitor of the Vertebrata probably possessed many vertebræ; the unknown progenitor of the Articulata, many segments; and the unknown progenitor of flowering plants, many leaves arranged in one or more spires. We have also formerly seen that parts many times repeated are eminently liable to vary, not only in number, but in form. Consequently such parts, being already present in considerable numbers, and being highly variable, would naturally afford the materials for adaptation to the most different purposes; yet they would generally retain, through the force of inheritance, plain traces of their original or fundamental resemblance. They would retain this resemblance all the more, as the variations, which afforded the basis for their subsequent modification through natural selection, would tend from the first to be similar; the parts being at an early stage of growth alike, and being subjected to nearly the same conditions. Such parts, whether more or less modified, unless their common origin became wholly obscured, would be serially homologous.

In the great class of molluscs, though the parts in distinct species can be shown to be homologous, only a few serial homologies, such as the valves of Chitons, can be indicated; that is, we are seldom enabled to say that one part is homologous with another part in the same individual. And we can understand this fact; for in molluscs, even in the lowest members of the class, we do not find nearly so much indefinite repetition of any one part as we find in the other great classes of the animal and vegetable kingdoms.

But morphology is a much more complex subject than it at first appears, as has lately been well shown in a remarkable paper by Mr. E. Ray Lankester, who has drawn an important distinction between certain classes of cases which have all been equally ranked by naturalists as homologous. He proposes to call the structures which resemble each other in distinct animals, owing to their descent from a common progenitor with subsequent modification, *homogenous;* and the resemblances which cannot thus be accounted for, he proposes to call *homoplastic.* For instance, he believes that the hearts of birds and mammals are as a whole homogenous,—that is, have been derived from a common progenitor; but that the four cavities of the heart in the two classes are homoplastic,—that is, have been independently developed. Mr. Lankester also adduces the close resemblance of the parts on the right and left sides of the body, and in the successive segments of the same individual animal; and here we have parts commonly called homologous, which bear no relation to the descent of distinct species from a common progenitor. Homoplastic structures are the same with those which I have classed, though in a very imperfect manner, as analogous modifications or resemblances. Their formation may be attributed in part to distinct organisms, or to distinct parts of the same organism, having varied in an analogous manner; and in part to similar modifications, having been preserved for the same general purpose or function,—of which many instances have been given.

Naturalists frequently speak of the skull as formed of metamorphosed vertebræ; the jaws of crabs as metamorphosed legs; the stamens and pistils in flowers as metamorphosed leaves; but it would in most cases be more correct, as Professor Huxley has remarked, to speak of both skull and vertebræ, jaws and legs, &c., as having been metamorphosed, not one from the other, as they now exist, but from some common and simpler element. Most naturalists, however, use such language only in a metaphorical sense; they are far from meaning that during a long course of descent, primordial organs of any kind—vertebræ in the one case and legs in the other—have actually been converted into skulls or jaws. Yet so strong is the appearance of this having occurred, that naturalists can hardly avoid employing language having this plain signification. According to the views here maintained, such language may be used literally; and the wonderful fact of the jaws, for instance, of a crab retaining numerous characters, which they probably would have retained through inheritance, if they had really been metamorphosed from true though extremely simple legs, is in part explained.

Development and Embryology.

This is one of the most important subjects in the whole round of natural history. The metamorphoses of insects, with which every one is familiar, are generally effected abruptly by a few stages; but the transformations are in reality numerous and gradual, though concealed. A certain ephemerous insect (Chlöeon) during its development, moults, as shown by Sir J. Lubbock, above twenty times, and each time undergoes a certain amount of change; and in this case we see the act of metamorphosis performed in a primary and gradual manner. Many insects, and especially certain crustaceans, show us what wonderful changes of structure can be effected during development. Such changes, however, reach their acme in the so-called alternate generations of some of the lower animals. It is, for instance, an astonishing fact that a delicate branching coralline, studded with polypi and attached to a submarine rock, should produce, first by budding and then by transverse division, a host of huge floating jelly-fishes; and that these should produce eggs, from which are hatched swimming animalcules, which attach themselves to rocks and become developed into branching corallines; and so on in an endless cycle. The belief in the essential identity of the process of alternate generation and of ordinary metamorphosis has been greatly strengthened by Wagner's discovery of the larva or maggot of a fly, namely the Cecidomyia, producing asexually other larvæ, and these others, which finally are developed into mature males and females, propagating their kind in the ordinary manner by eggs.

It may be worth notice that when Wagner's remarkable discovery was first announced, I was asked how was it possible to account for the larvæ of this fly having acquired the power of asexual reproduction. As long as the case remained unique no answer could be given. But already Grimm has shown that another fly, a Chironomus, reproduces itself in nearly the same manner, and he believes that this occurs frequently in the Order. It is the pupa, and not the larva, of the Chironomus which has this power; and Grimm further shows that this case, to a certain extent, " unites that of the Cecidomyia with the parthenogenesis of the Coccidæ ; "—the term parthenogenesis implying that the mature females of the Coccidæ are capable of producing fertile eggs without the concourse of the male. Certain animals belonging to several classes are now known to have the power of ordinary reproduction at an unusually early age; and we have only to accelerate parthenogenetic reproduction by gradual steps to an earlier and earlier age,—Chironomus showing us an almost exactly intermediate stage, viz., that of the pupa —and we can perhaps account for the marvellous case of the Cecidomyia.

It has already been stated that various parts in the same individual which are exactly alike during an early embryonic period, become widely different and serve for widely different purposes in the adult state. So again it has been shown that generally the embryos of the most distinct species belonging to the same class are closely similar, but become, when fully developed, widely dissimilar. A better proof of this latter fact cannot be given than the statement by Von Baer that "the embryos of mammalia, "of birds, lizards, and snakes, probably also of chelonia, "are in their earliest states exceedingly like one another, "both as a whole and in the mode of development of "their parts; so much so, in fact, that we can often "distinguish the embryos only by their size. In my "possession are two little embryos in spirit, whose "names I have omitted to attach, and at present I am "quite unable to say to what class they belong. They "may be lizards or small birds, or very young mam- "malia, so complete is the similarity in the mode "of formation of the head and trunk in these animals. "The extremities, however, are still absent in these "embryos. But even if they had existed in the earliest "stage of their development we should learn nothing, for "the feet of lizards and mammals, the wings and feet of "birds, no less than the hands and feet of man, all arise "from the same fundamental form." The larvæ of most crustaceans, at corresponding stages of development, closely resemble each other, however different the adults may become; and so it is with very many other animals. A trace of the law of embryonic resemblance occasionally lasts till a rather late age: thus birds of the same genus, and of allied genera, often resemble each other in their immature plumage; as we see in the spotted feathers in the young of the thrush group. In the cat tribe, most of the species when adult are striped or spotted in lines; and stripes or spots can be plainly distinguished in the whelp of the lion and the puma. We occasionally though rarely see something of the same kind in plants; thus the first leaves of the ulex or furze, and the first leaves of the phyllodineous acacias, are pinnate or divided like the ordinary leaves of the leguminosæ.

The points of structure, in which the embryos of widely different animals within the same class resemble each other, often have no direct relation to their conditions of existence. We cannot, for instance, suppose that in the embryos of the vertebrata the peculiar loop-like courses of the arteries near the branchial slits are related to similar conditions,—in the young mammal which is nourished in the womb of its mother, in the egg of the bird which is hatched in a nest, and in the spawn of a frog under water. We have no more reason to believe in such a relation, than we have to believe that the similar bones in the hand of a man, wing of a bat, and fin of a porpoise, are related to similar conditions of life. No one supposes that the stripes on the whelp of a lion, or the spots on the young blackbird, are of any use to these animals.

The case, however, is different when an animal during any part of its embryonic career is active, and has to provide for itself. The period of activity may come on earlier or later in life; but whenever it comes on, the adaptation of the larva to its conditions of life is just as perfect and as beautiful as in the adult animal. In how important a manner this has acted, has recently been well shown by Sir J. Lubbock in his remarks on the close similarity of the larvæ of some insects belonging to very different orders, and on the dissimilarity of the larvæ of other insects within the same order, according to their habits of life. Owing to such adaptations, the similarity of the larvæ of allied animals is sometimes greatly obscured; especially when there is a division of labour during the different stages of development, as when the same larva has during one stage to search for food, and during another stage has to search for a place of attachment.   Cases can even be given of the larvæ of allied species, or groups of species, differing more from each other than do the adults.   In most cases, however, the larvæ, though active, still obey, more or less closely, the law of common embryonic resemblance.   Cirripedes afford a good instance of this; even the illustrious Cuvier did not perceive that a barnacle was a crustacean : but a glance at the larva shows this in an unmistakable manner.   So again the two main divisions of cirripedes, the pedunculated and sessile, though differing widely in external appearance, have larvæ in all their stages barely distinguishable.

The embryo in the course of development generally rises in organisation; I use this expression, though I am aware that it is hardly possible to define clearly what is meant by the organisation being higher or lower.   But no one probably will dispute that the butterfly is higher than the caterpillar.   In some cases, however, the mature animal must be considered as lower in the scale than the larva, as with certain parasitic crustaceans. To refer once again to cirripedes : the larvæ in the first stage have three pairs of locomotive organs, a simple single eye, and a prosciformed mouth, with which they feed largely, for they increase much in size.   In the second stage, answering to the chrysalis stage of butterflies, they have six pairs of beautifully constructed natatory legs, a pair of magnificent compound eyes, and extremely complex antennæ; but they have a closed and imperfect mouth, and cannot feed : their function at this stage is, to search out by their well-developed organs of sense, and to reach by their active powers of swimming, a proper place on which to become attached and to undergo their final metamorphosis. When this is completed they are fixed for life: their legs are now converted into prehensile organs; they again obtain a well-constructed mouth; but they have no antennæ, and their two eyes are now reconverted into a minute, single, simple eye-spot. In this last and complete state, cirripedes may be considered as either more highly or more lowly organised than they were in the larval condition. But in some genera the larvæ become developed into hermaphrodites having the ordinary structure, and into what I have called complemental males; and in the latter the development has assuredly been retrograde, for the male is a mere sack, which lives for a short time and is destitute of mouth, stomach, and every other organ of importance, excepting those for reproduction.

We are so much accustomed to see a difference in structure between the embryo and the adult, that we are tempted to look at this difference as in some necessary manner contingent on growth. But there is no reason why, for instance, the wing of a bat, or the fin of a porpoise, should not have been sketched out with all their parts in proper proportion, as soon as any part became visible. In some whole groups of animals and in certain members of other groups this is the case, and the embryo does not at any period differ widely from the adult: thus Owen has remarked in regard to cuttle-fish, "there is no metamorphosis; the cephalopodic character is manifested long before the parts of the embryo are completed." Land-shells and fresh-water crustaceans are born having their proper forms, whilst the marine members of the same two great classes pass through considerable and often great changes during their development. Spiders, again, barely undergo any metamorphosis. The larvæ of most insects pass through a worm-like stage, whether they are active and adapted to diversified habits, or are inactive from being placed in the midst of proper nutriment or from being fed by their parents; but in some few cases, as in that of Aphis, if we look to the admirable drawings of the development of this insect, by Professor Huxley, we see hardly any trace of the vermiform stage.

Sometimes it is only the earlier developmental stages which fail. Thus Fritz Müller has made the remarkable discovery that certain shrimp-like crustaceans (allied to Penœus) first appear under the simple nauplius-form, and after passing through two or more zoea-stages, and then through the mysis-stage, finally acquire their mature structure: now in the whole great malacostracan order, to which these crustaceans belong, no other member is as yet known to be first developed under the nauplius-form, though many appear as zoeas; nevertheless Müller assigns reasons for his belief, that if there had been no suppression of development, all these crustaceans would have appeared as nauplii.

How, then, can we explain these several facts in embryology,—namely, the very general, though not universal, difference in structure between the embryo and the adult;—the various parts in the same individual embryo, which ultimately become very unlike and serve for diverse purposes, being at an early period of growth alike;—the common, but not invariable, resemblance between the embryos or larvæ of the most distinct species in the same class;—the embryo often retaining whilst within the egg or womb, structures which are of no service to it, either at that or at a later

2 R

period of life; on the other hand larvæ, which have to
provide for their own wants, being perfectly adapted
to the surrounding conditions;—and lastly the fact of
certain larvæ standing higher in the scale of organisation
than the mature animal into which they are developed?
I believe that all these facts can be explained, as follows.

It is commonly assumed, perhaps from monstrosities
affecting the embryo at a very early period, that slight
variations or individual differences necessarily appear
at an equally early period. We have little evidence on
this head, but what we have certainly points the other
way; for it is notorious that breeders of cattle, horses,
and various fancy animals, cannot positively tell, until
some time after birth, what will be the merits or
demerits of their young animals. We see this plainly
in our own children; we cannot tell whether a child
will be tall or short, or what its precise features will be.
The question is not, at what period of life each variation
may have been caused, but at what period the effects
are displayed. The cause may have acted, and I believe
often has acted, on one or both parents before the act of
generation. It deserves notice that it is of no importance to a very young animal, as long as it remains in
its mother's womb or in the egg, or as long as it is
nourished and protected by its parent, whether most of
its characters are acquired a little earlier or later in life.
It would not signify, for instance, to a bird which
obtained its food by having a much-curved beak
whether or not whilst young it possessed a beak of this
shape, as long as it was fed by its parents.

I have stated in the first chapter, that at whatever
age a variation first appears in the parent, it tends to
re-appear at a corresponding age in the offspring.

Certain variations can only appear at corresponding ages; for instance, peculiarities in the caterpillar, cocoon, or imago states of the silk-moth: or, again, in the full-grown horns of cattle. But variations, which, for all that we can see might have first appeared either earlier or later in life, likewise tend to reappear at a corresponding age in the offspring and parent. I am far from meaning that this is invariably the case, and I could give several exceptional cases of variations (taking the word in the largest sense) which have supervened at an earlier age in the child than in the parent.

These two principles, namely, that slight variations generally appear at a not very early period of life, and are inherited at a corresponding not early period, explain, as I believe, all the above specified leading facts in embryology. But first let us look to a few analogous cases in our domestic varieties. Some authors who have written on Dogs, maintain that the greyhound and bulldog, though so different, are really closely allied varieties, descended from the same wild stock; hence I was curious to see how far their puppies differed from each other: I was told by breeders that they differed just as much as their parents, and this, judging by the eye, seemed almost to be the case; but on actually measuring the old dogs and their six-days-old puppies, I found that the puppies had not acquired nearly their full amount of proportional difference. So, again, I was told that the foals of cart and race-horses —breeds which have been almost wholly formed by selection under domestication—differed as much as the full-grown animals; but having had careful measurements made of the dams and of three-days-old colts of race and heavy cart-horses, I find that this is by no means the case.

As we have conclusive evidence that the breeds of the Pigeon are descended from a single wild species, I compared the young within twelve hours after being hatched; I carefully measured the proportions (but will not here give the details) of the beak, width of mouth, length of nostril and of eyelid, size of feet and length of leg, in the wild parent-species, in pouters, fantails, runts, barbs, dragons, carriers, and tumblers. Now some of these birds, when mature, differ in so extraordinary a manner in the length and form of beak, and in other characters, that they would certainly have been ranked as distinct genera if found in a state of nature. But when the nestling birds of these several breeds were placed in a row, though most of them could just be distinguished, the proportional differences in the above specified points were incomparably less than in the full-grown birds. Some characteristic points of difference—for instance, that of the width of mouth— could hardly be detected in the young. But there was one remarkable exception to this rule, for the young of the short-faced tumbler differed from the young of the wild rock-pigeon and of the other breeds, in almost exactly the same proportions as in the adult state.

These facts are explained by the above two principles. Fanciers select their dogs, horses, pigeons, &c., for breeding, when nearly grown up: they are indifferent whether the desired qualities are acquired earlier or later in life, if the full-grown animal possesses them. And the cases just given, more especially that of the pigeons, show that the characteristic differences which have been accumulated by man's selection, and which

give value to his breeds, do not generally appear at a
very early period of life, and are inherited at a corresponding not early period. But the case of the short-
faced tumbler, which when twelve hours old possessed
its proper characters, proves that this is not the
universal rule; for here the characteristic differences
must either have appeared at an earlier period than
usual, or, if not so, the differences must have been
inherited, not at a corresponding, but at an earlier age.

Now let us apply these two principles to species in a
state of nature. Let us take a group of birds, descended
from some ancient form and modified through natural
selection for different habits. Then, from the many
slight successive variations having supervened in the
several species at a not early age, and having been
inherited at a corresponding age, the young will have
been but little modified, and they will still resemble
each other much more closely than do the adults,—just
as we have seen with the breeds of the pigeon. We
may extend this view to widely distinct structures and
to whole classes. The fore-limbs, for instance, which
once served as legs to a remote progenitor, may have
become, through a long course of modification, adapted
in one descendant to act as hands, in another as paddles,
in another as wings; but on the above two principles
the fore-limbs will not have been much modified in the
embryos of these several forms; although in each form
the fore-limb will differ greatly in the adult state.
Whatever influence long-continued use or disuse may
have had in modifying the limbs or other parts of any
species, this will chiefly or solely have affected it when
nearly mature, when it was compelled to use its full
powers to gain its own living; and the effects thus

produced will have been transmitted to the offspring at a corresponding nearly mature age. Thus the young will not be modified, or will be modified only in a slight degree, through the effects of the increased use or disuse of parts.

With some animals the successive variations may have supervened at a very early period of life, or the steps may have been inherited at an earlier age than that at which they first occurred. In either of these cases, the young or embryo will closely resemble the mature parent-form, as we have seen with the short-faced tumbler. And this is the rule of development in certain whole groups, or in certain sub-groups alone, as with cuttle-fish, land-shells, fresh-water crustaceans, spiders, and some members of the great class of insects. With respect to the final cause of the young in such groups not passing through any metamorphosis, we can see that this would follow from the following contingences; namely, from the young having to provide at a very early age for their own wants, and from their following the same habits of life with their parents; for in this case, it would be indispensable for their existence that they should be modified in the same manner as their parents. Again, with respect to the singular fact that many terrestrial and fresh-water animals do not undergo any metamorphosis, whilst marine members of the same groups pass through various transformations, Fritz Müller has suggested that the process of slowly modifying and adapting an animal to live on the land or in fresh water, instead of in the sea, would be greatly simplified by its not passing through any larval stage; for it is not probable that places well adapted for both the larval and mature stages, under such new and greatly changed habits of life, would commonly be found unoccupied or ill-occupied by other organisms. In this case the gradual acquirement at an earlier and earlier age of the adult structure would be favoured by natural selection; and all traces of former metamorphoses would finally be lost.

If, on the other hand, it profited the young of an animal to follow habits of life slightly different from those of the parent-form, and consequently to be constructed on a slightly different plan, or if it profited a larva already different from its parent to change still further, then, on the principle of inheritance at corresponding ages, the young or the larvæ might be rendered by natural selection more and more different from their parents to any conceivable extent. Differences in the larva might, also, become correlated with successive stages of its development; so that the larva, in the first stage, might come to differ greatly from the larva in the second stage, as is the case with many animals. The adult might also become fitted for sites or habits, in which organs of locomotion or of the senses, &c., would be useless; and in this case the metamorphosis would be retrograde.

From the remarks just made we can see how by changes of structure in the young, in conformity with changed habits of life, together with inheritance at corresponding ages, animals might come to pass through stages of development, perfectly distinct from the primordial condition of their adult progenitors. Most of our best authorities are now convinced that the various larval and pupal stages of insects have thus been acquired through adaptation, and not through inheritance from some ancient form. The curious case of Sitaris—a beetle which passes through certain unusual stages of development—will illustrate how this might occur. The first larval form is described by M. Fabre, as an active, minute insect, furnished with six legs, two long antennæ, and four eyes. These larvæ are hatched in the nests of bees; and when the male-bees emerge from their burrows, in the spring, which they do before the females, the larvæ spring on them, and afterwards crawl on to the females whilst paired with the males. As soon as the female bee deposits her eggs on the surface of the honey stored in the cells, the larvæ of the Sitaris leap on the eggs and devour them. Afterwards they undergo a complete change; their eyes disappear; their legs and antennæ become rudimentary, and they feed on honey; so that they now more closely resemble the ordinary larvæ of insects; ultimately they undergo a further transformation, and finally emerge as the perfect beetle. Now, if an insect, undergoing transformations like those of the Sitaris, were to become the progenitor of a whole new class of insects, the course of development of the new class would be widely different from that of our existing insects; and the first larval stage certainly would not represent the former condition of any adult and ancient form.

On the other hand it is highly probable that with many animals the embryonic or larval stages show us, more or less completely, the condition of the progenitor of the whole group in its adult state. In the great class of the Crustacea, forms wonderfully distinct from each other, namely, suctorial parasites, cirripedes, entomostraca, and even the malacostraca, appear at first as larvæ under the nauplius-form; and as these larvæ live and feed in the open sea, and are not adapted for any peculiar habits of life, and from other reasons assigned by Fritz Müller, it is probable that at some very remote period an independent adult animal, resembling the Nauplius, existed, and subsequently produced, along several divergent lines of descent, the above-named great Crustacean groups. So again it is probable, from what we know of the embryos of mammals, birds, fishes, and reptiles, that these animals are the modified descendants of some ancient progenitor, which was furnished in its adult state with branchiæ, a swim-bladder, four fin-like limbs, and a long tail, all fitted for an aquatic life.

As all the organic beings, extinct and recent, which have ever lived, can be arranged within a few great classes; and as all within each class have, according to our theory, been connected together by fine gradations, the best, and, if our collections were nearly perfect, the only possible arrangement, would be genealogical; descent being the hidden bond of connexion which naturalists have been seeking under the term of the Natural System. On this view we can understand how it is that, in the eyes of most naturalists, the structure of the embryo is even more important for classification than that of the adult. In two or more groups of animals, however much they may differ from each other in structure and habits in their adult condition, if they pass through closely similar embryonic stages, we may feel assured that they all are descended from one parent-form, and are therefore closely related. Thus, community in embryonic structure reveals community of descent; but dissimilarity in embryonic development does not prove discommunity of descent, for in one of two groups the developmental stages may have been suppressed, or may have been so greatly modified through adaptation to new habits of life, as to be no longer recognisable. Even in groups, in which the adults have been modified to an extreme degree, community of origin is often revealed by the structure of the larvæ; we have seen, for instance, that cirripedes, though externally so like shell-fish, are at once known by their larvæ to belong to the great class of crustaceans. As the embryo often shows us more or less plainly the structure of the less modified and ancient progenitor of the group, we can see why ancient and extinct forms so often resemble in their adult state the embryos of existing species of the same class. Agassiz believes this to be a universal law of nature; and we may hope hereafter to see the law proved true. It can, however, be proved true only in those cases in which the ancient state of the progenitor of the group has not been wholly obliterated, either by successive variations having supervened at a very early period of growth, or by such variations having been inherited at an earlier age than that at which they first appeared. It should also be borne in mind, that the law may be true, but yet, owing to the geological record not extending far enough back in time, may remain for a long period, or for ever, incapable of demonstration. The law will not strictly hold good in those cases in which an ancient form became adapted in its larvæ state to some special line of life, and transmitted the same larval state to a whole group of descendants; for such larval will not resemble any still more ancient form in its adult state.

Thus, as it seems to me, the leading facts in embryology, which are second to none in importance, are explained on the principle of variations in the many descendants from some one ancient progenitor, having appeared at a not very early period of life, and having been inherited at a corresponding period. Embryology rises greatly in interest, when we look at the embryo as a picture, more or less obscured, of the progenitor, either in its adult or larval state, of all the members of the same great class.

Rudimentary, Atrophied, and Aborted Organs.

Organs or parts in this strange condition, bearing the plain stamp of inutility, are extremely common, or even general, throughout nature. It would be impossible to name one of the higher animals in which some part or other is not in a rudimentary condition. In the mammalia, for instance, the males possess rudimentary mammæ; in snakes one lobe of the lungs is rudimentary; in birds the " bastard-wing " may safely be considered as a rudimentary digit, and in some species the whole wing is so far rudimentary that it cannot be used for flight. What can be more curious than the presence of teeth in fœtal whales, which when grown up have not a tooth in their heads; or the teeth, which never cut through the gums, in the upper jaws of unborn calves ?

Rudimentary organs plainly declare their origin and meaning in various ways. There are beetles belonging to closely allied species, or even to the same identical species, which have either full-sized and perfect wings, or mere rudiments of membrane, which not rarely lie under wing-covers firmly soldered together; and in these cases it is impossible to doubt, that the rudiments represent wings. Rudimentary organs sometimes retain their potentiality: this occasionally occurs with the mammæ of male mammals, which have been known to become well developed and to secrete milk. So again in the udders in the genus Bos, there are normally four developed and two rudimentary teats ; but the latter in our domestic cows sometimes become well developed and yield milk. In regard to plants the petals are sometimes rudimentary, and sometimes well-developed in the individuals of the same species. In certain plants having separated sexes Kölreuter found that by crossing a species, in which the male flowers included a rudiment of a pistil, with an hermaphrodite species, having of course a well-developed pistil, the rudiment in the hybrid offspring was much increased in size ; and this clearly shows that the rudimentary and perfect pistils are essentially alike in nature. An animal may possess various parts in a perfect state, and yet they may in one sense be rudimentary, for they are useless : thus the tadpole of the common Salamander or Water-newt, as Mr. G. H. Lewes remarks, " has gills, and " passes its existence in the water ; but the Salamandra " atra, which lives high up among the mountains, brings " forth its young full-formed. This animal never lives " in the water. Yet if we open a gravid female, we " find tadpoles inside her with exquisitely feathered " gills ; and when placed in water they swim about " like the tadpoles of the water-newt. Obviously this " aquatic organisation has no reference to the future " life of the animal, nor has it any adaptation to its " embryonic condition ; it has solely reference to ances- " tral adaptations, it repeats a phase in the development " of its progenitors."

An organ, serving for two purposes, may become rudimentary or utterly aborted for one, even the more important purpose, and remain perfectly efficient for the other. Thus in plants, the office of the pistil is to allow the pollen-tubes to reach the ovules within the ovarium. The pistil consists of a stigma supported on a style; but in some Compositæ, the male florets, which of course cannot be fecundated, have a rudimentary pistil, for it is not crowned with a stigma; but the style remains well developed and is clothed in the usual manner with hairs, which serve to brush the pollen out of the surrounding and conjoined anthers. Again, an organ may become rudimentary for its proper purpose, and be used for a distinct one: in certain fishes the swim-bladder seems to be rudimentary for its proper function of giving buoyancy, but has become converted into a nascent breathing organ or lung. Many similar instances could be given.

Useful organs, however little they may be developed, unless we have reason to suppose that they were formerly more highly developed, ought not to be considered as rudimentary. They may be in a nascent condition, and in progress towards further development. Rudimentary organs, on the other hand, are either quite useless, such as teeth which never cut through the gums, or almost useless, such as the wings of an ostrich, which serve merely as sails. As organs in this condition would formerly, when still less developed, have been of even less use than at present, they cannot formerly have been produced through variation and natural selection, which acts solely by the preservation of useful modifications. They have been partially retained by the power of inheritance, and relate to a former state of things. It is, however, often difficult to distinguish between rudimentary and nascent organs;

for we can judge only by analogy whether a part is
capable of further development, in which case alone it
deserves to be called nascent. Organs in this condition
will always be somewhat rare; for beings thus provided
will commonly have been supplanted by their successors
with the same organ in a more perfect state, and consequently will have become long ago extinct. The wing
of the penguin is of high service, acting as a fin; it
may, therefore, represent the nascent state of the wing:
not that I believe this to be the case; it is more probably a reduced organ, modified for a new function: the
wing of the Apteryx, on the other hand, is quite useless,
and is truly rudimentary. Owen considers the simple
filamentary limbs of the Lepidosiren as the "beginnings
of organs which attain full functional development in
higher vertebrates;" but, according to the view lately
advocated by Dr. Günther, they are probably remnants,
consisting of the persistent axis of a fin, with the lateral
rays or branches aborted. The mammary glands of the
Ornithorhynchus may be considered, in comparison
with the udders of a cow, as in a nascent condition.
The ovigerous frena of certain cirripedes, which have
ceased to give attachment to the ova and are feebly
developed, are nascent branchiæ.

Rudimentary organs in the individuals of the same
species are very liable to vary in the degree of their
development and in other respects. In closely allied
species, also, the extent to which the same organ has
been reduced occasionally differs much. This latter fact
is well exemplified in the state of the wings of female
moths belonging to the same family. Rudimentary
organs may be utterly aborted; and this implies, that
in certain animals or plants, parts are entirely absent

which analogy would lead us to expect to find in them, and which are occasionally found in monstrous individuals. Thus in most of the Scrophulariaceæ the fifth stamen is utterly aborted; yet we may conclude that a fifth stamen once existed, for a rudiment of it is found in many species of the family, and this rudiment occasionally becomes perfectly developed, as may sometimes be seen in the common snap-dragon. In tracing the homologies of any part in different members of the same class, nothing is more common, or, in order fully to understand the relations of the parts, more useful than the discovery of rudiments. This is well shown in the drawings given by Owen of the leg-bones of the horse, ox, and rhinoceros.

It is an important fact that rudimentary organs, such as teeth in the upper jaws of whales and ruminants, can often be detected in the embryo, but afterwards wholly disappear. It is also, I believe, a universal rule, that a rudimentary part is of greater size in the embryo relatively to the adjoining parts, than in the adult; so that the organ at this early age is less rudimentary, or even cannot be said to be in any degree rudimentary. Hence rudimentary organs in the adult are often said to have retained their embryonic condition.

I have now given the leading facts with respect to rudimentary organs. In reflecting on them, every one must be struck with astonishment; for the same reasoning power which tells us that most parts and organs are exquisitely adapted for certain purposes, tells us with equal plainness that these rudimentary or atrophied organs are imperfect and useless. In works on natural history, rudimentary organs are generally said to have been created " for the sake of symmetry," or in order "to complete the scheme of nature." But this is not an explanation, merely a re-statement of the fact. Nor is it consistent with itself: thus the boa-constrictor has rudiments of hind-limbs and of a pelvis, and if it be said that these bones have been retained "to complete the scheme of nature," why, as Professor Weismann asks, have they not been retained by other snakes, which do not possess even a vestige of these same bones? What would be thought of an astronomer who maintained that the satellites revolve in elliptic courses round their planets "for the sake of symmetry," because the planets thus revolve round the sun? An eminent physiologist accounts for the presence of rudimentary organs, by supposing that they serve to excrete matter in excess, or matter injurious to the system; but can we suppose that the minute papilla, which often represents the pistil in male flowers, and which is formed of mere cellular tissue, can thus act? Can we suppose that rudimentary teeth, which are subsequently absorbed, are beneficial to the rapidly growing embryonic calf by removing matter so precious as phosphate of lime? When a man's fingers have been amputated, imperfect nails have been known to appear on the stumps, and I could as soon believe that these vestiges of nails are developed in order to excrete horny matter, as that the rudimentary nails on the fin of the manatee have been developed for this same purpose.

On the view of descent with modification, the origin of rudimentary organs is comparatively simple; and we can understand to a large extent the laws governing their imperfect development. We have plenty of cases of rudimentary organs in our domestic productions,—as the stump of a tail in tailless breeds,—the vestige of an

ear in earless breeds of sheep,—the reappearance of
minute dangling horns in hornless breeds of cattle,
more especially, according to Youatt, in young animals,
—and the state of the whole flower in the cauliflower.
We often see rudiments of various parts in monsters;
but I doubt whether any of these cases throw light on
the origin of rudimentary organs in a state of nature,
further than by showing that rudiments can be produced; for the balance of evidence clearly indicates that
species under nature do not undergo great and abrupt
changes.   But we learn from the study of our domestic
productions that the disuse of parts leads to their
reduced size; and that the result is inherited.

It appears probable that disuse has been the main
agent in rendering organs rudimentary.   It would at
first lead by slow steps to the more and more complete
reduction of a part, until at last it became rudimentary,
—as in the case of the eyes of animals inhabiting dark
caverns, and of the wings of birds inhabiting oceanic
islands, which have seldom been forced by beasts of prey
to take flight, and have ultimately lost the power of
flying.   Again, an organ, useful under certain conditions,
might become injurious under others, as with the wings
of beetles living on small and exposed islands; and in
this case natural selection will have aided in reducing
the organ, until it was rendered harmless and rudimentary.

Any change in structure and function, which can be
effected by small stages, is within the power of natural
selection; so that an organ rendered, through changed
habits of life, useless or injurious for one purpose, might
be modified and used for another purpose.   An organ
might, also, be retained for one alone of its former func-

tions. Organs, originally formed by the aid of natural selection, when rendered useless may well be variable, for their variations can no longer be checked by natural selection. All this agrees well with what we see under nature. Moreover, at whatever period of life either disuse or selection reduces an organ, and this will generally be when the being has come to maturity and has to exert its full powers of action, the principle of inheritance at corresponding ages will tend to reproduce the organ in its reduced state at the same mature age, but will seldom affect it in the embryo. Thus we can understand the greater size of rudimentary organs in the embryo relatively to the adjoining parts, and their lesser relative size in the adult. If, for instance, the digit of an adult animal was used less and less during many generations, owing to some change of habits, or if an organ or gland was less and less functionally exercised, we may infer that it would become reduced in size in the adult descendants of this animal, but would retain nearly its original standard of development in the embryo.

There remains, however, this difficulty. After an organ has ceased being used, and has become in consequence much reduced, how can it be still further reduced in size until the merest vestige is left; and how can it be finally quite obliterated? It is scarcely possible that disuse can go on producing any further effect after the organ has once been rendered functionless. Some additional explanation is here requisite which I cannot give. If, for instance, it could be proved that every part of the organisation tends to vary in a greater degree towards diminution than towards augmentation of size, then we should be able to understand how an organ which has become useless would be rendered. independently of the effects of disuse, rudimentary and would at last be wholly suppressed; for the variations towards diminished size would no longer be checked by natural selection. The principle of the economy of growth, explained in a former chapter, by which the materials forming any part, if not useful to the possessor, are saved as far as is possible, will perhaps come into play in rendering a useless part rudimentary. But this principle will almost necessarily be confined to the earlier stages of the process of reduction; for we cannot suppose that a minute papilla, for instance, representing in a male flower the pistil of the female flower, and formed merely of cellular tissue, could be further reduced or absorbed for the sake of economising nutriment.

Finally, as rudimentary organs, by whatever steps they may have been degraded into their present useless condition, are the record of a former state of things, and have been retained solely through the power of inheritance,—we can understand, on the genealogical view of classification, how it is that systematists, in placing organisms in their proper places in the natural system, have often found rudimentary parts as useful as, or even sometimes more useful than, parts of high physiological importance. Rudimentary organs may be compared with the letters in a word, still retained in the spelling, but become useless in the pronunciation, but which serve as a clue for its derivation. On the view of descent with modification, we may conclude that the existence of organs in a rudimentary, imperfect, and useless condition, or quite aborted, far from presenting a strange difficulty, as they assuredly do on the old doctrine of creation, might even have been anticipated in accordance with the views here explained.

Summary.

In this chapter I have attempted to show, that the arrangement of all organic beings throughout all time in groups under groups—that the nature of the relationships by which all living and extinct organisms are united by complex, radiating, and circuitous lines of affinities into a few grand classes,—the rules followed and the difficulties encountered by naturalists in their classifications,—the value set upon characters, if constant and prevalent, whether of high or of the most trifling importance, or, as with rudimentary organs, of no importance,—the wide opposition in value between analogical or adaptive characters, and characters of true affinity; and other such rules;—all naturally follow if we admit the common parentage of allied forms, together with their modification through variation and natural selection, with the contingencies of extinction and divergence of character. In considering this view of classification, it should be borne in mind that the element of descent has been universally used in ranking together the sexes, ages, dimorphic forms, and acknowledged varieties of the same species, however much they may differ from each other in structure. If we extend the use of this element of descent,—the one certainly known cause of similarity in organic beings,—we shall understand what is meant by the Natural System: it is genealogical in its attempted arrangement, with the grades of acquired difference marked by the terms, varieties, species, genera, families, orders, and classes.

On this same view of descent with modification, most

of the great facts in Morphology become intelligible,—whether we look to the same pattern displayed by the different species of the same class in their homologous organs, to whatever purpose applied; or to the serial and lateral homologies in each individual animal and plant.

On the principle of successive slight variations, not necessarily or generally supervening at a very early period of life, and being inherited at a corresponding period, we can understand the leading facts in Embryology; namely, the close resemblance in the individual embryo of the parts which are homologous, and which when matured become widely different in structure and function; and the resemblance of the homologous parts or organs in allied though distinct species, though fitted in the adult state for habits as different as is possible. Larvæ are active embryos, which have been specially modified in a greater or less degree in relation to their habits of life, with their modifications inherited at a corresponding early age. On these same principles,—and bearing in mind that when organs are reduced in size, either from disuse or through natural selection, it will generally be at that period of life when the being has to provide for its own wants, and bearing in mind how strong is the force of inheritance—the occurrence of rudimentary organs might even have been anticipated. The importance of embryological characters and of rudimentary organs in classification is intelligible, on the view that a natural arrangement must be genealogical.

Finally, the several classes of facts which have been considered in this chapter, seem to me to proclaim so plainly, that the innumerable species, genera and families, with which this world is peopled, are all descended, each within its own class or group, from common parents, and have all been modified in the course of descent, that I should without hesitation adopt this view, even if it were unsupported by other facts or arguments.

CHAPTER XV.

RECAPITULATION AND CONCLUSION.

Recapitulation of the objections to the theory of Natural Selection
—Recapitulation of the general and special circumstances in its
favour—Causes of the general belief in the immutability of
species — How far the theory of Natural Selection may be
extended — Effects of its adoption on the study of Natural
History—Concluding remarks.

As this whole volume is one long argument, it may
be convenient to the reader to have the leading facts
and inferences briefly recapitulated.

That many and serious objections may be advanced
against the theory of descent with modification through
variation and natural selection, I do not deny. I have
endeavoured to give to them their full force. Nothing
at first can appear more difficult to believe than that
the more complex organs and instincts have been perfected, not by means superior to, though analogous with,
human reason, but by the accumulation of innumerable
slight variations, each good for the individual possessor.
Nevertheless, this difficulty, though appearing to our
imagination insuperably great, cannot be considered
real if we admit the following propositions, namely,
that all parts of the organisation and instincts offer, at
least, individual differences—that there is a struggle for
existence leading to the preservation of profitable deviations of structure or instinct—and, lastly, that gradations

in the state of perfection of each organ may have existed,
each good of its kind.   The truth of these propositions
cannot, I think, be disputed.

It is, no doubt, extremely difficult even to conjecture
by what gradations many structures have been perfected,
more especially amongst broken and failing groups of
organic beings, which have suffered much extinction;
but we see so many strange gradations in nature, that
we ought to be extremely cautious in saying that any
organ or instinct, or any whole structure, could not have
arrived at its present state by many graduated steps.
There are, it must be admitted, cases of special difficulty
opposed to the theory of natural selection; and one of
the most curious of these is the existence in the same
community of two or three defined castes of workers or
sterile female ants; but I have attempted to show how
these difficulties can be mastered.

With respect to the almost universal sterility of
species when first crossed, which forms so remarkable a
contrast with the almost universal fertility of varieties
when crossed, I must refer the reader to the recapitulation of the facts given at the end of the ninth chapter,
which seem to me conclusively to show that this sterility
is no more a special endowment than is the incapacity of
two distinct kinds of trees to be grafted together; but
that it is incidental on differences confined to the reproductive systems of the intercrossed species.   We see the
truth of this conclusion in the vast difference in the
results of crossing the same two species reciprocally,—
that is, when one species is first used as the father and
then as the mother.   Analogy from the consideration of
dimorphic and trimorphic plants clearly leads to the
same conclusion, for when the forms are illegitimately

united, they yield few or no seed, and their offspring are more or less sterile; and these forms belong to the same undoubted species, and differ from each other in no respect except in their reproductive organs and functions.

Although the fertility of varieties when intercrossed and of their mongrel offspring has been asserted by so many authors to be universal, this cannot be considered as quite correct after the facts given on the high authority of Gärtner and Kölreuter. Most of the varieties which have been experimented on have been produced under domestication; and as domestication (I do not mean mere confinement) almost certainly tends to eliminate that sterility which, judging from analogy, would have affected the parent-species if intercrossed, we ought not to expect that domestication would likewise induce sterility in their modified descendants when crossed. This elimination of sterility apparently follows from the same cause which allows our domestic animals to breed freely under diversified circumstances; and this again apparently follows from their having been gradually accustomed to frequent changes in their conditions of life.

A double and parallel series of facts seems to throw much light on the sterility of species, when first crossed, and of their hybrid offspring. On the one side, there is good reason to believe that slight changes in the conditions of life give vigour and fertility to all organic beings. We know also that a cross between the distinct individuals of the same variety, and between distinct varieties, increases the number of their offspring, and certainly gives to them increased size and vigour. This is chiefly owing to the forms which are crossed having been exposed to somewhat different conditions of life;

for I have ascertained by a laborious series of experiments that if all the individuals of the same variety be subjected during several generations to the same conditions, the good derived from crossing is often much diminished or wholly disappears. This is one side of the case.   On the other side, we know that species which have long been exposed to nearly uniform conditions, when they are subjected under confinement to new and greatly changed conditions, either perish, or if they survive, are rendered sterile, though retaining perfect health.   This does not occur, or only in a very slight degree, with our domesticated productions, which have long been exposed to fluctuating conditions. Hence when we find that hybrids produced by a cross between two distinct species are few in number, owing to their perishing soon after conception or at a very early age, or if surviving that they are rendered more or less sterile, it seems highly probable that this result is due to their having been in fact subjected to a great change in their conditions of life, from being compounded of two distinct organisations. He who will explain in a definite manner why, for instance, an elephant or a fox will not breed under confinement in its native country, whilst the domestic pig or dog will breed freely under the most diversified conditions, will at the same time be able to give a definite answer to the question why two distinct species, when crossed, as well as their hybrid offspring, are generally rendered more or less sterile, whilst two domesticated varieties when crossed and their mongrel offspring are perfectly fertile.

Turning to geographical distribution, the difficulties encountered on the theory of descent with modification are serious enough.   All the individuals of the same species, and all the species of the same genus, or even higher group, are descended from common parents; and therefore, in however distant and isolated parts of the world they may now be found, they must in the course of successive generations have travelled from some one point to all the others. We are often wholly unable even to conjecture how this could have been effected. Yet, as we have reason to believe that some species have retained the same specific form for very long periods of time, immensely long as measured by years, too much stress ought not to be laid on the occasional wide diffusion of the same species; for during very long periods there will always have been a good chance for wide migration by many means. A broken or interrupted range may often be accounted for by the extinction of the species in the intermediate regions. It cannot be denied that we are as yet very ignorant as to the full extent of the various climatal and geographical changes which have affected the earth during modern periods; and such changes will often have facilitated migration. As an example, I have attempted to show how potent has been the influence of the Glacial period on the distribution of the same and of allied species throughout the world. We are as yet profoundly ignorant of the many occasional means of transport. With respect to distinct species of the same genus inhabiting distant and isolated regions, as the process of modification has necessarily been slow, all the means of migration will have been possible during a very long period; and consequently the difficulty of the wide diffusion of the species of the same genus is in some degree lessened.

As according to the theory of natural selection an interminable number of intermediate forms must have existed, linking together all the species in each group by gradations as fine as are our existing varieties, it may be asked, Why do we not see these linking forms all around us? Why are not all organic beings blended together in an inextricable chaos? With respect to existing forms, we should remember that we have no right to expect (excepting in rare cases) to discover *directly* connecting links between them, but only between each and some extinct and supplanted form. Even on a wide area, which has during a long period remained continuous, and of which the climatic and other conditions of life change insensibly in proceeding from a district occupied by one species into another district occupied by a closely allied species, we have no just right to expect often to find intermediate varieties in the intermediate zones. For we have reason to believe that only a few species of a genus ever undergo change; the other species becoming utterly extinct and leaving no modified progeny. Of the species which do change, only a few within the same country change at the same time; and all modifications are slowly effected. I have also shown that the intermediate varieties which probably at first existed in the intermediate zones, would be liable to be supplanted by the allied forms on either hand; for the latter, from existing in greater numbers, would generally be modified and improved at a quicker rate than the intermediate varieties, which existed in lesser numbers; so that the intermediate varieties would, in the long run, be supplanted and exterminated.

On this doctrine of the extermination of an infinitude of connecting links, between the living and extinct inhabitants of the world, and at each successive period between the extinct and still older species, why is not every geological formation charged with such links? Why does not every collection of fossil remains afford plain evidence of the gradation and mutation of the forms of life? Although geological research has undoubtedly revealed the former existence of many links, bringing numerous forms of life much closer together, it does not yield the infinitely many fine gradations between past and present species required on the theory; and this is the most obvious of the many objections which may be urged against it. Why, again, do whole groups of allied species appear, though this appearance is often false, to have come in suddenly on the successive geological stages? Although we now know that organic beings appeared on this globe, at a period incalculably remote, long before the lowest bed of the Cambrian system was deposited, why do we not find beneath this system great piles of strata stored with the remains of the progenitors of the Cambrian fossils? For on the theory, such strata must somewhere have been deposited at these ancient and utterly unknown epochs of the world's history.

I can answer these questions and objections only on the supposition that the geological record is far more imperfect than most geologists believe. The number of specimens in all our museums is absolutely as nothing compared with the countless generations of countless species which have certainly existed. The parent-form of any two or more species would not be in all its characters directly intermediate between its modified offspring, any more than the rock-pigeon is directly intermediate in crop and tail between its descendants, the pouter and fantail pigeons. We should not be able to recognise a species as the parent of another and modified species, if we were to examine the two ever so closely, unless we possessed most of the intermediate links; and owing to the imperfection of the geological record, we have no just right to expect to find so many links. If two or three, or even more linking forms were discovered, they would simply be ranked by many naturalists as so many new species, more especially if found in different geological sub-stages, let their differences be ever so slight. Numerous existing doubtful forms could be named which are probably varieties; but who will pretend that in future ages so many fossil links will be discovered, that naturalists will be able to decide whether or not these doubtful forms ought to be called varieties? Only a small portion of the world has been geologically explored. Only organic beings of certain classes can be preserved in a fossil condition, at least in any great number. Many species when once formed never undergo any further change but become extinct without leaving modified descendants; and the periods, during which species have undergone modification, though long as measured by years, have probably been short in comparison with the periods during which they retained the same form. It is the dominant and widely ranging species which vary most frequently and vary most, and varieties are often at first local—both causes rendering the discovery of intermediate links in any one formation less likely. Local varieties will not spread into other and distant regions until they are considerably modified and improved; and when they have spread, and are discovered in a geological formation, they appear as if suddenly created there, and will be simply classed as new species. Most formations have been intermittent in their accumulation; and their duration has probably been shorter than the average duration of specific forms. Successive formations are in most cases separated from each other by blank intervals of time of great length; for fossiliferous formations thick enough to resist future degradation can as a general rule be accumulated only where much sediment is deposited on the subsiding bed of the sea. During the alternate periods of elevation and of stationary level the record will generally be blank. During these latter periods there will probably be more variability in the forms of life; during periods of subsidence, more extinction.

With respect to the absence of strata rich in fossils beneath the Cambrian formation, I can recur only to the hypothesis given in the tenth chapter; namely, that though our continents and oceans have endured for an enormous period in nearly their present relative positions, we have no reason to assume that this has always been the case; consequently formations much older than any now known may lie buried beneath the great oceans. With respect to the lapse of time not having been sufficient since our planet was consolidated for the assumed amount of organic change, and this objection, as urged by Sir William Thompson, is probably one of the gravest as yet advanced, I can only say, firstly, that we do not know at what rate species change as measured by years, and secondly, that many philosophers are not as yet willing to admit that we know enough of the constitution of the universe and of the interior of our globe to speculate with safety on its past duration.

That the geological record is imperfect all will admit; but that it is imperfect to the degree required by our theory, few will be inclined to admit. If we look to long enough intervals of time, geology plainly declares that species have all changed; and they have changed in the manner required by the theory, for they have changed slowly and in a graduated manner. We clearly see this in the fossil remains from consecutive formations invariably being much more closely related to each other, than are the fossils from widely separated formations.

Such is the sum of the several chief objections and difficulties which may be justly urged against the theory; and I have now briefly recapitulated the answers and explanations which, as far as I can see, may be given. I have felt these difficulties far too heavily during many years to doubt their weight. But it deserves especial notice that the more important objections relate to questions on which we are confessedly ignorant; nor do we know how ignorant we are. We do not know all the possible transitional gradations between the simplest and the most perfect organs; it cannot be pretended that we know all the varied means of Distribution during the long lapse of years, or that we know how imperfect is the Geological Record. Serious as these several objections are, in my judgment they are by no means sufficient to overthrow the theory of descent with subsequent modification.

Now let us turn to the other side of the argument. Under domestication we see much variability, caused, or at least excited, by changed conditions of life; but often in so obscure a manner, that we are tempted to consider the variations as spontaneous. Variability is governed by many complex laws,—by correlated growth, compensation, the increased use and disuse of parts, and the definite action of the surrounding conditions. There is much difficulty in ascertaining how largely our domestic productions have been modified; but we may safely infer that the amount has been large, and that modifications can be inherited for long periods. As long as the conditions of life remain the same, we have reason to believe that a modification, which has already been inherited for many generations, may continue to be inherited for an almost infinite number of generations. On the other hand, we have evidence that variability when it has once come into play, does not cease under domestication for a very long period; nor do we know that it ever ceases, for new varieties are still occasionally produced by our oldest domesticated productions.

Variability is not actually caused by man; he only unintentionally exposes organic beings to new conditions of life, and then nature acts on the organisation and causes it to vary. But man can and does select the variations given to him by nature, and thus accumulates them in any desired manner. He thus adapts animals and plants for his own benefit or pleasure. He may do this methodically, or he may do it unconsciously by preserving the individuals most useful or pleasing to him without any intention of altering the breed. It is certain that he can largely influence the character of a breed by selecting, in each successive generation, individual differences so slight as to be inappreciable except by an educated eye. This unconscious process of selection has been the great agency in the formation of the most distinct and useful domestic breeds. That many breeds produced by man have to a large extent the character of natural species, is shown by the inextricable doubts whether many of them are varieties or aboriginally distinct species.

There is no reason why the principles which have

2 T

acted so efficiently under domestication should not have
acted under nature. In the survival of favoured
individuals and races, during the constantly-recurrent
Struggle for Existence, we see a powerful and ever-
acting form of Selection. The struggle for existence
inevitably follows from the high geometrical ratio of
increase which is common to all organic beings. This
high rate of increase is proved by calculation,—by the
rapid increase of many animals and plants during a
succession of peculiar seasons, and when naturalised in
new countries. More individuals are born than can
possibly survive. A grain in the balance may determine
which individuals shall live and which shall die,—which
variety or species shall increase in number, and which
shall decrease, or finally become extinct. As the individuals of the same species come in all respects into the
closest competition with each other, the struggle will
generally be most severe between them; it will be
almost equally severe between the varieties of the same
species, and next in severity between the species of the
same genus. On the other hand the struggle will often
be severe between beings remote in the scale of nature.
The slightest advantage in certain individuals, at any
age or during any season, over those with which they
come into competition, or better adaptation in however
slight a degree to the surrounding physical conditions,
will, in the long run, turn the balance.

With animals having separated sexes, there will be
in most cases a struggle between the males for the possession of the females. The most vigorous males, or those
which have most successfully struggled with their conditions of life, will generally leave most progeny. But
success will often depend on the males having special

weapons, or means of defence, or charms; and a slight advantage will lead to victory.

As geology plainly proclaims that each land has undergone great physical changes, we might have expected to find that organic beings have varied under nature, in the same way as they have varied under domestication. And if there has been any variability under nature, it would be an unaccountable fact if natural selection had not come into play. It has often been asserted, but the assertion is incapable of proof, that the amount of variation under nature is a strictly limited quantity. Man, though acting on external characters alone and often capriciously, can produce within a short period a great result by adding up mere individual differences in his domestic productions; and every one admits that species present individual differences. But, besides such differences, all naturalists admit that natural varieties exist, which are considered sufficiently distinct to be worthy of record in systematic works. No one has drawn any clear distinction between individual differences and slight varieties; or between more plainly marked varieties and sub-species, and species. On separate continents, and on different parts of the same continent when divided by barriers of any kind, and on outlying islands, what a multitude of forms exist, which some experienced naturalists rank as varieties, others as geographical races or sub-species, and others as distinct, though closely allied species!

If then, animals and plants do vary, let it be ever so slightly or slowly, why should not variations or individual differences, which are in any way beneficial, be preserved and accumulated through natural selection, or the survival of the fittest? If man can by patience select variations useful to him, why, under changing and complex conditions of life, should not variations useful to nature's living products often arise, and be preserved or selected ? What limit can be put to this power, acting during long ages and rigidly scrutinising the whole constitution, structure, and habits of each creature,—favouring the good and rejecting the bad ? I can see no limit to this power, in slowly and beautifully adapting each form to the most complex relations of life. The theory of natural selection, even if we look no farther than this, seems to be in the highest degree probable. I have already recapitulated, as fairly as I could, the opposed difficulties and objections : now let us turn to the special facts and arguments in favour of the theory.

On the view that species are only strongly marked and permanent varieties, and that each species first existed as a variety, we can see why it is that no line of demarcation can be drawn between species, commonly supposed to have been produced by special acts of creation, and varieties which are acknowledged to have been produced by secondary laws. On this same view we can understand how it is that in a region where many species of a genus have been produced, and where they now flourish, these same species should present many varieties; for where the manufactory of species has been active, we might expect, as a general rule, to find it still in action; and this is the case if varieties be incipient species. Moreover, the species of the larger genera, which afford the greater number of varieties or incipient species, retain to a certain degree the character of varieties; for they differ from each other by a less amount of difference than do the species of smaller genera. The closely allied species also of the larger genera apparently have restricted ranges, and in their affinities they are clustered in little groups round other species—in both respects resembling varieties. These are strange relations on the view that each species was independently created, but are intelligible if each existed first as a variety.

As each species tends by its geometrical rate of reproduction to increase inordinately in number; and as the modified descendants of each species will be enabled to increase by as much as they become more diversified in habits and structure, so as to be able to seize on many and widely different places in the economy of nature, there will be a constant tendency in natural selection to preserve the most divergent offspring of any one species. Hence, during a long-continued course of modification, the slight differences characteristic of varieties of the same species, tend to be augmented into the greater differences characteristic of the species of the same genus. New and improved varieties will inevitably supplant and exterminate the older, less improved, and intermediate varieties; and thus species are rendered to a large extent defined and distinct objects. Dominant species belonging to the larger groups within each class tend to give birth to new and dominant forms; so that each large group tends to become still larger, and at the same time more divergent in character. But as all groups cannot thus go on increasing in size, for the world would not hold them, the more dominant groups beat the less dominant. This tendency in the large groups to go on increasing in size and diverging in character, together with the inevitable contingency of much extinction, explains the arrangement of all the forms of life in groups subordinate to groups, all within a few great classes, which has prevailed throughout all time.   This grand fact of the grouping of all organic beings under what is called the Natural System, is utterly inexplicable on the theory of creation.

As natural selection acts solely by accumulating slight, successive, favourable variations, it can produce no great or sudden modifications; it can act only by short and slow steps.   Hence, the canon of "Natura non facit saltum," which every fresh addition to our knowledge tends to confirm, is on this theory intelligible.   We can see why throughout nature the same general end is gained by an almost infinite diversity of means, for every peculiarity when once acquired is long inherited, and structures already modified in many different ways have to be adapted for the same general purpose.   We can, in short, see why nature is prodigal in variety, though niggard in innovation.   But why this should be a law of nature if each species has been independently created no man can explain.

Many other facts are, as it seems to me, explicable on this theory.   How strange it is that a bird, under the form of a woodpecker, should prey on insects on the ground; that upland geese which rarely or never swim, should possess webbed feet; that a thrush-like bird should dive and feed on sub-aquatic insects; and that a petrel should have the habits and structure fitting it for the life of an auk! and so in endless other cases.   But on the view of each species constantly trying to increase in number, with natural selection always ready to adapt the slowly varying descendants of each to any unoccupied or ill-occupied place in nature, these facts cease to be strange, or might even have been anticipated.

We can to a certain extent understand how it is that

there is so much beauty throughout nature; for this may be largely attributed to the agency of selection. That beauty, according to our sense of it, is not universal, must be admitted by every one who will look at some venomous snakes, at some fishes, and at certain hideous bats with a distorted resemblance to the human face. Sexual selection has given the most brilliant colours, elegant patterns, and other ornaments to the males, and sometimes to both sexes of many birds, butterflies, and other animals. With birds it has often rendered the voice of the male musical to the female, as well as to our ears. Flowers and fruit have been rendered conspicuous by brilliant colours in contrast with the green foliage, in order that the flowers may be easily seen, visited, and fertilised by insects, and the seeds disseminated by birds. How it comes that certain colours, sounds, and forms should give pleasure to man and the lower animals,—that is, how the sense of beauty in its simplest form was first acquired,—we do not know any more than how certain odours and flavours were first rendered agreeable.

As natural selection acts by competition, it adapts and improves the inhabitants of each country only in relation to their co-inhabitants; so that we need feel no surprise at the species of any one country, although on the ordinary view supposed to have been created and specially adapted for that country, being beaten and supplanted by the naturalised productions from another land. Nor ought we to marvel if all the contrivances in nature be not, as far as we can judge, absolutely perfect, as in the case even of the human eye; or if some of them be abhorrent to our ideas of fitness. We need not marvel at the sting of the bee, when used

against an enemy, causing the bee's own death; at
drones being produced in such great numbers for one
single act, and being then slaughtered by their sterile
sisters; at the astonishing waste of pollen by our fir-
trees; at the instinctive hatred of the queen-bee for her
own fertile daughters; at ichneumonidæ feeding within
the living bodies of caterpillars; or at other such cases.
The wonder indeed is, on the theory of natural selection,
that more cases of the want of absolute perfection have
not been detected.

The complex and little known laws governing the
production of varieties are the same, as far as we can
judge, with the laws which have governed the production
of distinct species. In both cases physical conditions
seem to have produced some direct and definite effect,
but how much we cannot say. Thus, when varieties
enter any new station, they occasionally assume some
of the characters proper to the species of that station.
With both varieties and species, use and disuse seem to
have produced a considerable effect; for it is impossible
to resist this conclusion when we look, for instance, at
the logger-headed duck, which has wings incapable of
flight, in nearly the same condition as in the domestic
duck; or when we look at the burrowing tucu-tucu,
which is occasionally blind, and then at certain moles,
which are habitually blind and have their eyes covered
with skin; or when we look at the blind animals inhabiting the dark caves of America and Europe. With
varieties and species, correlated variation seems to have
played an important part, so that when one part has
been modified other parts have been necessarily modified.
With both varieties and species, reversions to long-lost
characters occasionally occur. How inexplicable on the

theory of creation is the occasional appearance of stripes
on the shoulders and legs of the several species of the
horse-genus and of their hybrids! How simply is this
fact explained if we believe that these species are all
descended from a striped progenitor, in the same manner
as the several domestic breeds of the pigeon are descended
from the blue and barred rock-pigeon!

On the ordinary view of each species having been
independently created, why should specific characters,
or those by which the species of the same genus
differ from each other, be more variable than generic
characters in which they all agree? Why, for instance,
should the colour of a flower be more likely to vary in
any one species of a genus, if the other species possess
differently coloured flowers, than if all possessed the
same coloured flowers? If species are only well-marked
varieties, of which the characters have become in a high
degree permanent, we can understand this fact; for they
have already varied since they branched off from a
common progenitor in certain characters, by which they
have come to be specifically distinct from each other;
therefore these same characters would be more likely
again to vary than the generic characters which have
been inherited without change for an immense period.
It is inexplicable on the theory of creation why a part
developed in a very unusual manner in one species
alone of a genus, and therefore, as we may naturally infer,
of great importance to that species, should be eminently
liable to variation; but, on our view, this part has undergone, since the several species branched off from a
common progenitor, an unusual amount of variability
and modification, and therefore we might expect the
part generally to be still variable. But a part may be

developed in the most unusual manner, like the wing of a bat, and yet not be more variable than any other structure, if the part be common to many subordinate forms, that is, if it has been inherited for a very long period; for in this case it will have been rendered constant by long-continued natural selection.

Glancing at instincts, marvellous as some are, they offer no greater difficulty than do corporeal structures on the theory of the natural selection of successive, slight, but profitable modifications. We can thus understand why nature moves by graduated steps in endowing different animals of the same class with their several instincts. I have attempted to show how much light the principle of gradation throws on the admirable architectural powers of the hive-bee. Habit no doubt often comes into play in modifying instincts; but it certainly is not indispensable, as we see in the case of neuter insects, which leave no progeny to inherit the effects of long-continued habit. On the view of all the species of the same genus having descended from a common parent, and having inherited much in common, we can understand how it is that allied species, when placed under widely different conditions of life, yet follow nearly the same instincts; why the thrushes of tropical and temperate South America, for instance, line their nests with mud like our British species. On the view of instincts having been slowly acquired through natural selection, we need not marvel at some instincts being not perfect and liable to mistakes, and at many instincts causing other animals to suffer.

If species be only well-marked and permanent varieties, we can at once see why their crossed offspring should follow the same complex laws in their degrees and kinds of resemblance to their parents,—in being absorbed into each other by successive crosses, and in other such points,—as do the crossed offspring of acknowledged varieties. This similarity would be a strange fact, if species had been independently created and varieties had been produced through secondary laws.

If we admit that the geological record is imperfect to an extreme degree, then the facts, which the record does give, strongly support the theory of descent with modification. New species have come on the stage slowly and at successive intervals; and the amount of change, after equal intervals of time, is widely different in different groups. The extinction of species and of whole groups of species, which has played so conspicuous a part in the history of the organic world, almost inevitably follows from the principle of natural selection; for old forms are supplanted by new and improved forms. Neither single species nor groups of species reappear when the chain of ordinary generation is once broken. The gradual diffusion of dominant forms, with the slow modification of their descendants, causes the forms of life, after long intervals of time, to appear as if they had changed simultaneously throughout the world. The fact of the fossil remains of each formation being in some degree intermediate in character between the fossils in the formations above and below, is simply explained by their intermediate position in the chain of descent. The grand fact that all extinct beings can be classed with all recent beings, naturally follows from the living and the extinct being the offspring of common parents. As species have generally diverged in character during their long course of descent and modification, we can understand why it is that the more ancient forms, or early

progenitors of each group, so often occupy a position
in some degree intermediate between existing groups.
Recent forms are generally looked upon as being, on the
whole, higher in the scale of organisation than ancient
forms; and they must be higher, in so far as the later
and more improved forms have conquered the older and
less improved forms in the struggle for life; they have
also generally had their organs more specialised for different functions. This fact is perfectly compatible with
numerous beings still retaining simple and but little
improved structures, fitted for simple conditions of life;
it is likewise compatible with some forms having retrograded in organisation, by having become at each stage
of descent better fitted for new and degraded habits of
life. Lastly, the wonderful law of the long endurance
of allied forms on the same continent,—of marsupials
in Australia, of edentata in America, and other such
cases,—is intelligible, for within the same country the
existing and the extinct will be closely allied by descent.

Looking to geographical distribution, if we admit that
there has been during the long course of ages much migration from one part of the world to another, owing to
former climatal and geographical changes and to the
many occasional and unknown means of dispersal, then
we can understand, on the theory of descent with modification, most of the great leading facts in Distribution.
We can see why there should be so striking a parallelism
in the distribution of organic beings throughout space,
and in their geological succession throughout time; for
in both cases the beings have been connected by the
bond of ordinary generation, and the means of modification have been the same. We see the full meaning of
the wonderful fact, which has struck every traveller,

namely, that on the same continent, under the most diverse conditions, under heat and cold, on mountain and lowland, on deserts and marshes, most of the inhabitants within each great class are plainly related; for they are the descendants of the same progenitors and early colonists. On this same principle of former migration, combined in most cases with modification, we can understand, by the aid of the Glacial period, the identity of some few plants, and the close alliance of many others, on the most distant mountains, and in the northern and southern temperate zones; and likewise the close alliance of some of the inhabitants of the sea in the northern and southern temperate latitudes, though separated by the whole intertropical ocean. Although two countries may present physical conditions as closely similar as the same species ever require, we need feel no surprise at their inhabitants being widely different, if they have been for a long period completely sundered from each other; for as the relation of organism to organism is the most important of all relations, and as the two countries will have received colonists at various periods and in different proportions, from some other country or from each other, the course of modification in the two areas will inevitably have been different.

On this view of migration, with subsequent modification, we see why oceanic islands are inhabited by only few species, but of these, why many are peculiar or endemic forms. We clearly see why species belonging to those groups of animals which cannot cross wide spaces of the ocean, as frogs and terrestrial mammals, do not inhabit oceanic islands; and why, on the other hand, new and peculiar species of bats, animals which can traverse the ocean, are often found on islands far distant from any continent. Such cases as the presence of peculiar species of bats on oceanic islands and the absence of all other terrestrial mammals, are facts utterly inexplicable on the theory of independent acts of creation.

The existence of closely allied or representative species in any two areas, implies, on the theory of descent with modification, that the same parent-forms formerly inhabited both areas: and we almost invariably find that wherever many closely allied species inhabit two areas, some identical species are still common to both. Wherever many closely allied yet distinct species occur, doubtful forms and varieties belonging to the same groups likewise occur. It is a rule of high generality that the inhabitants of each area are related to the inhabitants of the nearest source whence immigrants might have been derived. We see this in the striking relation of nearly all the plants and animals of the Galapagos archipelago, of Juan Fernandez, and of the other American islands, to the plants and animals of the neighbouring American mainland; and of those of the Cape de Verde archipelago, and of the other African islands to the African mainland. It must be admitted that these facts receive no explanation on the theory of creation.

The fact, as we have seen, that all past and present organic beings can be arranged within a few great classes, in groups subordinate to groups, and with the extinct groups often falling in between the recent groups, is intelligible on the theory of natural selection with its contingencies of extinction and divergence of character. On these same principles we see how it is, that the mutual affinities of the forms within each class are so complex and circuitous. We see why certain characters are far more serviceable than others for classification;—why adaptive characters, though of paramount importance to the beings, are of hardly any importance in classification; why characters derived from rudimentary parts, though of no service to the beings, are often of high classificatory value; and why embryological characters are often the most valuable of all. The real affinities of all organic beings, in contradistinction to their adaptive resemblances, are due to inheritance or community of descent. The Natural System is a genealogical arrangement, with the acquired grades of difference, marked by the terms, varieties, species, genera, families, &c.; and we have to discover the lines of descent by the most permanent characters whatever they may be and of however slight vital importance.

The similar framework of bones in the hand of a man, wing of a bat, fin of the porpoise, and leg of the horse, —the same number of vertebræ forming the neck of the giraffe and of the elephant,—and innumerable other such facts, at once explain themselves on the theory of descent with slow and slight successive modifications. The similarity of pattern in the wing and in the leg of a bat, though used for such different purpose,—in the jaws and legs of a crab,—in the petals, stamens, and pistils of a flower, is likewise, to a large extent, intelligible on the view of the gradual modification of parts or organs, which were aboriginally alike in an early progenitor in each of these classes. On the principle of successive variations not always supervening at an early age, and being inherited at a corresponding not early period of life, we clearly see why the embryos of mammals, birds, reptiles, and fishes should be so closely similar, and so unlike the adult forms. We may cease marvelling at the embryo of an air-breathing mammal or bird having branchial slits and arteries running in loops, like those of a fish which has to breathe the air dissolved in water by the aid of well-developed branchiæ.

Disuse, aided sometimes by natural selection, will often have reduced organs when rendered useless under changed habits or conditions of life; and we can understand on this view the meaning of rudimentary organs. But disuse and selection will generally act on each creature, when it has come to maturity and has to play its full part in the struggle for existence, and will thus have little power on an organ during early life; hence the organ will not be reduced or rendered rudimentary at this early age. The calf, for instance, has inherited teeth, which never cut through the gums of the upper jaw, from an early progenitor having well-developed teeth; and we may believe, that the teeth in the mature animal were formerly reduced by disuse, owing to the tongue and palate, or lips, having become excellently fitted through natural selection to browse without their aid; whereas in the calf, the teeth have been left unaffected, and on the principle of inheritance at corresponding ages have been inherited from a remote period to the present day. On the view of each organism with all its separate parts having been specially created, how utterly inexplicable is it that organs bearing the plain stamp of inutility, such as the teeth in the embryonic calf or the shrivelled wings under the soldered wing-covers of many beetles, should so frequently occur. Nature may be said to have taken pains to reveal her scheme of modification, by means of rudimentary organs, of embryological and homologous structures, but we are too blind to understand her meaning.

I have now recapitulated the facts and considerations which have thoroughly convinced me that species have been modified, during a long course of descent. This has been effected chiefly through the natural selection of numerous successive, slight, favourable variations; aided in an important manner by the inherited effects of the use and disuse of parts; and in an unimportant manner, that is in relation to adaptive structures, whether past or present, by the direct action of external conditions, and by variations which seem to us in our ignorance to arise spontaneously. It appears that I formerly underrated the frequency and value of these latter forms of variation, as leading to permanent modifications of structure independently of natural selection. But as my conclusions have lately been much misrepresented, and it has been stated that I attribute the modification of species exclusively to natural selection, I may be permitted to remark that in the first edition of this work, and subsequently, I placed in a most conspicuous position—namely, at the close of the Introduction—the following words: "I am convinced that natural selection has been the main but not the exclusive means of modification." This has been of no avail. Great is the power of steady misrepresentation; but the history of science shows that fortunately this power does not long endure.

It can hardly be supposed that a false theory would explain, in so satisfactory a manner as does the theory of natural selection, the several large classes of facts above specified. It has recently been objected that this is an unsafe method of arguing; but it is a method used in judging of the common events of life, and has often been used by the greatest natural philosophers. The undulatory theory of light has thus been arrived at; and

2 U

the belief in the revolution of the earth on its own axis was until lately supported by hardly any direct evidence. It is no valid objection that science as yet throws no light on the far higher problem of the essence or origin of life. Who can explain what is the essence of the attraction of gravity? No one now objects to following out the results consequent on this unknown element of attraction; notwithstanding that Leibnitz formerly accused Newton of introducing "occult qualities and miracles into philosophy."

I see no good reason why the views given in this volume should shock the religious feelings of any one. It is satisfactory, as showing how transient such impressions are, to remember that the greatest discovery ever made by man, namely, the law of the attraction of gravity, was also attacked by Leibnitz, "as subversive of natural, and inferentially of revealed, religion." A celebrated author and divine has written to me that "he has gradu- "ally learnt to see that it is just as noble a conception "of the Deity to believe that He created a few original "forms capable of self-development into other and need- "ful forms, as to believe that He required a fresh act of "creation to supply the voids caused by the action of "His laws."

Why, it may be asked, until recently did nearly all the most eminent living naturalists and geologists disbelieve in the mutability of species. It cannot be asserted that organic beings in a state of nature are subject to no variation; it cannot be proved that the amount of variation in the course of long ages is a limited quantity; no clear distinction has been, or can be, drawn between species and well-marked varieties. It cannot be maintained that species when intercrossed are invari-

ably sterile, and varieties invariably fertile; or that
sterility is a special endowment and sign of creation.
The belief that species were immutable productions was
almost unavoidable as long as the history of the world
was thought to be of short duration; and now that we
have acquired some idea of the lapse of time, we are too
apt to assume, without proof, that the geological record
is so perfect that it would have afforded us plain evidence
of the mutation of species, if they had undergone
mutation.

But the chief cause of our natural unwillingness to
admit that one species has given birth to other and
distinct species, is that we are always slow in admitting
great changes of which we do not see the steps. The
difficulty is the same as that felt by so many geologists,
when Lyell first insisted that long lines of inland cliffs
had been formed, and great valleys excavated, by the
agencies which we see still at work. The mind cannot
possibly grasp the full meaning of the term of even a
million years; it cannot add up and perceive the full
effects of many slight variations, accumulated during an
almost infinite number of generations.

Although I am fully convinced of the truth of the
views given in this volume under the form of an abstract,
I by no means expect to convince experienced naturalists
whose minds are stocked with a multitude of facts all
viewed, during a long course of years, from a point of
view directly opposite to mine. It is so easy to hide
our ignorance under such expressions as the "plan of
creation," "unity of design," &c., and to think that we
give an explanation when we only re-state a fact. Any
one whose disposition leads him to attach more weight
to unexplained difficulties than to the explanation of a

certain number of facts will certainly reject the theory. A few naturalists, endowed with much flexibility of mind, and who have already begun to doubt the immutability of species, may be influenced by this volume; but I look with confidence to the future,—to young and rising naturalists, who will be able to view both sides of the question with impartiality. Whoever is led to believe that species are mutable will do good service by conscientiously expressing his conviction; for thus only can the load of prejudice by which this subject is overwhelmed be removed.

Several eminent naturalists have of late published their belief that a multitude of reputed species in each genus are not real species; but that other species are real, that is, have been independently created. This seems to me a strange conclusion to arrive at. They admit that a multitude of forms, which till lately they themselves thought were special creations, and which are still thus looked at by the majority of naturalists, and which consequently have all the external characteristic features of true species,—they admit that these have been produced by variation, but they refuse to extend the same view to other and slightly different forms. Nevertheless they do not pretend that they can define, or even conjecture, which are the created forms of life, and which are those produced by secondary laws. They admit variation as a *vera causa* in one case, they arbitrarily reject it in another, without assigning any distinction in the two cases. The day will come when this will be given as a curious illustration of the blindness of preconceived opinion. These authors seem no more startled at a miraculous act of creation than at an ordinary birth. But do they really believe that at innumerable periods in the earth's history certain elemental atoms have been commanded suddenly to flash into living tissues ?   Do they believe that at each supposed act of creation one individual or many were produced ? Were all the infinitely numerous kinds of animals and plants created as eggs or seed, or as full grown ? and in the case of mammals, were they created bearing the false marks of nourishment from the mother's womb ? Undoubtedly some of these same questions cannot be answered by those who believe in the appearance or creation of only a few forms of life, or of some one form alone.   It has been maintained by several authors that it is as easy to believe in the creation of a million beings as of one; but Maupertuis' philosophical axiom " of least action " leads the mind more willingly to admit the smaller number; and certainly we ought not to believe that innumerable beings within each great class have been created with plain, but deceptive, marks of descent from a single parent.

As a record of a former state of things, I have retained in the foregoing paragraphs, and elsewhere, several sentences which imply that naturalists believe in the separate creation of each species; and I have been much censured for having thus expressed myself.   But undoubtedly this was the general belief when the first edition of the present work appeared.   I formerly spoke to very many naturalists on the subject of evolution, and never once met with any sympathetic agreement. It is probable that some did then believe in evolution, but they were either silent, or expressed themselves so ambiguously that it was not easy to understand their meaning.   Now things are wholly changed, and almost every naturalist admits the great principle of evolution.

There are, however, some who still think that species
have suddenly given birth, through quite unexplained
means, to new and totally different forms: but, as I
have attempted to show, weighty evidence can be opposed to the admission of great and abrupt modifications.
Under a scientific point of view, and as leading to further
investigation, but little advantage is gained by believing
that new forms are suddenly developed in an inexplicable manner from old and widely different forms, over
the old belief in the creation of species from the dust of
the earth.

It may be asked how far I extend the doctrine of the
modification of species. The question is difficult to
answer, because the more distinct the forms are which
we consider, by so much the arguments in favour of
community of descent become fewer in number and
less in force. But some arguments of the greatest
weight extend very far. All the members of whole
classes are connected together by a chain of affinities,
and all can be classed on the same principle, in
groups subordinate to groups. Fossil remains sometimes tend to fill up very wide intervals between
existing orders.

Organs in a rudimentary condition plainly show that
an early progenitor had the organ in a fully developed
condition; and this in some cases implies an enormous
amount of modification in the descendants. Throughout
whole classes various structures are formed on the same
pattern, and at a very early age the embryos closely
resemble each other. Therefore I cannot doubt that the
theory of descent with modification embraces all the
members of the same great class or kingdom. I believe
that animals are descended from at most only four or

five progenitors, and plants from an equal or lesser number.

Analogy would lead me one step farther, namely, to the belief that all animals and plants are descended from some one prototype. But analogy may be a deceitful guide. Nevertheless all living things have much in common, in their chemical composition, their cellular structure, their laws of growth, and their liability to injurious influences. We see this even in so trifling a fact as that the same poison often similarly affects plants and animals; or that the poison secreted by the gall-fly produces monstrous growths on the wild rose or oak-tree. With all organic beings, excepting perhaps some of the very lowest, sexual reproduction seems to be essentially similar. With all, as far as is at present known, the germinal vesicle is the same; so that all organisms start from a common origin. If we look even to the two main divisions—namely, to the animal and vegetable kingdoms—certain low forms are so far intermediate in character that naturalists have disputed to which kingdom they should be referred. As Professor Asa Gray has remarked, " the spores and other repro- " ductive bodies of many of the lower algæ may claim " to have first a characteristically animal, and then an " unequivocally vegetable existence." Therefore, on the principle of natural selection with divergence of character, it does not seem incredible that, from some such low and intermediate form, both animals and plants may have been developed; and, if we admit this, we must likewise admit that all the organic beings which have ever lived on this earth may be descended from some one primordial form. But this inference is chiefly grounded on analogy, and it is immaterial whether or not it be accepted. No doubt it is possible, as Mr. G. H. Lewes has urged, that at the first commencement of life many different forms were evolved; but if so, we may conclude that only a very few have left modified descendants. For, as I have recently remarked in regard to the members of each great kingdom, such as the Vertebrata, Articulata, &c., we have distinct evidence in their embryological, homologous, and rudimentary structures, that within each kingdom all the members are descended from a single progenitor.

When the views advanced by me in this volume, and by Mr. Wallace, or when analogous views on the origin of species are generally admitted, we can dimly foresee that there will be a considerable revolution in natural history. Systematists will be able to pursue their labours as at present; but they will not be incessantly haunted by the shadowy doubt whether this or that form be a true species. This, I feel sure and I speak after experience, will be no slight relief. The endless disputes whether or not some fifty species of British brambles are good species will cease. Systematists will have only to decide (not that this will be easy) whether any form be sufficiently constant and distinct from other forms, to be capable of definition; and if definable, whether the differences be sufficiently important to deserve a specific name. This latter point will become a far more essential consideration than it is at present; for differences, however slight, between any two forms, if not blended by intermediate gradations, are looked at by most naturalists as sufficient to raise both forms to the rank of species.

Hereafter we shall be compelled to acknowledge that the only distinction between species and well-marked varieties is, that the latter are known, or believed, to be connected at the present day by intermediate gradations whereas species were formerly thus connected. Hence, without rejecting the consideration of the present existence of intermediate gradations between any two forms, we shall be led to weigh more carefully and to value higher the actual amount of difference between them. It is quite possible that forms now generally acknowledged to be merely varieties may hereafter be thought worthy of specific names; and in this case scientific and common language will come into accordance. In short, we shall have to treat species in the same manner as those naturalists treat genera, who admit that genera are merely artificial combinations made for convenience. This may not be a cheering prospect; but we shall at least be freed from the vain search for the undiscovered and undiscoverable essence of the term species.

The other and more general departments of natural history will rise greatly in interest. The terms used by naturalists, of affinity, relationship, community of type, paternity, morphology, adaptive characters, rudimentary and aborted organs, &c., will cease to be metaphorical, and will have a plain signification. When we no longer look at an organic being as a savage looks at a ship, as something wholly beyond his comprehension; when we regard every production of nature as one which has had a long history; when we contemplate every complex structure and instinct as the summing up of many contrivances, each useful to the possessor, in the same way as any great mechanical invention is the summing up of the labour, the experience, the reason, and even the blunders of numerous workmen; when we thus view each organic being, how far more interesting

—I speak from experience—does the study of natural history become!

A grand and almost untrodden field of inquiry will be opened, on the causes and laws of variation, on correlation, on the effects of use and disuse, on the direct action of external conditions, and so forth. The study of domestic productions will rise immensely in value. A new variety raised by man will be a more important and interesting subject for study than one more species added to the infinitude of already recorded species. Our classifications will come to be, as far as they can be so made, genealogies; and will then truly give what may be called the plan of creation. The rules for classifying will no doubt become simpler when we have a definite object in view. We possess no pedigrees or armorial bearings; and we have to discover and trace the many diverging lines of descent in our natural genealogies, by characters of any kind which have long been inherited. Rudimentary organs will speak infallibly with respect to the nature of long-lost structures. Species and groups of species which are called aberrant, and which may fancifully be called living fossils, will aid us in forming a picture of the ancient forms of life. Embryology will often reveal to us the structure, in some degree obscured, of the prototypes of each great class.

When we can feel assured that all the individuals of the same species, and all the closely allied species of most genera, have within a not very remote period descended from one parent, and have migrated from some one birth-place; and when we better know the many means of migration, then, by the light which geology now throws, and will continue to throw, on former changes of climate and of the level of the land, we shall surely be enabled to trace in an admirable manner the former migrations of the inhabitants of the whole world. Even at present, by comparing the differences between the inhabitants of the sea on the opposite sides of a continent, and the nature of the various inhabitants on that continent in relation to their apparent means of immigration, some light can be thrown on ancient geography.

The noble science of Geology loses glory from the extreme imperfection of the record. The crust of the earth with its imbedded remains must not be looked at as a well-filled museum, but as a poor collection made at hazard and at rare intervals. The accumulation of each great fossiliferous formation will be recognised as having depended on an unusual concurrence of favourable circumstances, and the blank intervals between the successive stages as having been of vast duration. But we shall be able to gauge with some security the duration of these intervals by a comparison of the preceding and succeeding organic forms. We must be cautious in attempting to correlate as strictly contemporaneous two formations, which do not include many identical species, by the general succession of the forms of life. As species are produced and exterminated by slowly acting and still existing causes, and not by miraculous acts of creation; and as the most important of all causes of organic change is one which is almost independent of altered and perhaps suddenly altered physical conditions, namely, the mutual relation of organism to organism,— the improvement of one organism entailing the improvement or the extermination of others; it follows, that the amount of organic change in the fossils of consecutive formations probably serves as a fair measure of the

relative, though not actual lapse of time. A number
of species, however, keeping in a body might remain for
a long period unchanged, whilst within the same period
several of these species by migrating into new countries
and coming into competition with foreign associates,
might become modified; so that we must not overrate
the accuracy of organic change as a measure of time.

In the future I see open fields for far more important
researches. Psychology will be securely based on the
foundation already well laid by Mr. Herbert Spencer,
that of the necessary acquirement of each mental power
and capacity by gradation. Much light will be thrown
on the origin of man and his history.

Authors of the highest eminence seem to be fully
satisfied with the view that each species has been
independently created. To my mind it accords better
with what we know of the laws impressed on matter by
the Creator, that the production and extinction of the
past and present inhabitants of the world should have
been due to secondary causes, like those determining the
birth and death of the individual. When I view all
beings not as special creations, but as the lineal descendants of some few beings which lived long before the first
bed of the Cambrian system was deposited, they seem
to me to become ennobled. Judging from the past, we
may safely infer that not one living species will transmit
its unaltered likeness to a distant futurity. And of the
species now living very few will transmit progeny of
any kind to a far distant futurity; for the manner in
which all organic beings are grouped, shows that the
greater number of species in each genus, and all the
species in many genera, have left no descendants, but
have become utterly extinct. We can so far take a

prophetic glance into futurity as to foretell that it will be the common and widely-spread species, belonging to the larger and dominant groups within each class, which will ultimately prevail and procreate new and dominant species. As all the living forms of life are the lineal descendants of those which lived long before the Cambrian epoch, we may feel certain that the ordinary succession by generation has never once been broken, and that no cataclysm has desolated the whole world. Hence we may look with some confidence to a secure future of great length. And as natural selection works solely by and for the good of each being, all corporeal and mental endowments will tend to progress towards perfection.

It is interesting to contemplate a tangled bank, clothed with many plants of many kinds, with birds singing on the bushes, with various insects flitting about, and with worms crawling through the damp earth, and to reflect that these elaborately constructed forms, so different from each other, and dependent upon each other in so complex a manner, have all been produced by laws acting around us. These laws, taken in the largest sense, being Growth with Reproduction; Inheritance which is almost implied by reproduction; Variability from the indirect and direct action of the conditions of life, and from use and disuse: a Ratio of Increase so high as to lead to a Struggle for Life, and as a consequence to Natural Selection, entailing Divergence of Character and the Extinction of less-improved forms. Thus, from the war of nature, from famine and death, the most exalted object which we are capable of conceiving, namely, the production of the higher animals, directly follows. There is grandeur in this view of life, with its several powers, having been originally breathed by the Creator into a few forms or into one; and that, whilst tnis planet has gone cycling on according to the fixed law of gravity, from so simple a beginning endless forms most beautiful and most wonderful have been, and are being evolved.

GLOSSARY

OF THE

PRINCIPAL SCIENTIFIC TERMS USED IN THE PRESENT VOLUME.*

———•◦•———

ABERRANT.—Forms or groups of animals or plants which deviate in important characters from their nearest allies, so as not to be easily included in the same group with them, are said to be aberrant.

ABERRATION (in Optics).—In the refraction of light by a convex lens the rays passing through different parts of the lens are brought to a focus at slightly different distances,—this is called *spherical aberration*; at the same time the coloured rays are separated by the prismatic action of the lens and likewise brought to a focus at different distances,—this is *chromatic aberration*.

ABNORMAL.—Contrary to the general rule.

ABORTED.—An organ is said to be aborted, when its development has been arrested at a very early stage.

ALBINISM.—Albinos are animals in which the usual colouring matters characteristic of the species have not been produced in the skin and its appendages. Albinism is the state of being an albino.

ALGÆ.—A class of plants including the ordinary sea-weeds and the filamentous fresh-water weeds.

————

* I am indebted to the kindness of Mr. W. S. Dallas for this Glossary, which has been given because several readers have complained to me that some of the terms used were unintelligible to them. Mr. Dallas has endeavoured to give the explanations of the terms in as popular a form as possible.

ALTERNATION OF GENERATIONS.—This term is applied to a peculiar mode of reproduction which prevails among many of the lower animals, in which the egg produces a living form quite different from its parent, but from which the parent-form is reproduced by a process of budding, or by the division of the substance of the first product of the egg.

AMMONITES.—A group of fossil, spiral, chambered shells, allied to the existing pearly Nautilus, but having the partitions between the chambers waved in complicated patterns at their junction with the outer wall of the shell.

ANALOGY.—That resemblance of structures which depends upon similarity of function, as in the wings of insects and birds. Such structures are said to be *analogous*, and to be *analogues* of each other.

ANIMALCULE.—A minute animal : generally applied to those visible only by the microscope.

ANNELIDS.—A class of worms in which the surface of the body exhibits a more or less distinct division into rings or segments, generally provided with appendages for locomotion and with gills. It includes the ordinary marine worms, the earthworms, and the leeches.

ANTENNÆ.—Jointed organs appended to the head in Insects, Crustacea and Centipedes, and not belonging to the mouth.

ANTHERS.—The summits of the stamens of flowers, in which the pollen or fertilising dust is produced.

APLACENTALIA, APLACENTATA or Aplacental Mammals. See *Mammalia*.

ARCHETYPAL.—Of or belonging to the Archetype, or ideal primitive form upon which all the beings of a group seem to be organised.

ARTICULATA.—A great division of the Animal Kingdom characterised generally by having the surface of the body divided into rings called segments, a greater or less number of which are furnished with jointed legs (such as Insects, Crustaceans and Centipedes).

ASYMMETRICAL.—Having the two sides unlike.

ATROPHIED.—Arrested in development at a very early stage.

BALANUS.—The genus including the common Acorn-shells which live in abundance on the rocks of the sea-coast.

BATRACHIANS.—A class of animals allied to the Reptiles, but

undergoing a peculiar metamorphosis, in which the young animal is generally aquatic and breathes by gills. (*Examples*, Frogs, Toads, and Newts.)

BOULDERS.—Large transported blocks of stone generally imbedded in clays or gravels.

BRACHIOPODA.—A class of marine Mollusca, or soft-bodied animals, furnished with a bivalve shell, attached to submarine objects by a stalk which passes through an aperture in one of the valves, and furnished with fringed arms, by the action of which food is carried to the mouth.

BRANCHIÆ.—Gills or organs for respiration in water.

BRANCHIAL.—Pertaining to gills or branchiæ.

CAMBRIAN SYSTEM.—A Series of very ancient Palæozoic rocks, between the Laurentian and the Silurian. Until recently these were regarded as the oldest fossiliferous rocks.

CANIDÆ.—The Dog-family, including the Dog, Wolf, Fox, Jackal, &c.

CARAPACE.—The shell enveloping the anterior part of the body in Crustaceans generally; applied also to the hard shelly pieces of the Cirripedes.

CARBONIFEROUS.—This term is applied to the great formation which includes, among other rocks, the coal-measures. It belongs to the oldest, or Palæozoic, system of formations.

CAUDAL.—Of or belonging to the tail.

CEPHALOPODS.—The highest class of the Mollusca, or soft-bodied animals, characterised by having the mouth surrounded by a greater or less number of fleshy arms or tentacles, which, in most living species, are furnished with sucking-cups. (*Examples*, Cuttle-fish, Nautilus.)

CETACEA.—An order of Mammalia, including the Whales, Dolphins, &c., having the form of the body fish-like, the skin naked, and only the fore-limbs developed.

CHELONIA.—An order of Reptiles including the Turtles, Tortoises, &c.

CIRRIPEDES.—An order of Crustaceans including the Barnacles and Acorn-shells. Their young resemble those of many other Crustaceans in form; but when mature they are always attached to other objects, either directly or by means of a stalk, and their bodies are enclosed by a calcareous shell composed of several pieces, two of which can open to give issue to a bunch of curled, jointed tentacles, which represent the limbs.

2 X

Coccus.—The genus of Insects including the Cochineal. In these the male is a minute, winged fly, and the female generally a motionless, berry-like mass.

Cocoon.—A case usually of silky material, in which insects are frequently enveloped during the second or resting-stage (pupa) of their existence. The term "cocoon-stage" is here used as equivalent to "pupa-stage."

Cœlospermous.—A term applied to those fruits of the Umbelliferæ which have the seed hollowed on the inner face.

Coleoptera.—Beetles, an order of Insects, having a biting mouth and the first pair of wings more or less horny, forming sheaths for the second pair, and usually meeting in a straight line down the middle of the back.

Column.—A peculiar organ in the flowers of Orchids, in which the stamens, style and stigma (or the reproductive parts) are united.

Compositæ or Compositous Plants.—Plants in which the inflorescence consists of numerous small flowers (florets) brought together into a dense head, the base of which is enclosed by a common envelope. (*Examples*, the Daisy, Dandelion, &c.)

Confervæ.—The filamentous weeds of fresh water.

Conglomerate.—A rock made up of fragments of rock or pebbles, cemented together by some other material.

Corolla.—The second envelope of a flower usually composed of coloured, leaf-like organs (petals), which may be united by their edges either in the basal part or throughout.

Correlation.—The normal coincidence of one phenomenon, character, &c., with another.

Corymb.—A bunch of flowers in which those springing from the lower part of the flower stalk are supported on long stalks so as to be nearly on a level with the upper ones.

Cotyledons.—The first or seed-leaves of plants.

Crustaceans.—A class of articulated animals, having the skin of the body generally more or less hardened by the deposition of calcareous matter, breathing by means of gills. (*Examples*, Crab, Lobster, Shrimp, &c.)

Curculio.—The old generic term for the Beetles known as Weevils, characterised by their four-jointed feet, and by the head being produced into a sort of beak, upon the sides of which the antennæ are inserted.

Cutaneous.—Of or belonging to the skin.

DEGRADATION.—The wearing down of land by the action of the sea or of meteoric agencies.

DENUDATION.—The wearing away of the surface of the land by water.

DEVONIAN SYSTEM or formation.—A series of Palæozoic rocks, including the Old Red Sandstone.

DICOTYLEDONS or DICOTYLEDONOUS PLANTS.—A class of plants characterised by having two seed-leaves, by the formation of new wood between the bark and the old wood (exogenous growth) and by the reticulation of the veins of the leaves. The parts of the flowers are generally in multiples of five.

DIFFERENTIATION.—The separation or discrimination of parts or organs which in simpler forms of life are more or less united.

DIMORPHIC.—Having two distinct forms.—Dimorphism is the condition of the appearance of the same species under two dissimilar forms.

DIŒCIOUS.—Having the organs of the sexes upon distinct individuals.

DIORITE.—A peculiar form of Greenstone.

DORSAL.—Of or belonging to the back.

EDENTATA.—A peculiar order of Quadrupeds, characterised by the absence of at least the middle incisor (front) teeth in both jaws. (*Examples*, the Sloths and Armadillos.)

ELYTRA.—The hardened fore-wings of Beetles, serving as sheaths for the membranous hind-wings, which constitute the true organs of flight.

EMBRYO.—The young animal undergoing development within the egg or womb.

EMBRYOLOGY.—The study of the development of the embryo.

ENDEMIC.—Peculiar to a given locality.

ENTOMOSTRACA.—A division of the class Crustacea, having all the segments of the body usually distinct, gills attached to the feet or organs of the mouth, and the feet fringed with fine hairs. They are generally of small size.

EOCENE.—The earliest of the three divisions of the Tertiary epoch of geologists. Rocks of this age contain a small proportion of shells identical with species now living.

EPHEMEROUS INSECTS.—Insects allied to the May-fly.

FAUNA.—The totality of the animals naturally inhabiting a certain country or region, or which have lived during a given geological period.

FELIDÆ.—The Cat-family.

FERAL.—Having become wild from a state of cultivation or domestication.

FLORA.—The totality of the plants growing naturally in a country, or during a given geological period.

FLORETS.—Flowers imperfectly developed in some respects, and collected into a dense spike or head, as in the Grasses, the Dandelion, &c.

FŒTAL.—Of or belonging to the fœtus, or embryo in course of development.

FORAMINIFERA.—A class of animals of very low organisation, and generally of small size, having a jelly-like body, from the surface of which delicate filaments can be given off and retracted for the prehension of external objects, and having a calcareous or sandy shell, usually divided into chambers, and perforated with small apertures.

FOSSILIFEROUS.—Containing fossils.

FOSSORIAL.—Having a faculty of digging. The Fossorial Hymenoptera are a group of Wasp-like Insects, which burrow in sandy soil to make nests for their young.

FRENUM (pl. FRENA).—A small band or fold of skin.

FUNGI (sing. FUNGUS).—A class of cellular plants, of which Mushrooms, Toadstools, and Moulds, are familiar examples.

FURCULA.—The forked bone formed by the union of the collar-bones in many birds, such as the common Fowl.

GALLINACEOUS BIRDS.—An order of Birds of which the common Fowl, Turkey, and Pheasant, are well-known examples.

GALLUS.—The genus of birds which includes the common Fowl.

GANGLION.—A swelling or knot from which nerves are given off as from a centre.

GANOID FISHES.—Fishes covered with peculiar enamelled bony scales. Most of them are extinct.

GERMINAL VESICLE.—A minute vesicle in the eggs of animals, from which the development of the embryo proceeds.

GLACIAL PERIOD.—A period of great cold and of enormous extension of ice upon the surface of the earth. It is believed that glacial periods have occurred repeatedly during the geological history of the earth, but the term is generally applied to the close of the Tertiary epoch, when nearly the whole of Europe was subjected to an arctic climate.

GLAND.—An organ which secretes or separates some peculiar product from the blood or sap of animals or plants.

GLOTTIS.—The opening of the windpipe into the œsophagus or gullet.

GNEISS.—A rock approaching granite in composition, but more or less laminated, and really produced by the alteration of a sedimentary deposit after its consolidation.

GRALLATORES.—The so-called Wading-birds (Storks, Cranes, Snipes, &c.), which are generally furnished with long legs, bare of feathers above the heel, and have no membranes between the toes.

GRANITE.—A rock consisting essentially of crystals of felspar and mica in a mass of quartz.

HABITAT.—The locality in which a plant or animal naturally lives.

HEMIPTERA.—An order or sub-order of Insects, characterised by the possession of a jointed beak or rostrum, and by having the forewings horny in the basal portion and membranous at the extremity, where they cross each other. This group includes the various species of Bugs.

HERMAPHRODITE.—Possessing the organs of both sexes.

HOMOLOGY.—That relation between parts which results from their development from corresponding embryonic parts, either in different animals, as in the case of the arm of man, the fore-leg of a quadruped, and the wing of a bird; or in the same individual, as in the case of the fore and hind legs in quadrupeds, and the segments or rings and their appendages of which the body of a worm, a centipede, &c., is composed. The latter is called *serial homology*. The parts which stand in such a relation to each other are said to be *homologous*, and one such part or organ is called the *homologue* of the other. In different plants the parts of the flower are homologous, and in general these parts are regarded as homologous with leaves.

HOMOPTERA.—An order or sub-order of Insects having (like the Hemiptera) a jointed beak, but in which the fore-wings are either wholly membranous or wholly leathery. The *Cicadæ*, Frog-hoppers, and *Aphides*, are well-known examples.

HYBRID.—The offspring of the union of two distinct species.

HYMENOPTERA.—An order of Insects possessing biting jaws and usually four membranous wings in which there are a few veins. Bees and Wasps are familiar examples of this group.

HYPERTROPHIED.—Excessively developed.

ICHNEUMONIDÆ.—A family of Hymenopterous insects, the members of which lay their eggs in the bodies or eggs of other insects.

IMAGO.—The perfect (generally winged) reproductive state of an insect.

INDIGENS. — The aboriginal animal or vegetable inhabitants of a country or region.

INFLORESCENCE.—The mode of arrangement of the flowers of plants.

INFUSORIA.—A class of microscopic Animalcules, so called from their having originally been observed in infusions of vegetable matters. They consist of a gelatinous material enclosed in a delicate membrane, the whole or part of which is furnished with short vibrating hairs (called cilia), by means of which the animalcules swim through the water or convey the minute particles of their food to the orifice of the mouth.

INSECTIVOROUS.—Feeding on Insects.

INVERTEBRATA, or INVERTEBRATE ANIMALS.—Those animals which do not possess a backbone or spinal column.

LACUNÆ.—Spaces left among the tissues in some of the lower animals, and serving in place of vessels for the circulation of the fluids of the body.

LAMELLATED.—Furnished with lamellæ or little plates.

LARVA (pl. LARVÆ).—The first condition of an insect at its issuing from the egg, when it is usually in the form of a grub, caterpillar, or maggot.

LARYNX.—The upper part of the windpipe opening into the gullet.

LAURENTIAN.—A group of greatly altered and very ancient rocks, which is greatly developed along the course of the St. Laurence, whence the name. It is in these that the earliest known traces of organic bodies have been found.

LEGUMINOSÆ.—An order of plants represented by the common Peas and Beans, having an irregular flower in which one petal stands up like a wing, and the stamens and pistil are enclosed in a sheath formed by two other petals. The fruit is a pod (or legume).

LEMURIDÆ.—A group of four-handed animals, distinct from the Monkeys and approaching the Insectivorous Quadrupeds in some of their characters and habits. Its members have the nostrils curved or twisted, and a claw instead of a nail upon the first finger of the hind hands.

LEPIDOPTERA.—An order of Insects, characterised by the possession of a spiral proboscis, and of four large more or less scaly wings. It includes the well-known Butterflies and Moths.

LITTORAL.—Inhabiting the seashore.

LOESS.—A marly deposit of recent (Post-Tertiary) date, which occupies a great part of the valley of the Rhine.

MALACOSTRACA.—The higher division of the Crustacea, including the ordinary Crabs, Lobsters, Shrimps, &c., together with the Woodlice and Sand-hoppers.

MAMMALIA.—The highest class of animals, including the ordinary hairy quadrupeds, the Whales, and Man, and characterised by the production of living young which are nourished after birth by milk from the teats (*Mammæ*, *Mammary glands*) of the mother. A striking difference in embryonic development has led to the division of this class into two great groups; in one of these, when the embryo has attained a certain stage, a vascular connection, called the *placenta*, is formed between the embryo and the mother; in the other this is wanting, and the young are produced in a very incomplete state. The former, including the greater part of the class, are called *Placental mammals;* the latter, or *Aplacental mammals*, include the Marsupials and Monotremes (*Ornithorhynchus*).

MAMMIFEROUS. Having mammæ or teats (see MAMMALIA).

MANDIBLES, in Insects.—The first or uppermost pair of jaws, which are generally solid, horny, biting organs. In Birds the term is applied to both jaws with their horny coverings. In Quadrupeds the mandible is properly the lower jaw.

MARSUPIALS.—An order of Mammalia in which the young are born in a very incomplete state of development, and carried by the mother, while sucking, in a ventral pouch (marsupium), such as the Kangaroos, Opossums, &c. (see MAMMALIA).

MAXILLÆ, in Insects.—The second or lower pair of jaws, which are composed of several joints and furnished with peculiar jointed appendages called palpi, or feelers.

MELANISM.—The opposite of albinism; an undue development of colouring material in the skin and its appendages.

METAMORPHIC ROCKS.—Sedimentary rocks which have undergone alteration, generally by the action of heat, subsequently to their deposition and consolidation.

MOLLUSCA.—One of the great divisions of the Animal Kingdom, including those animals which have a soft body, usually furnished with a shell, and in which the nervous ganglia, or centres, present no definite general arrangement. They are generally known under the denomination of "shell-fish;" the cuttle-fish, and the common snails, whelks, oysters, mussels, and cockles, may serve as examples of them.

MONOCOTYLEDONS, or MONOCOTYLEDONOUS PLANTS.— Plants in which the seed sends up only a single seed-leaf (or cotyledon); characterised by the absence of consecutive layers of wood in the stem (endogenous growth), by the veins of the leaves being generally straight, and by the parts of the flowers being generally in multiples of three. (*Examples*, Grasses, Lilies, Orchids, Palms, &c.)

MORAINES.—The accumulations of fragments of rock brought down by glaciers.

MORPHOLOGY.—The law of form or structure independent of function.

MYSIS-STAGE.—A stage in the development of certain Crustaceans (Prawns), in which they closely resemble the adults of a genus (*Mysis*) belonging to a slightly lower group.

NASCENT.—Commencing development.

NATATORY.—Adapted for the purpose of swimming.

NAUPLIUS-FORM.—The earliest stage in the development of many Crustacea, especially belonging to the lower groups. In this stage the animal has a short body, with indistinct indications of a division into segments, and three pairs of fringed limbs. This form of the common fresh-water *Cyclops* was described as a distinct genus under the name of *Nauplius*.

NEURATION.—The arrangement of the veins or nervures in the wings of Insects.

NEUTERS.—Imperfectly developed females of certain social insects (such as Ants and Bees), which perform all the labours of the community. Hence they are also called *workers*.

NICTITATING MEMBRANE.—A semi-transparent membrane, which can be drawn across the eye in Birds and Reptiles, either to moderate the effects of a strong light or to sweep particles of dust, &c., from the surface of the eye.

OCELLI.—The simple eyes or stemmata of Insects, usually situated on the crown of the head between the great compound eyes.

ŒSOPHAGUS.—The gullet.

OOLITIC.—A great series of secondary rocks, so called from the texture of some of its members, which appear to be made up of a mass of small *egg-like* calcareous bodies.

OPERCULUM.—A calcareous plate employed by many Mollusca to close the aperture of their shell. The *opercular valves* of Cirripedes are those which close the aperture of the shell.

ORBIT.—The bony cavity for the reception of the eye.

ORGANISM.—An organised being, whether plant or animal.

ORTHOSPERMOUS.—A term applied to those fruits of the Umbelliferæ which have the seed straight.

OSCULANT.—Forms or groups apparently intermediate between and connecting other groups are said to be osculant.

OVA.—Eggs.

OVARIUM or OVARY (in plants).— The lower part of the pistil or female organ of the flower, containing the ovules or incipient seeds; by growth after the other organs of the flower have fallen, it usually becomes converted into the fruit.

OVIGEROUS.—Egg-bearing.

OVULES (of plants).—The seeds in the earliest condition.

PACHYDERMS.—A group of Mammalia, so called from their thick skins, and including the Elephant, Rhinoceros, Hippopotamus, &c.

PALÆOZOIC.—The oldest system of fossiliferous rocks.

PALPI.—Jointed appendages to some of the organs of the mouth in Insects and Crustacea.

PAPILIONACEÆ.—An order of Plants (see LEGUMINOSÆ).—The flowers of these plants are called *papilionaceous*, or butterfly-like, from the fancied resemblance of the expanded superior petals to the wings of a butterfly.

PARASITE.—An animal or plant living upon or in, and at the expense of, another organism.

PARTHENOGENESIS.—The production of living organisms from unimpregnated eggs or seeds.

PEDUNCULATED.—Supported upon a stem or stalk. The pedunculated oak has its acorns borne upon a footstool.

PELORIA or PELORISM.—The appearance of regularity of structure in the flowers of plants which normally bear irregular flowers.

PELVIS.—The bony arch to which the hind limbs of vertebrate animals are articulated.

PETALS.—The leaves of the corolla, or second circle of organs in a flower. They are usually of delicate texture and brightly coloured.

PHYLLODINEOUS.—Having flattened, leaf-like twigs or leafstalks instead of true leaves.

PIGMENT.—The colouring material produced generally in the superficial parts of animals. The cells secreting it are called *pigment-cells*.

PINNATE.—Bearing leaflets on each side of a central stalk.

PISTILS.—The female organs of a flower, which occupy a position in the centre of the other floral organs. The pistil is generally divisible into the ovary or germen, the style and the stigma.

PLACENTALIA, PLACENTATA, or Placental Mammals.—See MAMMALIA.

PLANTIGRADES.—Quadrupeds which walk upon the whole sole of the foot, like the Bears.

PLASTIC.—Readily capable of change.

PLEISTOCENE PERIOD.—The latest portion of the Tertiary epoch.

PLUMULE (in plants).—The minute bud between the seed-leaves of newly-germinated plants.

PLUTONIC ROCKS.—Rocks supposed to have been produced by igneous action in the depths of the earth.

POLLEN.—The male element in flowering plants; usually a fine dust produced by the anthers, which, by contact with the stigma effects the fecundation of the seeds. This impregnation is brought about by means of tubes (*pollen-tubes*) which issue from the pollen-grains adhering to the stigma, and penetrate through the tissues until they reach the ovary.

POLYANDROUS (flowers).—Flowers having many stamens.

POLYGAMOUS PLANTS.—Plants in which some flowers are unisexual and others hermaphrodite. The unisexual (male and female) flowers, may be on the same or on different plants.

POLYMORPHIC.—Presenting many forms.

POLYZOARY.—The common structure formed by the cells of the Polyzoa, such as the well-known Sea-mats.

PREHENSILE.—Capable of grasping.

PREPOTENT.—Having a superiority of power.

PRIMARIES.—The feathers forming the tip of the wing of a bird, and inserted upon that part which represents the hand of man.

PROCESSES.—Projecting portions of bones, usually for the attachment of muscles, ligaments, &c.

PROPOLIS.—A resinous material collected by the Hive-Bees from the opening buds of various trees.

PROTEAN.—Exceedingly variable.

PROTOZOA.—The lowest great division of the Animal Kingdom. These animals are composed of a gelatinous material, and show scarcely any trace of distinct organs. The Infusoria, Foraminifera, and Sponges, with some other forms, belong to this division.

PUPA (pl. PUPÆ).—The second stage in the development of an Insect, from which it emerges in the perfect (winged) reproductive form. In most insects the *pupal stage* is passed in perfect repose. The *chrysalis* is the pupal state of butterflies.

RADICLE.—The minute root of an embryo plant.

RAMUS.—One half of the lower jaw in the Mammalia. The portion which rises to articulate with the skull is called the *ascending ramus*.

RANGE.—The extent of country over which a plant or animal is naturally spread. *Range in time* expresses the distribution of a species or group through the fossiliferous beds of the earth's crust.

RETINA.—The delicate inner coat of the eye, formed by nervous filaments spreading from the optic nerve, and serving for the perception of the impressions produced by light.

RETROGRESSION.—Backward development. When an animal, as it approaches maturity, becomes less perfectly organised than might be expected from its early stages and known relationships, it is said to undergo a *retrograde development* or *metamorphosis*.

RHIZOPODS.—A class of lowly organised animals (Protozoa), having a gelatinous body, the surface of which can be protruded in the form of root-like processes or filaments, which serve for locomotion and the prehension of food. The most important order is that of the Foraminifera.

RODENTS.—The gnawing Mammalia, such as the Rats, Rabbits, and Squirrels. They are especially characterised by the possession of a single pair of chisel-like cutting teeth in each jaw, between which and the grinding teeth there is a great gap.

RUBUS.—The Bramble Genus.

RUDIMENTARY.—Very imperfectly developed.

RUMINANTS.—The group of Quadrupeds which ruminate or chew

the cud, such as oxen, sheep, and deer. They have divided hoofs, and are destitute of front teeth in the upper jaw.

SACRAL.—Belonging to the sacrum, or the bone composed usually of two or more united vertebræ to which the sides of the pelvis in vertebrate animals are attached.

SARCODE.—The gelatinous material of which the bodies of the lowest animals (Protozoa) are composed.

SCUTELLÆ.—The horny plates with which the feet of birds are generally more or less covered, especially in front.

SEDIMENTARY FORMATIONS.—Rocks deposited as sediments from water.

SEGMENTS.—The transverse rings of which the body of an articulate animal or Annelid is composed.

SEPALS.—The leaves or segments of the calyx, or outermost envelope of an ordinary flower. They are usually green, but sometimes brightly coloured.

SERRATURES.—Teeth like those of a saw.

SESSILE.—Not supported on a stem or footstalk.

SILURIAN SYSTEM.—A very ancient system of fossiliferous rocks belonging to the earlier part of the Palæozoic series.

SPECIALISATION.—The setting apart of a particular organ for the performance of a particular function.

SPINAL CHORD.—The central portion of the nervous system in the Vertebrata, which descends from the brain through the arches of the vertebræ, and gives off nearly all the nerves to the various organs of the body.

STAMENS.—The male organs of flowering plants, standing in a circle within the petals. They usually consist of a filament and an anther, the anther being the essential part in which the pollen, or fecundating dust, is formed.

STERNUM.—The breast-bone.

STIGMA.—The apical portion of the pistil in flowering plants.

STIPULES.—Small leafy organs placed at the base of the footstalks of the leaves in many plants.

STYLE.—The middle portion of the perfect pistil, which rises like a column from the ovary and supports the stigma at its summit.

SUBCUTANEOUS.—Situated beneath the skin.

SUCTORIAL.—Adapted for sucking.

SUTURES (in the skull).—The lines of junction of the bones of which the skull is composed.

Tarsus (pl. **Tarsi**).—The jointed feet of articulate animals, such as Insects.

Teleostean Fishes.—Fishes of the kind familiar to us in the present day, having the skeleton usually completely ossified and the scales horny.

Tentacula or **Tentacles.**—Delicate fleshy organs of prehension or touch possessed by many of the lower animals.

Tertiary.—The latest geological epoch, immediately preceding the establishment of the present order of things.

Trachea.—The windpipe or passage for the admission of air to the lungs.

Tridactyle.—Three-fingered, or composed of three movable parts attached to a common base.

Trilobites.—A peculiar group of extinct Crustaceans, somewhat resembling the Woodlice in external form, and, like some of them, capable of rolling themselves up into a ball. Their remains are found only in the Palæozoic rocks, and most abundantly in those of Silurian age.

Trimorphic.—Presenting three distinct forms.

Umbelliferæ.—An order of plants in which the flowers, which contain five stamens and a pistil with two styles, are supported upon footstalks which spring from the top of the flower stem and spread out like the wires of an umbrella, so as to bring all the flowers in the same head (*umbel*) nearly to the same level. (*Examples*, Parsley and Carrot.)

Ungulata.—Hoofed quadrupeds.

Unicellular.—Consisting of a single cell.

Vascular.—Containing blood-vessels.

Vermiform.—Like a worm.

Vertebrata; or **Vertebrate Animals.**—The highest division of the animal kingdom, so called from the presence in most cases of a backbone composed of numerous joints or *vertebræ*, which constitutes the centre of the skeleton and at the same time supports and protects the central parts of the nervous system.

Whorls.—The circles or spiral lines in which the parts of plants are arranged upon the axis of growth.

Workers.—See Neuters.

ZoËA-STAGE.—The earliest stage in the development of many of
the higher Crustacea, so called from the name of *Zoëa* applied to
these young animals when they were supposed to constitute a
peculiar genus.

ZOOIDS.—In many of the lower animals (such as the Corals, Medusæ,
&c.) reproduction takes place in two ways, namely, by means of
eggs and by a process of budding with or without separation
from the parent of the product of the latter, which is often very
different from that of the egg. The individuality of the species
is represented by the whole of the form produced between two
sexual reproductions; and these forms, which are apparently
individual animals, have been called *zooids*.

INDEX.

A.

THE END.